Pushpendra K. Jain
and John S. Nkoma

INTRODUCTION TO CLASSICAL MECHANICS

Kinematics, Newtonian and Lagrangian

MKUKI NA NYOTA
DAR—ES—SALAAM

PUBLISHED BY

Mkuki na Nyota Publishers Ltd
P. O. Box 4246
Dar es Salaam, Tanzania
www.mkukinanyota.com

First published 2004 by Bay Publishers, Gaborone
Second Edition 2019 by Mkuki na Nyota, Dar es Salaam

© Pushpendra K. Jain and John S. Nkoma, 2019

ISBN 978-9987-08-370-1

Typesetting by John S. Nkoma

Visit www.mkukinanyota.com to read more about and to purchase any of Mkuki na Nyota books. You will also find featured authors, interviews and news about other publisher/author events. Sign up for our e-newsletters for updates on new releases and other announcements.

Distributed world wide outside Africa by African Books Collective.
www.africanbookscollective.com

Contents

CONTENTS

The Authors

Professor P K Jain holds a PhD degree in physics from the University of Connecticut, USA, and MSc and BSc degrees from India. He is a Chartered Physicist (CPhys) and a Fellow of the Institute of Physics (FInstP), UK. Prof. Jain has over 50 years of teaching and research experience in India, USA, Zambia including past 31 years in Botswana. He has been external examiner at several universities in Africa. Prof. Jain has made over 150 contributions to research in materials science, renewable energy, and general physics, has lectured internationally, and has received several awards and honours. Prof. Jain has been the UB Coordinator (2008-2016) of the African Materials Science and Engineering Network: AMSEN a Carnegie-IAS-RISE network of six African universities with the aim to train manpower at MSc and PhD degree levels, to build supervisory capacity and to develop research infrastructure in materials science and engineering in member universities. He can be contacted at his email address: jainpk@mopipi.ub.bw

Professor J S Nkoma obtained his BSc in Physics and Mathematics from the University of Dar es Salaam, Tanzania, and later his MSc and PhD in Physics from the University of Essex, UK. Prof. Nkoma has over 30 years experience of university teaching and research in Tanzania, Botswana, Italy and UK, and has been external examiner to several universities. His research interests are in condensed matter physics, materials science and ICT Regulation in telecommunications/internet, broadcasting and postal communications. He is a Fellow of the African Academy of Sciences. He also served as the Director General/CEO of the Tanzania Communications Regulatory Authority (TCRA) during 2004 to 2015. He recently published the title: Introduction to Basic Concepts for Engineers and Scientists: Electromagnetic, Quantum, Statistical and Relativistic Concepts, John. S. Nkoma, Mkuki na Nyota Publishers (2018). He can be contacted at his emailaddress: jsnkoma@hotmail.com

Professors P K Jain and J S Nkoma have co-authored these two books:
1. Introduction to Optics: Geometrical, Physical and Quantum, J S Nkoma and P K Jain, Bay Publishers, Gaborone ISBN 99912-511-6-2 (2003) and Second Edition, Mkuki na Nyota Publishers, Dar es Salaam (2019), and
2. Introduction to Mechanics: Kinematics, Newtonian and Lagrangian, P K Jain and J S Nkoma, Bay Publishers, Gaborone ISBN 99912-561-4-8 (2004) and Second Edition, Introduction to Classical Mechanics: Kinematics, Newtonian and Lagrangian, Mkuki na Nyota Publishers, Dar es Salaam (2019).

Preface to the Second Edition

Following the success of the first edition of our book *Introduction to Mechanics: Kinematics, Newtonian and Lagrangian* which was published fourteen years ago by Bay Publishing (Pty) Ltd. in Gaborone, Botswana, we bring the second edition, *Introduction to Classical Mechanics: Kinematics, Newtonian and Lagrangian* published by Mkuki na Nyota Publishers in Dar es Salaam, Tanzania. The Foreword, Preface and Content of the first edition are still relevant and are reproduced herein in this second edition.

Why the second edition? First, there is the addition of the word Classical to the title for the reason of providing more clarity, classical mechanics, as opposed to other branches of mechanics such as quantum mechanics, statistical mechanics, relativistic mechanics and others. Secondly, we have fixed a few typos as is usually done in such undertakings, and the contents of each chapter have been updated, and noting that problem solving is important for students, additional recipes have been added. Thirdly, the current publisher, Mkuki na Nyota has ably redesigned the book in a new format with an attractive cover representative of the subject matter, and overall they have produced this second edition with vision, excellence and professionalism.

In this second edition, the authors still acknowledge colleagues mentioned in the preface of the first edition as well as our many students over the years. We also thank staff of Mkuki na Nyota Publishers for several discussions during manuscript preparation and for finally producing the book with an excellent quality. Patience and understanding of our families and friends is gratefully acknowledged.

Pushpendra K. Jain
Professor of Physics
Department of Physics
University of Botswana
Gaborone, Botswana

October 2018

John S. Nkoma
Professor of Physics and
Ex-Director General/CEO, TCRA
Mbezi Beach
Dar es Salaam, Tanzania

October 2018

Foreword to the First Edition

Scientists and technocrats are key role players in the industrial and economic development of any society. In most countries of Africa, there is a dire shortage of such indigenous trained manpower. Some countries, even after several decades of independence, continue to face the shortage of science teachers, thus stopping the science based development in its tracks. Several factors can be attributed to this phenomenon; the high cost of training in science based careers being the prime cause, as well as the high cost of the infrastructure required for such training. The exorbitantly high cost of imported text books and teaching materials coupled with weak national currencies, and lack of foreign currency compound the problem even further. To overcome some of the barriers, there is an express need for the development of indigenous quality text books written by mature and experienced African academics and professionals, and that such textbooks are produced locally at an affordable cost for the African market. The book *Introduction to Mechanics: Kinematics, Newtonian and Lagrangian* by Professors P K Jain and J S Nkoma is a most welcome contribution in this direction of Physics education. The authors are not only highly experienced academics, but also they gained most of their experience working at various universities in Africa, and are well conversant with the status science in the continent. However, I must add that the book shall be found useful by students globally.

The book takes into account the varied academic background of entry level students, and the language barrier that some students may initially face. Some of the unique features of the book are its simple language so that it is easy to understand; a clear, pointwise and elaborate discussion of difficult to comprehend concepts; the recipes to guide students to problem solving; solved examples that illustrate the application of recipes followed by exercises for practice; and a large number of problems at the end of each chapter. An appendix on common mathematical formulas adds further value to the book. I commend Professors Jain and Nkoma for their initiative and valuable contribution to the development of a science and technology base in Africa through the use of indigenous resources.

Juma Shabani
Professor of Mathematical Physics, and
Director, UNESCO Harare Cluster Office
July 2004

Preface to the First Edition

Contemporary physics programmes are under increasing pressure to provide a balance between coverage of several core branches of Physics such as mechanics, optics, electromagnetism, thermodynamics, quantum physics, relativity, and statistical physics on the one hand, and to expose students to new and emerging areas of research on the other hand. At the same time, it is also important to provide an indepth background to certain core areas of physics to students who may not become professional physicists but would need them in their chosen profession. Mechanics - the science of motion and equilibrium is one such area of physics. Concepts of mechanics such as vector and scalar quantities, forces, the laws of motion, work, energy, the conservation laws, gravitation, circular, orbital and oscillatory motions cut across not only most branches of physics such as electromagnetism, atomic, molecular, nuclear, astro and space physics, but are also applied to most branches of engineering and technology. This makes mechanics an important component of physics which students must master well at an early stage before branching to various career options. The book, *Introduction to Physics: Kinematics, Newtonian and Lagrangian* is written with a view to meet the needs of both groups of University undergraduate students namely those in pure physical sciences and those proceeding to programmes in applied areas of engineering and technology. In some universities, mechanics may be offered as a a one semester course or may take more than one semester, while in others it may be offered as modules within a general physics courses. With a choice of appropriate topics, the book can be adapted to teach mechanics curricula in both of these approaches.

This book covers three crucial subareas of mechanics namely Kinematics, Newtonian mechanics and Lagrangian mechanics. Chapter 1 covers introductory aspects. Kinematics is discussed in chapter 2. Newton's laws of motion are introduced in chapter 3. Chapter 4 deals with the conservation of linear momentum. Work, energy and power are covered in chapter 5. Circular motion, Gravitation and planetary motion, and oscillations are covered in chapters 6, 7 and 8 respectively. Chapter 9 presents the aspects of rigid body dynamics, and Lagrangian mechanics is introduced in chapter 10, which lays a foundation for advanced courses in mechanics. Towards the end of the book there is a comprehensive bibliography that includes both recent references and a number of old classic books that are still regarded as the authoritative writing on the subject, and can be found in most well stocked libraries. Self motivated students with a keen interest in the subject shall find it advantageous to consult these references for indepth self-study of the topics of interest.

It is well acknowledged that the language of physics is universal, and the book is suited to students globally. However, the book recognizes and addresses the specific needs of students in African Universities. There is a marked heterogeneity in the background of students attending the same courses.

This ranges from well prepared to not so well prepared students. Those with stronger academic backgrounds need a bit of a challenge in their work, whereas the others need to be taught in fair detail to makeup for their deficiencies and to maintain their interest in the subject. By the end of the course all students should have achieved near equal competence in the subject. The book meets the needs of all students with varying academic preparedness. It presents detailed explanations of difficult-to-grasp topics with the help of simple but clearly drawn and labeled diagrams. The discussions and conclusions are presented pointwise, and key words, definitions, laws etc. are highlighted. A number of tasks similar to those discussed within the text are assigned for students to take up a challenge. A unique feature of the book is a number of *'Recipes'* which give students tailor-made guidance to problems solving. Applications of the recipes are illustrated by solved examples, followed by similar exercises for students to practice. There are a large number of problems and exercises at the end of each chapter to further sharpen the students' skills. The book distills over 60 years of combined teaching experience of two authors in Botswana, India, Tanzania, UK, USA and Zambia, and the experience of having been external examiners at many Universities in Southern and Eastern Africa, as well as serving in other capacities on matters associated with Science and Technology.

Acknowledgments

The authors express their sincere gratitude to a number of colleagues: Professor E M Lungu, Dr S Chimidza, Dr P V C Luhanga, Dr J Prakash, Dr L K Sharma, Dr T S Verma and Dr P Winkoun for carefully reading through the manuscript and for making many valuable suggestions. We also thank our publisher for having produced the book so well and for discussions on manuscript preparation. Patience and understanding of our families of us having stayed away for extended hours and over weekends during the course of this work are gratefully acknowledged. Lastly, but not the least, we thank over 25,000 students over the years who gave us the opportunity to enhance our understanding of physics as reflected in the book.

P K Jain and J S Nkoma
Department of Physics
University of Botswana
Gaborone
July 2004

Chapter 1

Introduction: Units, Dimensions, Scalars and Vectors

There are two distinct aspects of Physics. The first aspect concerns measurements which is the experimental aspect. Linked with measurement is the use of appropriate units of the measured (or mathematically calculated) physical quantities. Secondly, physics relies heavily on mathematics which is the theoretical aspect of physics. Amongst the mathematical aspect, vectors are of particular interest as they play a central role in all branches of physics. A number of quantities such as work, torque, angular momentum, force on a current carrying conductor in a magnetic field, force on a moving charge in a magnetic field etc. are expressed in terms of related vector quantities. Students of a basic physics course are generally found to lack skills in both these aspects, namely the use of units, and the handling of vector quantities. These deficiencies not only hamper their understanding of the principles of physics but also affect their performance adversely. On the units front, the difficulties faced by students include the use of no or wrong units, conversion of units, and mixing of incompatible units. Amongst the use of vectors, it is the vector algebra namely the addition, subtraction and multiplication of vector quantities that is at the root of the problem. In this chapter both these issues are addressed. First we deal with units and dimensions where the classification of units, systems of units, primary standards of units, and conversion of units are discussed. Dimensions, dimensional analysis, its applications in physics and limitations are introduced. Use of very large and very small numbers as powers of tens, and prefixes for powers of tens are presented. In the later part of the chapter scalar and vector quantities are distinguished, representations of vectors, and vector algebra for the addition, subtraction and multiplication of vectors are discussed. By the end of this chapter students are expected to have acquired the skills in the use of units, and in handling of vector quantities which both cut across the entire physics.

1.1 Units

All physical quantities, measured or calculated, *with the exception of a few* are expressed in terms of the units in which they are measured. Magnitude of such quantities stated without the use of appropriate units is not only incomplete, it is incomprehensible. For example, let us say the distance from Gaborone to Francistown is 420 km, then stating the distance as 420 only, without the use of *km* shall convey only the partial information, and could not be interpreted correctly. Likewise, if Mpho is 18 *years* old, then years must be included along with his age, without which he could be 18 *hours*, 18 *days* or 18 *months* old. One may define *unit* in the following manner:

Definition 1.1: *Unit is defined as a measure of the measurement in terms of which a physical quantity is expressed.*

The quantities that do not have units are known as the *unitless*, or more commonly as *dimensionless* quantities. Examples of dimensionless quantities are, the relative density and the refractive index, which are defined as follows:

$$\text{Relative density of a substance} = \frac{\text{Density of the substance}}{\text{Density of water at } 4\,^{\circ}C}$$

$$\text{Refractive index of a medium} = \frac{\text{Speed of light in vacuum}}{\text{Speed of light in the medium}}$$

From these examples one notes, that dimensionless quantities are expressed as the ratio of similar physical quantities which have the same units, so that on taking the ratio, the units from the numerator and denominator cancel out. Other examples of dimensionless quantities are efficiency, magnification, plane and solid angles, resolving power of optical instruments, and trigonometric, logarithmic and exponential functions.

1.1.1   Classification of Units

There are innumerable measured and calculated physical quantities and physical constants in the universe each of which must be assigned a unit. It will be a daunting task if one were to define units for every such quantity. Fortunately it is not so, because all physical quantities and constants can be expressed in terms of just a few physical quantities. By assigning appropriate units to those few quantities only, one can obtain the units of all known quantities and constants. Thus units are classified into two categories: *(i) Primary units* or *Fundamental units* or *Basic units*, and *(ii) Secondary units* or *Derived Units*.

Primary units:

Definition 1.2: *Primary units are a set of independent units that can not be expressed in terms of each other or any other unit by any means.*

There are only seven quantities which constitute the primary units. These are: *mass, length, time, electric current, temperature, amount of substance,* and *luminous intensity.* In addition, in some literature, plane angle, and solid angle with their units *radian (rad)* and *steradian (sr)* respectively are treated as primary units, making a total of nine primary units. Strictly speaking, both the plane angle and the solid angle are dimensionless quantities. The plane angle in radians is expresses as the ratio of the length of the arc of the circle to its radius, and the solid angle in steradian is the ratio of the surface area of a conic segment of a sphere to the square of its radius. Thus, both radian and steradian are dimensionless.

Secondary units:

Definition 1.3: *Secondary units are the units which are expressed in terms of the primary units. Units of all physical quantities and constants other than the primary quantities constitute the secondary units.*

If one can express a physical quantity or a constant in terms of the primary quantities, then its unit is obtained simply by substituting the units of the constituent primary quantities. For example velocity is the ratio of the distance moved to the time interval, and hence the unit of velocity is the unit of length divided by the unit of time. Likewise, the unit of force which is mass times acceleration is the product of the unit of mass and that of acceleration, noting that the unit of acceleration itself is a secondary unit.

1.1.2 Systems of Units

A consistent system of units depends on the choice of the units of primary quantities from which all other units are derived. There are three main systems of units that are used in sciences, commerce, and in everyday life. They are:

- Systéme International (SI) units. It is also the modified MKS (meter, kilogram and seconds) or the metric system. This system is discussed in detail in section 1.1.3.

- cgs (centimeter, gram and second) or the Gaussian system. In this system the unit of length is $centimeter$ ($= 10^{-2}m$), the unit of mass is $gram$ ($= 10^{-3}kg$), and the unit of time is *second*. Although, the units of length, mass and time in cgs system are related to the corresponding units in the SI system, but the units of electric and magnetic quantities are quite different in the two systems.

- British, or the Imperial or the FPS (foot, pound and second) system. In this system mass is not the basic unit, rather the unit of force, *Pound*, is the basic unit. The unit of mass, *slug*, is derived from the unit of force, and $1\ slug = 14.59\ kg$. Commonly, it is the Pound (mass), denoted by *lb* that is used as the unit of mass where *1 slug = 32.17 Pound mass (lb)*, and $1\ lb = 453.53\ g$ where $g = 9.81\ ms^{-2}$. To distinguish the force and mass expressed in Pound, one simply calls them as Pound-force, and Pound-mass respectively. The unit of length is *foot* *(= 30 inches, 1 inch = 2.54 cm)*, unit of time is *second*, and the unit of temperature is *degree Fahrenheit*. Large distances in this system are measured in *miles* where *(1 mile = 1760 yards, and 1 yard = 3 feet)*

All three systems of units are used for various applications in different parts of the world. For example, the British system of units still continues to be the most commonly used units throughout the United States of America even though most of the world has changed to the SI units. The precious commodities such as gold are measured and traded in grams rather than in kilograms. A commonly used unit for the speed of ships and planes is *knot*, where $1\ knot = 1.151\ mile\ per\ hour = 1.832\ kmh^{-1}$

In sciences also all three systems of units have been in use through various periods of time. Most of the pre 1950's scientific literature is found in the Gaussian and the British units. In order to unify the scientific work and for a clearer understanding of scientific literature globally, an international committee on weights and measures recommended the use of SI system of units for all scientific work and publications. However, in certain branches of sciences, certain specialized units defined in terms of SI units are used for the sake of convenience of presentation and computation. Examples of some such specialized units are:

- In optics: Wavelength of radiation is expressed in Angstrom (Å): $1 \overset{\circ}{A} = 1 \times 10^{-10} m$

- In atomic/ nuclear Physics: Energy is expressed in electron volt (eV):
  $1 \, eV = e \, Joules = 1.6 \times 10^{-19} \, J$, where e is the electronic charge.

- In nuclear Physics: Radius of the nucleus is expressed in Fermi (F): $1 \, F = 1 \times 10^{-15} \, m$. Mass of elementary particles is expressed in Million eV (MeV): $1 \, MeV = e \times \frac{10^6}{c^2} \, kg$, where c is the speed of light in vacuum in ms^{-1}

In this book we shall use SI system of units. Students are also strongly advised to make use of these units only. Whenever some parameter or a quantity is given in units other than SI units, they must first convert it to SI units before proceeding to solve a problem. This shall have two advantages: The final result obtained shall automatically be in SI units, and it shall ensure that no mistakes are made due to the wrong use or mixing of units.

1.1.3   SI Units and the Fundamental SI Standards of Basic Quantities

Table 1.1 gives the primary SI units of the seven plus two fundamental quantities. In mechanics, mass, length, and time are the three basic quantities, and the units of everything else encountered in mechanics can be expressed in terms of the units of these quantities. In this book we shall define the fundamental standards of these three quantities only. The fundamental standards of other primary units are the subject of discussion for textbooks in respective areas of physics where those units are most commonly used.

Table 1.1: Primary units in the SI system of units.

Primary Quantity	SI Unit (Symbol)
Length	meter (m)
Mass	kilogram (kg)
Time	second (s)
Electric current	Ampere (A)
Temperature	Kelvin (K)
Amount of substance	mole (mol)
Luminous intensity	candela (cd)
Plane angle	radian (rad)
Solid angle	steradian (sr)

Standards for the SI units of mass, length and time are based on three fundamental standards. Historically they were originally adopted arbitrarily. In recent years two of them have been defined formally for the sake of permanency in case the original standards preserved in museums of weights and measures are destroyed.

Fundamental standard of mass: Originally, the SI unit of mass, *kilogram (kg)*, was defined as the mass of 1 $liter(= 10^{-3}\ m^3)$ of pure water at 4 ^oC. In 1887, the standard of mass was adopted as the mass of a platinum-iridium alloy cylinder preserved at the International Bureau of Weights and Measures, Sévres, France. Platinum-iridium alloy is used because it is a stable alloy and is completely free from deterioration due to corrosion etc. This is the only standard that has not been changed since its first adoption, and still continues to be expressed in terms of an artifact. An exact replica of the kg-cylinder in France is kept at the National Bureau of Standards in the USA, which has an accuracy of 1 part in 10^8. Replicas of the kg-mass are produced by using an equal-arm, beam balance.

Fundamental standard of length: In 1799, the SI unit of length, *meter (m)*, was defined in France as one ten-millionth of the distance from the North Pole to the equator along a particular longitude line passing through Paris. Later, 1 *m* distance was defined as the distance between the two end-marks on a platinum-iridium alloy rod stored at a controlled temperature, which continued to be in use until the 1960s. In 1960s the meter length was defined as equal to 1,650,763.73 wavelengths of the orange-red light in vacuum emitted from a krypton-86 lamp. In 1983 1 meter length was redefined as the distance traveled by light in vacuum during a time interval of {1/ (299,792,458)} second. This precisely establishes the speed of light in vacuum as 299, 792, 458 ms^{-1}. The new standard of meter length cannot be destroyed, is independent of small variations in temperature, can be reproduced anywhere, and is more accurate-free from the error resulting from the thickness of the marks themselves on the historic alloy-rod standard.

Fundamental standard of time: Originally, 1 *second* was defined as $\frac{1}{60} \times \frac{1}{60} \times \frac{1}{24}$ of a mean solar day. Mean solar day is the time interval between two successive appearance of sun overhead, averaged over a year. In 1956, 1 s was redefined to be equal to the fraction {1/ (31,556,925.9747)} of the tropical year 1900. In 1967, the standard of time was adopted in terms of a cesium atomic clock as the time interval of 9,912,631,770 vibrations of radiation from the cesium-133 atom. The cesium clock has an accuracy of 1 part in 10^{11}, and two synchronized cesium clocks may differ by only 1 s in 5000 $years$. The atomic standard of time has the same advantages as the spectral standard of length.

Arbitrariness in the definition of fundamental standards of units is prominently reflected from the above examples. It is, therefore, not surprising to note that amongst rural communities throughout the world sometimes informal units are used for trading and commerce locally, for example, trading a tin, or a basket or a cup-full of produce or an arm length of fabric for the specified amount of money. Keeping in line with such an informality in defining the standards of units, it shall be an interesting and amusing exercise for students to devise a personal system of units in which the mass of the students is the standard of mass, height is the unit of length, and the unit of time is the time interval of a lecture period. Give these personal units some name, for example, personal-mass (pm), personal-length (pl), and personal-time (pt), and convert some known quantities to the personal system of units, for example mass of the earth and the sun, distance between two major cities, and the day length. Also express the speed of light in $(pl)/(pt)$, density of water in $(pm)/(pl)^3$, personal units of force (personal-force: pf) in terms of Newton, and the personal unit of energy (personal-energy: pe) in terms of Joule. *HINT:* Conversion of units discussed in section 1.1.5 may be useful here.

1.1.4 Secondary SI Units and Special Nomenclature

Examples of some secondary or derived SI units are given in Table 1.2. The first column lists the quantities, in second column their relationship to primary quantities are given, from which the SI units given in the third column are derived.

Table 1.2: Some Secondary SI Units and Dimensions

Physical Quantity	Relationships	SI Units	Special Name	Dimensions
Speed	(Distance)/(Time)	ms^{-1}		ML^{-1}
Acceleration	(Velocity)/ (Time)	ms^{-2}		ML^{-2}
Density	(Mass)/ (Volume)	kgm^{-3}		ML^{-3}
Force	(Mass) × (Acceleration)	$kgms^{-2}$	Newton (N)	MLT^{-2}
Work (Energy)	(Force) × (Displacement)	kgm^2s^{-2}	$Nm \Rightarrow Joule(J)$	ML^2T^{-2}
Pressure	(Force)/ (Area)	$kgm^{-1}s^{-2}$	$Nm^{-2} \Rightarrow Pascal(Pa)$	$ML^{-1}T^{-2}$
Power	(Work or Energy)/ (Time)	kgm^2s^{-3}	$Js^{-1} \Rightarrow Watt(W)$	ML^2T^{-3}

Some units have been given special names in honour of the scientists who made the most noteworthy contributions to the subarea of science where these units are most frequently used. Examples of special names of the units in the SI system are given in the fourth column of the table. The special units are exactly at par with the units expressed in terms of the primary quantities, for example, $1N = 1kg\,m\,s^{-2}$, and they are used more frequently than the full-blown expression for these units because of the inherent convenience. From the fourth column, it is also interesting to note that certain units that are expressed in terms of the special units of another quantity, are in turn given a special name for themselves. For example the units of work (energy) and pressure which can be expressed in terms of Newton (N) as Nm and Nm^{-2} respectively are given names as Joule (J) and Pascal (Pa) respectively, and Power $\Rightarrow N\,m\,s^{-1} = J\,s^{-1} = Watt\,(W)$. The dimensions in the last column of the table are discussed in section 1.3.

1.1.5 Conversion of Units

Conversion of units from one system to another is an essential skill which one needs to apply quite frequently. Students are often found to make mistakes in conversion of units which are carried through the problem solving leading to wrong results. Here we give a simple, fool proof method for the conversion of units, which when applied systematically shall yield error free results. The method is based on a simple principle: *"When a physical quantity is multiplied with a dimensionless ones, it remains unchanged"*. The appropriate dimensionless ones are expressed in terms of the conversion factor of the corresponding units. Some examples of dimensionless ones are given below.

$$1 = \frac{1\,kg}{1000\,g} = \frac{1000\,g}{1\,kg}$$

$$1 \quad = \quad \frac{1\ m}{100\ cm} = \frac{100\ cm}{1\ m} = \frac{1000\ m}{1\ km} = \frac{1\ km}{1000\ m}$$

$$1 \quad = \quad \frac{1\ min.}{60\ s} = \frac{60\ s}{1\ min} = \frac{1\ hr.}{3600\ s} = \frac{3600\ s}{1\ hr.}$$

In this manner one may define dimensionless one for any pair of primary units, and the quantity in given units can be converted to any desired units using the following simple recipe.

Recipe 1.1: Conversion of Units.

- Define appropriate dimensionless ones for the desired conversion.

- Multiply the given quantity along with its units with the chosen dimensionless ones as many times as necessary, so that the given units cancel out and replaced with the new units.

- Simplify to obtain the result in new units.

Example 1.1: Convert the density of mercury from $13.6\ g\,cm^{-3}$ to SI units.

Solution: SI unit of density is $kg\,m^{-3}$. Therefore,

$$\begin{aligned} \text{Density of Mercury} \quad &= \quad 13.6 \times \frac{g}{cm^3} \times \frac{1\ kg}{1000\ g} \times \left(\frac{100\ cm}{1\ m}\right)^3 \\ &= \quad 13,600\ kg\,m^{-3} \end{aligned}$$

Exercise 1.1: A car is moving at 50 miles per hour (mph). Convert its speed into kmh^{-1}, and into SI Units. (Conversion factors for miles to metric system are given in section 1.1.2 under the FPS units system.)

1.2 Prefixes to Units

Physics is the science of nature, and in nature one encounters very large and very small numbers involving up to 20 or more zeros on either side of the decimal point. Such numbers are expressed in terms of the integer powers of ten which are positive for large numbers and negative for small numbers. General guide line to express a number as a power of ten are:

- Express the number with only one digit to the right of the decimal point, and multiply it with appropriate power of ten, for example: $x = y.zzz... \times 10^n$, where n could be a positive or a negative integer.

- Powers of ten with $|n| < 3$ are generally not used.

Table 1.3 gives examples of very large, intermediate and very small quantities in terms of mass, length and time encountered in nature and in physics. Examples of physical constants are also included.

Table 1.3: Examples of very large, intermediate and very small quantities in mass, length, time and physical constants

Quantity	Magnitude
Mass:	
Estimated mass of the visible universe	$\sim 10^{52}\ kg$
Mass of a human being	$\sim 80\ kg$
Mass of an electron	$9.1 \times 10^{-31}\ kg$
Length:	
Distance from the earth to the most remote known quasar	$1.4 \times 10^{26}\ m$
Length of a football field	$91\ m$
Diameter of a proton	$\sim 10^{-15}\ m$
Time:	
Age of the universe	$5 \times 10^{17}\ s$
Interval between normal heart beats	$0.8\ s$
Time for light to pass across a proton	$\sim 10^{-24}\ s$
Physical constants:	
Avogadro's number	$6.022 \times 10^{26}\ (kmol)^{-1}$
Average acceleration due to gravity	$9.8\ ms^{-2}$
Planck's constant	$6.626 \times 10^{-34}\ Js^{-1}$

Table 1.4: Prefixes for the powers of 10

10^n	Name	Symbol	Example of the area of usage
10^2	hecto	h	Not commonly used
10^1	deka	da	Not commonly used
10^{-1}	deci	d	Not commonly used
10^{-2}	centi	c	$1\ cm = 10^{-2}\ m$
10^{24}	Yotta	Y	Not commonly used
10^{21}	Zetta	Z	Not commonly used
10^{18}	Exa	E	Energy release in nuclear explosion
10^{15}	Peta	P	Annual energy consumption of the world
10^{12}	Tera	T	Annual energy consumption of a small nation
10^9	Giga	G	Transmission frequencies
10^6	Mega (Million)	M	Generating capacity of a power station
10^3	kilo	k	Distances in kilometer (km)
10^{-3}	milli	m	small lengths (mm)/ masses (mg)
10^{-6}	micro	μ	Value of an inductance in an electrical circuit
10^{-9}	nano	n	Wavelength of light
10^{-12}	pico	p	value of a capacitance in an electrical circuit
10^{-15}	femto	f	Diameter of the nucleus
10^{-18}	atto	a	Time interval in nuclear processes
10^{-21}	zepto	z	Not commonly used
10^{-24}	yocto	y	Not commonly used

Powers of ten in multiples of 3 are the most commonly used, and they are designated with prefixes to the units. Table 1.4 gives such prefixes to units and examples of where they are used.

Students are also found to have difficulty in handling algebra involving numbers with powers of ten. Although it is not the objective of this book to train students in such a basic mathematical skill, the following simple rules may be found helpful by those who wish to practice on their own.

- In multiplication, the powers of ten get added algebraically, for example $10^{22} \times 10^{-14} = 10^8$.

- In division, sign of the power of ten in the denominator is changed and is added algebraically to the power of ten in the numerator, for example $\frac{10^{-7}}{10^{-11}} = 10^{-7} \times 10^{11} = 10^4$

- For addition or subtraction factorize the largest or the smallest power of ten, and simplify, for example $2.4 \times 10^7 \pm 4.7 \times 10^9 = 10^7 \times (2.4 \pm 470) = 472.4 \times 10^7 = 4.724 \times 10^9$

1.3 Dimensions

Definition 1.4: *Dimensions of a physical quantity tell us how it is related to the constituent fundamental quantities*

Once again we shall restrict ourselves only to the dimensions of physical quantities encountered in mechanics, and discuss the applications of dimensional analysis to mechanical systems. Dimensions of the three fundamental quantities in mechanics, namely mass, length and time are denoted as M, L, and T respectively. The dimension of a dimensionless quantity is a dimensionless-1. Examples of the dimensions of some of the quantities in mechanics are given in the last column of Table 1.2. These are obtained by replacing mass with M, distance (length) with L, and time with T in the second column of the table. It is to be noted that dimensions are not the same as units, a common mistake made by students. However, units in a given system can be obtained from the dimensional formula, by replaying M, L, and T with appropriate units in the chosen system. For example from the dimensions of force as MLT^{-2}, the unit of force in SI system is $kg\,m\,s^{-2}$, in cgs system is $g\,cm\,s^{-2}$, and in FPS system is $slug\,ft\,s^{-2}$.

1.3.1 Dimensional Analysis and Applications

A straight forward application of dimensional formula is to obtain the units of a physical quantity in any system of units as has been demonstrated above. In addition, there are two important applications of dimensional analysis:
(i) To verify the validity of a mathematical equation, and
(ii) to derive a mathematical formula for a physical quantity in terms of the parameters on which it depends.
Both of these applications depend on the following principle of dimensions: *In a given mathematical equation or a formula, the dimensions of quantities to the left of the equal sign are equal to the dimensions of quantities on the right.*

(i) **To check the validity of a mathematical equation** one follows Recipe 1.2 as illustrated by Example 1.2.

Recipe 1.2:

- Write down the dimensions of the left side of the equation and simplify.

- Repeat the process for the right side of the equation.

- If the dimensions of both the sides are equal, the equation is dimensionally valid.

Limitation: From this method one can not verify the validity of multiplicative dimensionless constants in the equation. Explain why?

Example 1.2: Is the given equation: $s = k_1 u t + k_2 a t^2$ dimensionally correct, where k_1 and k_2 are dimensionless constants, and rest of the symbols have their usual meaning? What are the limitations of this method, and why?

Solution: The dimensions of the left side of the equation are:

$$s \Rightarrow L \qquad (1.1)$$

The dimensions of the right side of the equation are:

$$k_1 u t + k_2 a t^2 \Rightarrow 1(LT^{-1})T + 1(LT^{-2})T^2 = L + L = 2L \Rightarrow L \qquad (1.2)$$

Comparing equations (1.1) and (1.2) we note that the dimensions of both sides of the equation are the same, hence it is dimensionally correct.

The limitation of the method is that we cannot check the validity of the dimensionless constants in the equation. They can be given any value, and the equation shall still be dimensionally correct. The reason for this limitation is that the dimensions of all dimensionless constants are the same, *i.e.* 1.

Exercise 1.2: Check the dimensional correctness of the expression for the total energy (E) of an oscillating spring mass system given as: $E = a m v^2 + b k x^2$, where a and b are dimensionless constants, v is the velocity of the mass m, x is its displacement from the equilibrium position, and k is the force constant (force per unit extension or compression) of the spring. Can you determine the constants a and b by this method? Justify your answer.

(ii) **To derive a mathematical formula:** The method is given as Recipe 1.3 and illustrated by Example 1.3.

Recipe 1.3:

- Identify the parameters on which the required quantity depends.

- Express the formula as the product of the powers of the quantities identified, multiplied with a dimensionless constant.

- Write down the dimensions for both sides of the formula.

- By equating the dimensions on both sides, determine the unknown powers.

- Using the powers so determined, write down the final formula.

Limitation: From this method one can not determine the multiplicative dimensionless constant. Explain why.

Example 1.3: Find an expression for the time period of a simple pendulum using dimensional analysis.

Solution: A simple pendulum consists of a small mass, freely suspended with a long, massless, unstreachable string. When the mass is displaced by a small distance to one side from its equilibrium position and let go, it oscillates back and forth under the influence of the gravitational force. Hence, the only parameters on which the time period of the pendulum could possibly depend on are the mass m, the length of the pendulum l, and the acceleration due to gravity g. We can, therefore, write the expression for the time period of the oscillator as:

$$t = k\, m^a\, l^b\, g^c \quad \text{where k is a dimensionless constant} \tag{1.3}$$

Taking dimensions of the left and the right side of equation (1.3)

$$\text{Left side}: t \;\Rightarrow\; T = M^0\, L^0\, T^1 \tag{1.4}$$
$$\text{Right side}: k\, m^a\, l^b\, g^c \;\Rightarrow\; 1 M^a\, L^b\, (LT^{-2})^c = M^a\, L^{(b+c)}\, T^{-2c} \tag{1.5}$$

Equating the dimensions of the left and right side from Equations (1.4) and (1.5) respectively, one gets:

$$M^0\, L^0\, T^1 = M^a\, L^{(b+c)}\, T^{-2c} \tag{1.6}$$

Equation (1.6) gives:

$$
\begin{aligned}
a &= 0 \\
b + c &= 0 \\
(-)2c &= 1
\end{aligned}
\tag{1.7}
$$

Solving Equations (1.7) for $a, b,$ and c, gives $a = 0$, $b = \frac{1}{2}$, and $c = (-)\frac{1}{2}$. Substituting these in Equation (1.3), the expression for time period of a simple pendulum is found to be:

$$t = k\,\sqrt{\frac{l}{g}} \tag{1.8}$$

We can not determine the dimensionless constant k by this method, which from exact derivation is found to be 2π. Note that the time period of a simple pendulum does not depend on its mass.

Exercise 1.3: Velocity of transverse waves on a stretched string depends on the tension (force) in the string, mass of the string, and the length of the string. By method of dimensions find an expression for the velocity of the waves. What are the limitations of this method?

From the above discussion on the derivation of a mathematical formula by dimensional analysis, one is left with an impression that one can only obtain formulas of the form: $f = k\, X^a\, Y^b\, Z^c$ by this technique, and not the formulas involving linear combination of terms such as: $f = k_1\, W^a\, X^b\, Y^c + k_2\, X^d\, Y^e\, Z^g$. Formulas of this nature can be derived by dimensional analysis with a bit of ingenuity, and well-reasoned scientific logic. We illustrate this with the example of the formula for the displacement in uniformly accelerated motion.

Displacement of an object moving with uniform acceleration clearly depends on the initial velocity (u), the acceleration (a) and the duration of motion t. As a first step towards the derivation of the formula, one would be tempted to express the displacement as: $s = k\, u^b\, a^c\, t^d$. On careful consideration this can not be the correct form of the displacement because if the initial velocity is zero, the displacement shall be zero (or ∞ if the power b turns out to be negative) irrespective of the value of the acceleration. Similar reason applies if the acceleration is zero. Hence the initial velocity, u, and the acceleration a can not be included together within the same expression. However, since the displacement in any case shall depend on the duration of the motion, t, we must consider the combinations of initial velocity u and t, and acceleration a and t separately. Thus the possible expressions for the displacement are: $s_1 = k_1\, u^b\, t^c$, and $s_2 = k_2\, a^d\, t^e$. After going through dimensional analysis of both the expressions, one finds: $s_1 = k_1\, u\, t$, and $s_2 = k_2\, a\, t^2$. Since both are dimensionally correct expressions for the displacement, a linear combination of these shall also be a valid, and a more general expression for s. Therefore, the general expression for the displacement is: $s = s_1 + s_2 = k_1\, u\, t + k_2\, a\, t^2$.

Although we have already ruled out the combination of u and a as a possible expression for s, it is illustrative to persue this combination further. If we express: $s_3 = k_3\, u^g\, a^h$, dimensional analysis gives: $s_3 = k_3\, u^2\, a^{-1}$. Although, the expression for s_3 is dimensionally valid, it can not be a physically valid expression for many reasons amongst which the notable ones are: *(i)* If the velocity is constant, *i.e.* if $a = 0$, the displacement becomes ∞ which is nonsense. *(ii)* Displacement must depend on the duration of the motion. Lastly, one may imagine displacement to depend on the final velocity as given by the expression $v^2 = u^2 + 2\, a\, s$. But the final velocity is not an independent parameter because it depends on the initial velocity, acceleration and the duration of motion. Thus, the application of dimensional analysis requires more of logic and imagination than simply to be treated as a mathematical exercise.

From these applications of dimensional analysis one interestingly notes that the method has certain limitations. A question arises naturally: What is the advantage of dimensional analysis? Well, if one is working on a new concept to derive a mathematical expression, by using dimensional analysis one can either before hand determine the anticipated form of the expression, or one can check the validity of the derived expression by dimensional analysis. It can also be very helpful for students to decide on the correct form of a formula when one may be confused between the various possible forms under the stressful examination conditions.

1.4 Scalars and Vectors

Most physical quantities and measurables can be classified into two categories: *scalars* and *vectors*. The exceptions are some physical properties in elasticity, optics, electricity, magnetism, and mechanics that involve *tensors*, for example the stress tensor, polarizability tensor, electrical conductivity tensor, susceptibility tensor, and the moment of inertia tensor. In fact, a scalar is a tensor of rank zero, and a vector is a tensor of rank one. In this section we shall restrict ourselves only to scalars and vectors that are defined as follows.

Definition 1.5: *Scalars are the physical quantities that have magnitude only, (and no direction).*

Scalar quantities may or may not have units. Dimensionless quantities are all scalars. Examples of scalars are: mass, length, volume, distance, speed, work, energy, power, electrostatic potential, relative density and refractive index. The scalar quantities follow the simple algebraic rules, for example a 5 kg mass added to a 3 kg mass gives an 8 kg mass, and if a distance of 1.5 km is traversed 3 times is 4.5 km.

Definition 1.6: *Vector quantities have both magnitude and direction.*

Vector quantities must always be specified with units. The examples of vectors are displacement, velocity, acceleration, force, electric field, magnetic field. Specifying a vector without stating its direction not only conveys partial information, but also the partial information may not be useful for practical application or analysis. Under a cause-effect situation the effect of a vector quantity on the system depends on its direction. For example, when two forces of 4 N and 3 N are applied simultaneously to an object which is free to move, the combined effect of the forces on the motion of the object depends besides their magnitudes, on their directions as well. The vector quantities do not follow the simple algebraic rules, for example the resultant of adding two forces of 4 N and 3 N may have any value between 1 N and 7 N, the magnitude and direction of which depends on the relative direction of the two forces. The vector algebra is discussed in Section 1.5.

1.4.1 Representation of Vectors

Vector quantities, being distinctly different in terms of the algebraic rules they follow and their effect on the system with which they interact, must be clearly distinguished from the scalar quantities in their representation. This involves the use of symbols used to denote them and their graphical and analytical representations.

(A) Vector notation

In the printed text, vectors are denoted by bold face letters for example **F** for the force, **E** for the electric field, and **B** for the magnetic induction. A vector represented in normal type face (without bold) represents the magnitude of the vector quantity, for examples F, E and B represent the magnitudes of the corresponding vectors. In the hand written text (script) they are represented with an arrow sign above the symbol such as $\vec{F}$, $\vec{E}$ and $\vec{B}$. Individual styles may also be adopted, such as the use of a bar below ($\underline{F}$, $\underline{E}$ and $\underline{B}$), or above the symbol ($\bar{F}$, $\bar{E}$ and $\bar{B}$), so long as there is consistency.

(B) Vectors representation

The vector quantities can be represented in two ways: graphically or analytically.

(i) Graphical representation of vectors can only be used conveniently for vectors in a plane (in two dimensions). A vector is represented graphically by a straight line drawn to scale along the direction of the vector, with an arrow head at one end of the line pointing in the direction of the vector. The end of the line with the arrow head is known as the tip, while the other end is known as the tail. Figure 1.1 shows four force vectors, F_1, F_2, F_3, and F_4 graphically. This representation is used in diagrams, maps, graphs, sketches, and has limited use in analytical work.

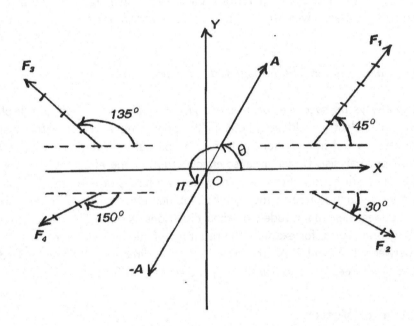

Figure 1.1: Graphical representation of vectors

The negative of a vector is a vector of the same magnitude, directed along the diametrically opposite direction. Figure 1.1 also shows a vector **A** and the negative vector - **A**.

(ii) Analytical representation of vectors: Analytically a vector is represented either in terms of its magnitude, and angular directions or in terms of its components in an orthogonal coordinates system. In order to specify a vector in one-, two and three- dimensions one requires one, two and three parameters respectively as discussed below.

(a) Vectors in one dimension: In one dimension a vector can have only two possible directions opposite to each other. The two directions are distinguished from each other by adopting a sign

convention such that the vector along one of the direction is represented as a positive quantity, and along the opposite direction it is given by a negative quantity, for example if a vector **A** = A in magnitude, then -**A** = -A. The acceleration due to gravity, which acts vertically downwards is generally taken to be negative while the velocity of an object thrown vertically upwards is taken to be positive. Its application shall be seen in the chapter on free motion under gravitational field, and the two-dimensional motion in a vertical plane, known as the projectile motion.

(b) Vectors in two dimensions: Two parameters are required to represent a vector in two dimensions. These parameters could either be the magnitude and its angular direction measured from a reference direction, or its two orthogonal components.

- While representing a vector in 2-D in terms of the magnitude and the direction, the direction is measured from the positive direction of the $x-$axis, such that the angles measured in the counter-clockwise direction are positive, and the angles measured in the clockwise direction are negative. For example, the four forces shown in Figure 1.1 can be expressed analytically as: $F_1 = (5N, \ 45^o), F_2 = (3.5N, \ -30^o), F_3 = (4N, \ 135^o)$, and $F_4 = (2.5N, \ 210^o) = (2.5N, \ -150^o)$. Sometimes in place of the angular directions geographical directions, North (N), South (S), East (E) and West (W), are used. In terms of the geographical directions the four forces in Figure 1.1 can be written as: $F_1 = (5N, \ NE), F_2 = (3.5N, \ 30^oSE), F_3 = (4N, \ NW)$, and $F_4 = (2.5N, \ 30^oSW) = (2.5N, \ 60^oWS)$.

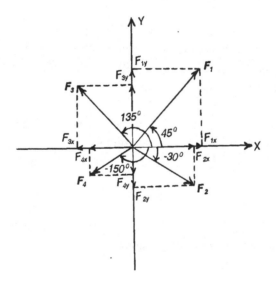

Figure 1.2: Components of vectors in two dimensions.

- To represent a vector in terms of the components, the components of the vector are given by its projections along two orthogonal axes in the plane of the vector. These axes are denoted as the $x-$ axis along the horizontal, and the $y-$ axis along the vertical. If A_x and A_y are the components of a vector **A** = (A, θ), then in terms of its components the vector is expressed as: **A** = $(A_x, \ A_y)$, where $A_x = A\cos \theta$ and $A_y = A\sin \theta$. Figure 1.2 shows the components of the four forces $\mathbf{F_1}$, $\mathbf{F_2}$, $\mathbf{F_3}$, and $\mathbf{F_4}$, and their values are given in Table 1.5. Students are assigned to verify these values.

Table 1.5: Components of the forces shown in Figure 1.1

$\mathbf{F} = (F, \theta)$	$F_x = F\cos\theta$	$F_y = F\sin\theta$	$\mathbf{F} = (F_x, F_y)$
$\mathbf{F_1} = (5N, 45°)$	$F_{1x} = 5\cos45 = 3.54N$	$F_{1y} = 5\sin45 = 3.54N$	$\mathbf{F_1} = (3.54N, 3.54N)$
$\mathbf{F_2} = (3.5N, -30°)$	$F_{2x} = 3.5\cos(-30) = 3.03N$	$F_{2y} = 3.5\sin(-30) = -1.75N$	$\mathbf{F_2} = (3.03N, -1.75N)$
$\mathbf{F_3} = (4N, 135°)$	$F_{3x} = 4\cos135 = -2.83N$	$F_{3y} = 4\sin135 = 2.83N$	$\mathbf{F_3} = (-2.83N, 2.83N)$
$\mathbf{F_4} = (2.5N, -150°)$	$F_{4x} = 2.5\cos(-150) = -2.17N$	$F_{4y} = 2.5\sin(-150) = -1.25N$	$\mathbf{F_4} = (-2.17N, -1.25N)$
$\mathbf{F_4} = (2.5N, 210°)$	$F_{4x} = 2.5\cos(210) = -2.17N$	$F_{4y} = 2.5\sin(210) = -1.25N$	$\mathbf{F_4} = (-2.17N, -1.25N)$

The components of force F_4 in Table 1.5 are obtained by using the two possible angles, measured in anticlockwise and clockwise directions respectively, and we note that the components remain the same. The components being vectors in one dimension, they can either be along the positive direction of the axis or in the negative direction. Accordingly the components are positive or negative. The signs of the components are automatically obtained correctly once the angular direction of the vector is used correctly.

Here we introduce unit vectors $\mathbf{i}$ and $\mathbf{j}$, *i.e.*, vectors of unit magnitude along the positive directions of x- and y- axes respectively. In terms of the unit vectors and the components, a vector $\mathbf{F}$ can be written as: $\mathbf{F} = (F_x, F_y) = \mathbf{F_x} + \mathbf{F_y} = \mathbf{i}\, F_x + \mathbf{j}\, F_y$.

In the above discussion we have obtained components of a vector from its magnitude and angular direction. One can also carry out the reverse transformation to obtain the magnitude and direction of the vectors from the given components as: $F = \sqrt{F_x^2 + F_y^2}$ and $\theta = \tan^{-1}(F_y/F_x)$, the verification is left as an exercise for the students.

The negative of a vector in two dimensions, is a vector directed diametrically opposite to the given vector. If we have vector $\mathbf{A} = (A, \theta) = (A_x, A_y)$ then the negative vector, $-\mathbf{A} = (A, (\pi + \theta)) = (-A_x, -A_y)$

Recipe 1.4:

- Given a vector $\mathbf{A} = (A, \theta) = (A_x, A_y)$, the conversion of the vector from one representation to another is done by using the following expressions.

$$A_x = A\cos\theta, \qquad A_y = A\sin\theta \qquad (1.9)$$

and

$$A = \sqrt{A_x^2 + A_y^2}, \qquad \theta = \tan^{-1}\left(\frac{A_y}{A_x}\right) \qquad (1.10)$$

- **NOTE:** The angles are measured from the positive x- axis as per the convention. If angles are measured from any other reference direction, say from the y- axis, then corrections to the recipe should be made accordingly. *This is one of the common mistakes made by students when dealing with this aspect of vectors.*

(c) Vectors in three dimensions: From what has been discussed about vectors in one and two dimensions, and obviously from the basic rule, one requires three independent parameters to specify a vector in three dimensions. These parameters could either be the magnitude of the vector and two independent angular directions, or the three orthogonal components as shown in Figure 1.3, and described below.

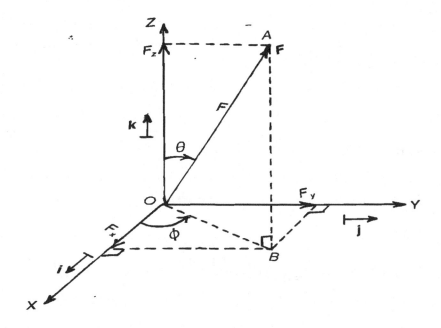

Figure 1.3: Representation of vectors in three dimensions.

- To specify a vector in terms of the magnitude and two angular directions, one follows the spherical-polar coordinate system, shown in Figure 1.3. Let **F** be a vector in three dimensions, then it can be expressed as: $\mathbf{F} = (F, \theta, \phi)$ where the various parameters are defined as:

  F is the magnitude of the vector. In the coordinate system shown in Figure 1.3, it is the length of the radius vector from the origin, drawn to scale along the direction of the vector. It can take any positive value from > 0 to ∞.

  θ is the azimuthal angle, measured from the positive z- axis as shown in the figure. It can take values from 0 to π

  The polar angle ϕ is the angle of the projection OB of the vector in the x - y- plane measured from the positive x- axis towards the positive y- axis. It can take values from 0 to 2π

  NOTE that there are no negative angles involved here, as the convention for the reference direction from which the angles are measured is strictly adhered to.

- In Cartesian coordinates system, the three orthogonal components of the vector are F_x, F_y and F_z along the x-, y-, and z- axes respectively, so that one can express the vector as: $\mathbf{F} = (F_x, F_y, F_z) = \mathbf{F_x} + \mathbf{F_y} + \mathbf{F_z} = \mathbf{i}\,F_x + \mathbf{j}\,F_y + \mathbf{k}\,F_z$, where $\mathbf{i, j, k}$ are the unit vectors along the x-, y-, z- axes respectively. The relationships between the two representations of vectors in three dimensions is given in Recipe 1.5, and its proof from Figure 1.3 is left as an exercise for students.

The negative of a vector $\mathbf{F} = (F,\ \theta,\ \phi) = (F_x,\ F_y,\ F_z)$ in three dimensions is given by: $-\mathbf{F} = (F,\ (\pi - \theta),\ (\pi + \phi)) = (-F_x,\ -F_y,\ -F_z)$

Recipe 1.5:

- Given a vector $\mathbf{F} = (F,\ \theta,\ \phi) = (F_x,\ F_y,\ F_z)$. The relationships between the two representations are given by the following equations.

$$F_x = F \sin\theta \cos\phi, \qquad F_y = F \sin\theta \sin\phi, \qquad F_z = F \cos\theta \qquad (1.11)$$

and

$$F = \sqrt{F_x^2 + F_y^2 + F_z^2}, \qquad \theta = \tan^{-1}\left(\frac{\sqrt{F_x^2 + F_y^2}}{F_z}\right) \qquad \phi = \tan^{-1}\left(\frac{F_y}{F_x}\right) \qquad (1.12)$$

One can also represent a vector in three dimensions in terms of its components in other orthogonal coordinates systems, for example in the spherical polar coordinates as: $\mathbf{F} = (F_r, F_\theta, F_\phi)$ or in circular cylindrical coordinates as: $\mathbf{F} = (F_\rho, F_\phi, F_z)$. These are outside the scope of this book.

From the above discussion on representation on vectors in two and three dimensions, one notes that a vector expressed in terms of its magnitude and angular direction(s) gives a *visual feel* of the actual magnitude and direction of the vector, whereas such a feel is not explicit when it is represented in terms of its components.

1.5   Vector Algebra

In this section we shall only consider the addition, subtraction and multiplication of vectors, whereby subtraction is just a special case of addition of the negative of the vector to be subtracted. *In vector algebra, there is no division of two vectors, that is, division of two vectors is not defined.*

1.5.1   Addition of Vectors

Needless to say that vector addition (subtraction) can be applied only to vectors representing the same physical quantity, such as the force vectors, or the displacement vectors, or the electric field vectors *etc.* Vectors addition (subtraction) can be carried out in two ways: Graphically and Analytically.

(A) Graphical addition of vectors

In this method one works with the magnitude and the angular direction of the vectors. The method can be used only for vectors lying in the same plane (in two dimensions), and not for vectors in three dimensions. This method of addition of vectors is covered in the school-physics course, and therefore, without going into its details, we give below the method as Recipe 1.6 to refresh students memory. Students are assigned to add the four forces given in Table 1.5 graphically to determine the resultant force as a revision exercise.

Recipe 1.6:

- Choose a largest possible and convenient scale to graphically represent the vectors. Choice of the largest possible scale has two implications: one, to fit the graphical drawing within the maximum area of the available space, secondly, to have the maximum accuracy of graphical measurements. A convenient scale shall be the one which is easily divisible by the given magnitudes of the vectors, for example a scale with 3 or 7 vector units to 1 scale unit may not be convenient, compared to 2, 4, or 5 vector units to 1 unit of scale.

- Draw to scale all the given vectors, tail to tip along the given directions, *i.e.* tail of the next vector rests at the tip of the previous vector and each vector points along the given direction. It does not matter in which order the vectors are plotted, *i.e* the first listed vector may be drawn the last and the last may be drawn first, or in any other order.

- Join the tail of the first drawn vector to the tip of the last drawn vector. This gives the resultant vector directed from the first tail to the last tip. Measure its magnitude and the direction.

Accuracy of results by this method depends on the choice of the scale, care with which the vectors are drawn and the resultant is measured, and is restricted by the limitations of the graphing instruments. The graphical addition of vectors is, therefore, not used for analytical work and one uses the analytical method of addition instead. However, with careful working it is possible to achieve an accuracy of $\pm 1\%$.

In order to subtract two vectors graphically, one first defines the negative of the vector to be subtracted, and then add it to the other vector in the usual manner, *i.e.*, $\mathbf{S} = \mathbf{A} - \mathbf{B} = \mathbf{A} + (\mathbf{-B})$. As an exercise, find $(\mathbf{F_1} - \mathbf{F_2} + \mathbf{F_3} - \mathbf{F_4})$ graphically for the vectors given in Table 1.5.

(B) Analytical addition of vectors

For analytical addition of vectors one works with the orthogonal components of the vectors. Here we shall use the components in the Cartesian coordinates system only. If we are to add four forces analytically given in Table 1.5 , *i.e.*, to determine the resultant vector: $\mathbf{R} = (\mathbf{F_1} + \mathbf{F_2} + \mathbf{F_3} + \mathbf{F_4})$, or any other set of *n* vectors in 3-D, the following recipe is followed.

Recipe 1.7:

- Determine the components of the vectors to be added.

- Determine the components of the resultant vector **R** algebraically (with positive and negative sign preserved) adding the corresponding components of the given vectors, *i.e.*, $R_x = \sum_{i=1}^{n} F_{ix}$, $R_y = \sum_{i=1}^{n} F_{iy}$ and $R_z = \sum_{i=1}^{n} F_{iz}$ for vectors in three dimensions.

- One may then express the answer either in terms of the component itself, or may determine the magnitude and angular directions of the resultant vector following Recipes 1.4 and 1.5 for vectors in two and three dimensions respectively.

If Subtraction of vectors is involved, one follows the same recipe except that one uses the negative of the vector to be subtracted. Determine $(\mathbf{F_1} + \mathbf{F_2} + \mathbf{F_3} + \mathbf{F_4})$ and $(\mathbf{F_1} - \mathbf{F_2} + \mathbf{F_3} - \mathbf{F_4})$ analytically for the forces given in Table 1.5, and compare the analytical results to the results obtained by the graphical method.

Example 1.3: A particle is subjected to three forces (in Newton) given as: $\mathbf{F_1} = 23\,\mathbf{i} + 30\,\mathbf{j} + 12\,\mathbf{k}$, $\mathbf{F_2} = -13\,\mathbf{i} + 15\,\mathbf{j} - 5\,\mathbf{k}$, and $\mathbf{F_3} = 15\,\mathbf{i} - 14\,\mathbf{j}$. Find the resultant force, and express it in terms of the magnitude and angular directions.

Solution: Let the resultant force: $\mathbf{F_R} = \mathbf{F_1} + \mathbf{F_2} + \mathbf{F_3} = F_{Rx}\,\mathbf{i} + F_{Ry}\,\mathbf{j} + F_{Rz}\,\mathbf{k}$, where the components F_{Rx}, F_{Ry}, and F_{Rz} of the resultant force are given by:

$F_{Rx} = F_{1x} + F_{2x} + F_{3x} = 23 + (\text{-}13) + 15 = 25\ (N)$
$F_{Ry} = F_{1y} + F_{2y} + F_{3y} = 30 + 15 + (\text{-}14) = 31\ (N)$
$F_{Rz} = F_{1z} + F_{2z} + F_{3z} = 12 + (\text{-}5) = 7\ (N)$, and the resultant force is:

$$\mathbf{F_R} = 25\,\mathbf{i} + 31\,\mathbf{j} + 7\,\mathbf{k}\ (N)$$

The magnitude, $F_R = \sqrt{25^2 + 31^2 + 7^2} = 40.4\ N$

The azimuthal angle, $\theta = \tan^{-1}\frac{\sqrt{25^2 + 31^2}}{7} = 80.0^o$, and

The polar angle, $\phi = \tan^{-1}\frac{25}{31} = 38.9^o$

Thus, $\mathbf{F_R} = (25N,\ 31N,\ 7N) = (40.4N,\ 80.0^o,\ 38.9^o)$

Exercise 1.3: From the forces given in Exercise 1.3, evaluate $\mathbf{F_1} - \mathbf{F_2} + 2 \times \mathbf{F_3}$. Express the result in terms of both, the components, and the magnitude and angular directions.

1.5.2   Relative Velocity and Addition of Velocities

The relative velocity and the addition of velocities are the direct application of addition and subtraction of vectors which are of practical importance. For example consider the following scenarios:

- Two cars *A* and *B* are driving along a straight highway in the same direction. Car *B* with larger (smaller) velocity is following car *A*. As seen by the driver of car *B*, car *A* would appear to approach (going away from) him. Likewise, a passenger in car *A* shall see car *B* approaching (going away from) him.

- A boat in a flowing river is being rowed with a

- certain velocity. It shall be pushed by water along the direction of the flow, and the path of the boat as seen by the boat man and by a person on the shore shall be different.

- If the wind is blowing, a weather balloon when released straight up from the ground shall ascend while at the same time shall be pushed along the direction of the wind.

All these examples demonstrate the principle of relative velocity, *i.e.*, the velocity of moving objects as seen by an observer from another moving or stationary frame of reference. The relative velocity can be found by subtraction (addition) of vectors. Since most often, only two vectors are involved in such exercises, the graphical representation of vectors together with analytical calculations is found to be much easier and a very illustrative approach.

Relative velocity of two vehicles driving along a straight high way: Let us consider two cars *A* and *B* driving along a straight highway along the positive *x*- axis with velocities $80 km h^{-1}$ and $60 km h^{-1}$ respectively. Car *B* is following car *A*. The following observations shall be made by observers in the two cars:

(i) Observer in car *B* shall see car *A* going away from him with a velocity of 20 $km\ h^{-1}$ along the positive *x*- axis.

(ii) Observer in car *A* shall see car *B* going away from him with a velocity of 20 $km\ h^{-1}$ along the negative *x*- axis.

If the two cars are driving towards each other, car *B* along the positive *x*- axis and car *A* along the negative *x*- axis, then the following observations shall be made:

(iii) Observer in car *B* shall see car *A* approaching him with a velocity of 140 $km\ h^{-1}$ along the negative *x*- axis.

(iv) Observer in car *A* shall see car *B* approaching him with a velocity of 140 $km\ h^{-1}$ along the positive *x*- axis.

All these observations can be expressed in vector notation as in equation (1.13), noting that vectors in one dimension follow a sign convention as discussed earlier.

$$\mathbf{V_{AB}} = \mathbf{V_A} - \mathbf{V_B}, \qquad and \qquad \mathbf{V_{BA}} = \mathbf{V_B} - \mathbf{V_A} \qquad (1.13)$$

where $\mathbf{V_{AB}}$ is the relative velocity of *A* as seen from *B*, and $\mathbf{V_{BA}}$ is the relative velocity of *B* as seen from *A*. Equation (1.13) is a general vector equation which applies to any general directions of the velocities that are not necessarily along the same straight line. Thus, the relative velocity of an object *1* as seen by an observer from a moving object *2* is obtained from the following vector subtraction expression:

$$\mathbf{V_{12}} = \mathbf{V_1} - \mathbf{V_2} \qquad (1.14)$$

The river-boat problem and the addition of velocities Let us consider a river flowing west to east with velocity V_w, and a (motor) boat is rowed (driven) with velocity V_b, which is the velocity of the boat in still water. In the flowing water as the boat is rowed in a particular direction, it shall be simultaneously pushed along the flow of the river with velocity V_w. As a consequence if the boat is to cross the river from south bank to reach an exactly opposite point on the north bank, it must be rowed in the north-westerly direction. Instead, if the boat is rowed directly towards the opposite point on the north bank, it shall arrive at a point down stream due to the flow of water. The resultant velocity V_r of the boat in the flowing water is the addition of the velocities of the boat and the river as follows:

$$\mathbf{V_r} = \mathbf{V_B} + \mathbf{V_w} \qquad (1.15)$$

Application of equation (1.15) is illustrated in Example 1.4.

Example 1.4: A *1 km* wide river flows west to east at a velocity of $3\ km\ h^{-1}$, and a boat is rowed at a speed of $4\ km\ h^{-1}$ in still water. *(a)* The boat crosses the river from the south bank, and arrives at an exactly opposite point on the north bank. Calculate *(i)* the direction in which the boat must be rowed, *(ii)* the resultant speed of the boat and its direction as seen by an observer on the south bank of the river, and *(iii)* time taken by the boat to cross the river. *(b)* If the boat is rowed towards the opposite point on the north bank, calculate: *(i)* the resultant speed and direction of motion of the boat as seen by an observer on the south bank, *(ii)* time taken by the boat to cross the river, and *(iii)* how far and in which direction from the exactly opposite intended destination shall the boat dock at the north bank. *(c)* Compare and comment of the results in parts *(a)* and *(b)*.

Solution: *(a)* For the boat to reach the opposite point on the north bank, it should be rowed in the north-westerly direction at an angle θ_1, so that when the boat's velocity is added to the velocity of the water, the resultant velocity is directed straight northwards. The addition of velocities is shown graphically in Figure 1.4 (b), where $AB = V_b = 4\ km\ h^{-1}$, and $BC = V_w = 3\ km\ h^{-1}$.

(i) From the figure, the angle θ_1 along which the boat is rowed $= \sin^{-1}\frac{3}{4} = 48.6^o$

(ii) Resultant velocity of the boat as seen from the south bank $= AC = V_{r1} = \sqrt{4^2 - 3^2} = 2.65\ km\ h^{-1}$ and it is due north.

(iii) Time taken by the boat to cross the river $= t_1 = \frac{1\ km}{2.65\ km\ h^{-1}} = 0.38\ hrs. = 22.68\ minutes$. One can also calculate the time by calculating the effective distance covered by the boat along direction AB at a speed of $4\ km\ h^{-1}$, which shall turn out to be the same. It is left as an exercise for students.

(b) As shown in Figure 1.4 (c) if the boat is rowed directly northwards at a speed of $4\ km\ h^{-1}$ along DE, water shall carry it along EF with a speed of $3\ km\ h^{-1}$.

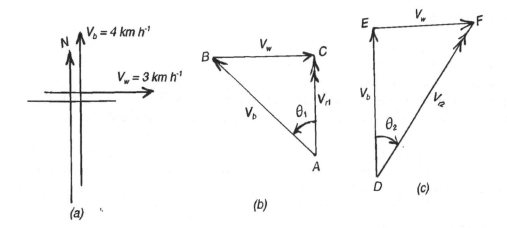

Figure 1.4: Addition of velocities for the river-boat problem.

(i) The resultant velocity V_{r2} is then DE at an angle θ_2. From the figure: $V_{r2} = \sqrt{4^2 + 3^2} = 5 \ km \ h^{-1}$ and the direction $\theta_2 = \tan^{-1}\frac{3}{4} = 36.87^\circ$.

(ii) The boat is effectively traveling due north at a speed of $4 \ km \ h^{-1}$. Hence the time taken by the boat to cross the river $= t_2 = \frac{1 \ km}{4 \ km \ h^{-1}} = 0.25 \ hour = 15 \ minutes$. Students are assigned to verify this time from the actual distance traveled by the boat along DF at a speed of $5 \ km \ h^{-1}$

(iii) Distance of the boat from the intended destination due north = distance pushed by the flow of water during the river crossing $= 0.25 \ hour \times 3 \ km \ h^{-1} = 0.75 \ km$ down stream (East).

(c) Crossing of the river in two cases is not symmetrical as one might have casually expected. Neither the time taken to cross the river nor the direction of the boat are the same in two cases. This can be explained in terms of how the velocities of the boat and water are added in the two cases as shown in Figure 1.4 (b) and (c). The effective speed of the boat due north is smaller than the true speed of the boat in the first case and hence the time taken is larger, whereas in the second case the speed of the boat due north remains the same as its speed in still water, and consequently it takes less time. Only thing is it is pushed along the water flow as it is rowed to the opposite bank.

Exercise 1.4 A weather balloon rises vertically up at a speed of $5 \ m \ s^{-1}$ in still air. Easterly wind is blowing at $3 \ m \ s^{-1}$ relative to earth's surface. Assuming these velocities to be uniform with altitude, calculate the following: *Part (a) (i)* In which direction relative to earth's surface should the balloon be released so that it shall rise vertically up. *(ii)* What is the resultant velocity of rise of the balloon, and how long will it take to rise to a height of $5 \ km$. *Part (b)* If the balloon is released vertically, calculate: *(i)* Direction and velocity of the rise of the balloon, *(ii)* time it takes to rise to a 5 km height, and *(iii)* its horizontal distance and direction from the point on ground from where it was launched.

1.5.3   Vector Products

Before we venture into the product of two vectors, let us first consider the trivial case of the product of a scalar s and a vector **V**, *i.e.*, s**V**. The result is a vector along the same direction as vector **V**, or opposite to it if the scalar quantity is negative, and magnitude is s times the magnitude of the vector. An application of this is the Newton's second law of motion: **F** $= m$ **a**, where m is the mass, **a** is the acceleration, and **F** is the force along the same direction as the acceleration.

When two vectors are multiplied, their product is defined in two ways, which are distinctly different from each other in terms of their results and applications. These two methods of vector multiplication are known as the:

- Dot product or the Scalar product.

- Cross product or the Vector product.

Dot product or the Scalar product

Let **A** and **B** be two vectors, and let θ be the angle between their directions. NOTE THAT two vectors always lie in a plane with an angle between them. Their dot product or the scalar product is expressed as **A.B**, (read as:A *dot* B). The result of the multiplication is a scalar quantity as the name suggest, and hence has magnitude only and not the direction. The magnitude is given by:

$$\mathbf{A.B} = A\,B\,\cos\theta \tag{1.16}$$

where A and B are the magnitudes of the vectors, and the magnitude of the dot product could be either positive or negative depending on the sign of $cos\theta$. If one interchanged the order of the two vectors, *i.e.* considered the dot product **B.A**, then from equation (1.16) it is the same as **A.B**. Thus *the dot product of two vectors is commutative.*

The dot products of the unit Cartesian vectors can be readily obtained from equation (1.16) as:

$$\begin{aligned}\mathbf{i.i} &= i\,i\cos(0) = 1 = \mathbf{j.j} = \mathbf{k.k}, \quad \text{and}\\ \mathbf{i.j} &= i\,j\cos(90^\circ) = 0 = \mathbf{j.k} = \mathbf{k.i}\end{aligned} \tag{1.17}$$

Using equation (1.17) the scalar product can be expressed in terms of the components of the vectors as:

$$\mathbf{A.B} = A_x\,B_x + A_y\,B_y + A_z\,B_z \tag{1.18}$$

where the components in equation (1.18) are employed algebraically, *i.e.*, along with their signs. Combining equations (1.16) and (1.18) one obtains the angle between the two vectors as:

$$\theta = \cos^{-1}\left(\frac{A_x\,B_x + A_y\,B_y + A_z\,B_z}{A\,B}\right). \tag{1.19}$$

Equation (1.19) gives two possible values of the angle θ. The scalar product, therefore, alone can not be used to determine the angle between two vectors. One must also find another set of two possible values of the angle from the cross product (discussed later), and the angle which is common to both the sets is the correct angle between the vectors. This is illustrated by the example following the discussion of the cross product.

Physical significance of dot product and its application in Physics: Figure 1.5 shows two vectors **A** and **B** and their components whose dot product is given by equation (1.16). The components of vector **A** parallel and perpendicular to vector **B** are $A\cos\theta$ and $A\sin\theta$ respectively. Likewise, the components of vector **B** parallel and perpendicular to vector **A** are $B\cos\theta$ and $B\sin\theta$ respectively. From equation (1.16) the dot product of the two vectors can be expressed as:

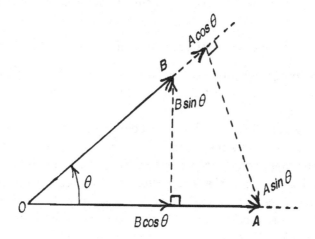

Figure 1.5: Two vectors in a plane and their components parallel and perpendiculat to each other.

$$
\begin{aligned}
\mathbf{A}.\mathbf{B} &= A\,B\cos\theta = A\,(B\cos\theta) \\
&= \text{(Magnitude of vector A)} \times \text{(Component of vector B parallel to A)} \\
&= B\,(A\cos\theta) \\
&= \text{(Magnitude of vector B)} \times \text{(Component of vector A parallel to B)} \tag{1.20}
\end{aligned}
$$

Thus from equation (1.20) one can state that the dot product of two vectors is equal to: *(Magnitude of one of the vector)* × *(Component of the other vector parallel to the first one)*.

A well known application of the dot product of two vectors is the work done by force **F** during displacement **s**. Both the quantities involved in work done are vectors, whereas the work itself is a scalar quantity. Furthermore, work is defined as the product of the force and displacement parallel to the direction of the force. Hence one can express the work done *(W)* in vector notation as:

$$W = \mathbf{F}.\mathbf{s} = \mathbf{s}.\mathbf{F} \tag{1.21}$$

Another example is the instantaneous power (P) delivered by a force $\mathbf{F}$ to an object moving with velocity $\mathbf{v}$ given by a dot product as: $P = \mathbf{F} . \mathbf{v} = \mathbf{v} . \mathbf{F}$.

Cross product or the Vector product

Cross product also known as the vector product of two vectors **A** and **B** shown in Figure 1.5 is denoted as $\mathbf{A} \times \mathbf{B}$ (read as: *A cross B*). As the alternate name suggests, the product is a vector quantity for which both magnitude and direction must be determined. The magnitude is given as:

$$
\begin{aligned}
\mathbf{A} \times \mathbf{B} \;=\;& A\,B \sin\theta \\
=\;& A\,(B \sin\theta) \\
=\;& \text{(Magnitude of vector A)} \times \text{(Component of vector B perpendicular to A)} \\
=\;& B\,(A \sin\theta) \\
=\;& \text{(Magnitude of vector B)} \times \text{(Component of vector A perpendicular to B)} \tag{1.22}
\end{aligned}
$$

Thus from equation (1.22), the magnitude of the cross product of two vectors is equal to the *(Magnitude of one of the vector)* × *(Component of the other vector perpendicular to the first one)*.

The direction of the product is perpendicular to the plane containing the two vectors (perpendicular to both the vectors), and is determined by the *right hand screw rule*. We designate the vector to the left of the " × " sign as the first vector and the one to the right as the second vector, and the right hand screw rule is stated as : *"Turn the first vector into the second vector with your right hand through the obtuse angle between them as if turning a screw. The direction in which the screw advances is the direction of the cross product of the two vectors"*. Applying the rule to the vectors in Figure 1.5, the cross product $\mathbf{A} \times \mathbf{B}$ is along the perpendicular directed out of the plane of the paper, and the cross product $\mathbf{B} \times \mathbf{A}$ is directed into the plane of the paper. Thus the cross products of two vectors does not commute, and:

$$\mathbf{A} \times \mathbf{B} = -\mathbf{B} \times \mathbf{A} \tag{1.23}$$

From equation (1.22), and using the right hand screw rule, the cross products of unit Cartesian vectors are:

$$
\begin{aligned}
\mathbf{i} \times \mathbf{i} \;=\;& i\,i \sin(0) = 0 = \mathbf{j} \times \mathbf{j} = \mathbf{k} \times \mathbf{k}, \\
\mathbf{i} \times \mathbf{j} \;=\;& \mathbf{k} = -\mathbf{j} \times \mathbf{i}, \\
\mathbf{j} \times \mathbf{k} \;=\;& \mathbf{i} = -\mathbf{k} \times \mathbf{j}, \quad \text{and} \\
\mathbf{k} \times \mathbf{i} \;=\;& \mathbf{j} = -\mathbf{i} \times \mathbf{k} \tag{1.24}
\end{aligned}
$$

Thus the cross products of unit vectors, **i, j, k** follow the cyclic order, *i.e.*, $\mathbf{i} \times \mathbf{j} = \mathbf{k}$, and the cross product in the reverse order results in a negative sign,*i.e.*, $\mathbf{j} \times \mathbf{i} = -\mathbf{k}$.

One may use the cross products of the unit vectors, to determine the cross product of two vectors in terms of their Cartesian components by taking the cross product of each component of the first vector with each component of the second vector as follows:

$$
\begin{aligned}
\mathbf{A} \times \mathbf{B} &= (\mathbf{i}A_x + \mathbf{j}A_y + \mathbf{k}A_z) \times (\mathbf{i}B_x + \mathbf{j}B_y + \mathbf{k}B_z) \\
&= (\mathbf{i} \times \mathbf{i}A_xB_x + \mathbf{i} \times \mathbf{j}A_xB_y + \mathbf{i} \times \mathbf{k}A_xB_z) + \dots \\
&= \mathbf{i}(A_yB_z - A_zB_y) + \mathbf{j}(A_zB_x - A_xB_z) + \mathbf{k}(A_xB_y - A_yB_x)
\end{aligned}
\tag{1.25}
$$

A proof of equation (1.25) by filling in the gaps is left as an exercise for students. Equation (1.25) for the cross product in terms of components can be obtained more easily by expressing the products in the determinant form as follows:

$$
\begin{aligned}
\mathbf{A} \times \mathbf{B} &= \begin{vmatrix} \mathbf{i} & \mathbf{j} & \mathbf{k} \\ A_x & A_y & A_z \\ B_x & B_y & B_z \end{vmatrix} \\
&= \mathbf{i}(A_yB_z - A_zB_y) + \mathbf{j}(A_zB_x - A_xB_z) + \mathbf{k}(A_xB_y - A_yB_x)
\end{aligned}
\tag{1.26}
$$

Equation (1.25) is a complete specification of the cross prooduct in terms of the components of the product, from which one can obtain its magnitude and angular directions using Recipe 1.5. The magnitude is of particular interest if one needs to determine the angle between the two vectors **A** and **B** as we have done using the dot product (Equation 1.19). The magnitude of the cross product from equation (1.25) is:

$$
|\mathbf{A} \times \mathbf{B}| = \sqrt{(A_yB_z - A_zB_y)^2 + (A_zB_x - A_xB_z)^2 + (A_xB_y - A_yB_x)^2}
\tag{1.27}
$$

Comparing the magnitude of the cross product from equations (1.22) and (1.27), the angle θ between vectors **A** and **B** is given as:

$$
\theta = \sin^{-1}\left(\frac{\sqrt{(A_yB_z - A_zB_y)^2 + (A_zB_x - A_xB_z)^2 + (A_xB_y - A_yB_x)^2}}{AB} \right)
\tag{1.28}
$$

Equation (1.28), like equation (1.19) gives two values of the angle θ. The common angle obtained from equations (1.19) and equation (1.28) is the correct angle between the two vectors.

Applications of the cross product of vectors in Physics are much more numerous and common compared to the applications of the dot product. Some examples are: *(i)* torque $\tau = \mathbf{d} \times \mathbf{F}$, where **d** is the displacement vector and **F** is the force, *(ii)* angular momentum $\mathbf{L} = \mathbf{r} \times \mathbf{P}$, where **r** is the radius vector and **P** is the linear momentum, *(iii)* force on a charge q moving with velocity **v** in a magnetic field **B** is $\mathbf{F} = q\mathbf{v} \times \mathbf{B}$, and *(iv)* force on a current (**I**) carrying conductor of length L in a magnetic field **B** is $\mathbf{F} = \mathbf{I}L \times \mathbf{B}$. NOTE that in all applications of cross product, the order in which the constituent vectors are multiplied is important to obtain the correct direction of the resultant quantity because the cross product of vectors does not commute.

Example 1.5: Given two vectors: $\mathbf{A} = 23\,\mathbf{i} + 30\,\mathbf{j} + 12\,\mathbf{k}$, and $\mathbf{B} = -13\,\mathbf{i} + 15\,\mathbf{j} - 5\,\mathbf{k}$. Determine the angle between them.

Solution: *(i)* Calculate the magnitude of both the vectors using equation (1.12).

$$
\begin{aligned}
A &= \sqrt{23^2 + 30^2 + 12^2} = 39.66, \quad and \\
B &= \sqrt{(-13)^2 + 15^2 + (-5)^2} = 20.47
\end{aligned}
$$

(ii) Calculate the dot product and the angle between the vectors using equations (1.16), (1.18) and (1.19).

$$
\begin{aligned}
\mathbf{A.B} &= 39.66 \times 20.47 \cos\theta = 811.84 \cos\theta \\
&= 23 \times (-13) + 30 \times 15 + 12 \times (-5) = 91, \quad and \\
\theta &= \cos^{-1}\frac{91}{811.84} = 83.6^o \quad or \quad -83.6^o
\end{aligned}
$$

(iii) Calculate the cross product and the angle between the vectors using equations (1.22), (1.27) and (1.28).

$$
\begin{aligned}
\mathbf{A} \times \mathbf{B} &= 39.66 \times 20.47 \sin\theta = 811.84 \sin\theta \\
&= \sqrt{(30 \times (-5) - 12 \times 15)^2 + (12 \times (-13) - 23 \times (-5))^2 + (23 \times 15 - 30 \times (-13))^2}, \\
&= 806.73 \quad and \\
\theta &= \sin^{-1}\frac{806.73}{811.84} = 83.6^o \quad or \quad 96.4^o
\end{aligned}
$$

The angle common to both the calculations is 83.6^o, and hence is the angle between the given vectors.

Exercise 1.5: Given two forces: $\mathbf{F_1} = 3\mathbf{i} - 2\mathbf{j} + 4\mathbf{k}$, and $\mathbf{F_2} = 2\mathbf{i} - 3\mathbf{j} - 2\mathbf{k}$. Determine the angle between the forces.

1.6   Problems

1.1. Convert the following units: (i) density of gold from $19.3 \times 10^3\ kg\ m^{-3}$ to $g\ cc^{-1}$, (ii) Dimensions and area of a plot of land from $20\ m \times 30\ m$ to units in the British system, (iii) 1 US gallon to liters (1 US gallon $= 231\ (inch)^3$), (iv) acceleration due to gravity $= 9.8\ ms^{-2}$ to fts^{-2} and kmh^{-2}, and (v) 1 kWh of energy to Joules.

1.2. Express the following numbers using the nearest prefix for the power of ten: $10^{-14}\ m, 5 \times 10^{11}\ s$, $1.66 \times 10^{-27}\ kg$, $2300\ kWh$, $6.022 \times 10^{26}\ (kmol)^{-1}$, and $1.6 \times 10^{13}\ s$

1.3. Verify which ones of the following equations are dimensionally correct and which ones are not. All the symbols used have their usual meaning.

$$(i) \quad \mathsf{E} = mc^2$$
$$(ii) \quad \mathsf{F} = \frac{m\,v^2}{R}$$
$$(iii) \quad \mathsf{KE} = m\,r^2\,\omega^2$$
$$(iv) \quad (\text{Work done})_{friction} = \mu\,R\,d$$
$$(v) \quad \omega^2 = \omega_o^2 + 2\sqrt{\alpha\,\theta}$$

1.4. The force of fluid friction on a sphere moving through a viscous fluid depends on the radius and velocity of the sphere, and the coefficient of viscosity η of the fluid. The SI units of η are $Nm^{-2}s$. Derive an expression for the viscous force on the sphere in terms of the given quantities by method of dimensional analysis.

1.5. The centripetal acceleration of an object moving along a circular trajectory depends on the velocity of the object and the radius of the trajectory. By the method of dimensions, verify if the acceleration also depends on the mass of the object or not, and derive an expression for the acceleration.

1.6. The speed of an object as a function of time is given as: $v = At^3 + Bt^2 - C\,a\,t$, where a is the acceleration. Determine the dimensions and SI units of constants A, B, and C

1.7. Four co-planar forces, $\mathbf{F_1} = (150N, 37^o)$, $\mathbf{F_2} = (70N, 120^o)$, $\mathbf{F_3} = (110N, -135^o)$, and $\mathbf{F_4} = (50N, 300^o)$ are applied to an object simultaneously. Represent each of the individual forces graphically in a single diagram. Determine the resultant force on the object graphically.

1.8. Find the resultant of the forces in Problem 1.7 analytically, and compare the result to the result obtained graphically. A fifth force $\mathbf{F_5}$ is applied to keep the object in equilibrium (net force on the object is zero). Determine the force $\mathbf{F_5}$.

1.9. Determine the resultant force $(\mathbf{F_1} - \mathbf{F_2} + \mathbf{F_3} - \mathbf{F_4})$ for the forces given in Problem 1.7 graphically and analytically.

1.10. A box is pulled along a horizontal surface by applying three forces in a horizontal plane. Force $\mathbf{F_1} = 100N$ is applied along the x- axis, force $\mathbf{F_2} = 80N$ is applied at 150° from the x- axis. In which direction should a third force, $\mathbf{F_3} = 60N$ be applied so that the box moves along the y- axis?

1.11. An object under goes four successive displacements in the 3-D space given in terms of their components or in spherical polar coordinates as: $\mathbf{d_1} = (2m, -5m, 7m)$, $\mathbf{d_2} = (20m, 60^o, 110^o)$, $\mathbf{d_3} = (-6m, 8m, -11m)$, and $\mathbf{d_4} = (15m, 150^o, 75^o)$. Express the displacements in representations other than those given here. Find the resultant displacement, and express it in terms of both, the components and the spherical polar coordinates.

1.12. For the displacements given in Problem 1.11, determine the resultant of displacements $(-\mathbf{d_1} + \mathbf{d_2} - \mathbf{d_3} + \mathbf{d_4})$.

1.13 Solve Example 1.4 for the case when the boat crosses the river in the opposite direction, *i.e.*, from the north bank to the south keeping the values of all the given parameters unchanged.

1.14 A parachuter is released from an altitude of 3 km and descends vertically at a constant speed of 5 m s^{-1} if there in no wind. Consider that there is wind blowing north to south at 3 m s^{-1}. Assuming these velocities to be uniform with altitude, calculate the following: *Part (a) (i)* In which direction should the parachuter steer himself so that he descends vertically down. *(ii)* what is the resultant velocity of descent, and how long will it take for the person to touch ground. *Part (b)* If the person does not correct for the wind direction and continues to descend freely, calculate: *(iii)* Direction and velocity of the descent of the person, *(iv)* time it takes to touch ground, and *(v)* horizontal distance and its direction along which the person drifts due to the wind on lending.

1.15. A car A is traveling north at 80 km h^{-1} and another car B behind it is also traveling north at 120 km h^{-1}. Calculate the relative velocity of each car with respect to the other, (i) before B passes A, and (ii) after B passes A.

1.16 A helicopter is flying due east at 300 km h^{-1} with respect to the still air, and wind is blowing it due north at 80 km h^{-1}. What is the resultant velocity of the helicopter? If the pilot wants to go due east, in which direction should he fly, and what shall be the resultant velocity in that case? What shall be the displacement vector of the helicopter with respect to earth after 3 hours of flying in both cases?

1.17. Determine the angles between the forces, $(\mathbf{F_1}, \mathbf{F_2})$, $(\mathbf{F_1}, \mathbf{F_3})$, $(\mathbf{F_1}, \mathbf{F_4})$, $(\mathbf{F_2}, \mathbf{F_3})$, $(\mathbf{F_2}, \mathbf{F_4})$ and $(\mathbf{F_3}, \mathbf{F_4})$ given in Problem 1.7.

1.18. Determine the angles between the displacements, $(\mathbf{d_1}, \mathbf{d_2})$, $(\mathbf{d_1}, \mathbf{d_3})$, $(\mathbf{d_1}, \mathbf{d_4})$, $(\mathbf{d_2}, \mathbf{d_3})$, $(\mathbf{d_2}, \mathbf{d_4})$ and $(\mathbf{d_3}, \mathbf{d_4})$ given in Problem 1.11

Chapter 2

Kinematics

Kinematics is a branch of mechanics dealing with the study of motion in which the only parameters considered are displacement, velocity and acceleration. Other parameteres such as forces, masses, momentum etc *are not* considered in kinematics, and are considered in another branch of mechanics known as dynamics. In this chapter we shall apply kinematics to study

- One Dimensional ($1d$) Motion, and

- Two Dimensional ($2d$) Motion

of bodies.

2.1   Definitions

Before we analyze the motion of objects, it is pertinent that we carefully understand and define the kinematic quantities, namely, displacement and distance, velocity and speed, and acceleration.

2.1.1   Displacement and distance

Displacement is defined as the vector separation between two points along the straight line joining the points. If r_i is the position vector of an initial position of an object, and r_f is the position vector of the final position of the object, then the displacement of the object in going from the initial position to the final position is given as:

$$\mathbf{r} = \mathbf{r_f} - \mathbf{r_i} \tag{2.1}$$

The displacement being a vector can be expressed in terms of components in the chosen set of coordinates axes.

Distance is a scalar defined as the physical separation between two points, or the physical length covered by a moving object which may not be along a straight line. As a scalar, distance can not be expressed in terms of components. However, under certain situations distance can be calculated from the components of displacements. For examples distance between two fixed points is equal in

31

magnitude to the displacement between the points, and one can calculate it from the components of the displacement. For a moving object the distance traveled by the object depends on the trajectory of the motion and not on the initial and final position of the object. In that case the distance is not necessary equal in magnitude to the displacement. This is illustrated by the following examples:

- A student leaves his seat in the class room to get a book from the library which is $500\,m$ away and after some time returns to his seat in the room. The displacement of the students over the period of his absence from the class is zero, but the distance moved by him/her during the same time is $1\,km$.

- A player kicks a football on the ground that rises up and lands into the goal post $100\,m$ away. The displacement of the ball is $100\,m$ along the horizontal, $0\,m$ along the vertical, and the distance traveled by the ball is equal to the length of the trajectory along which it moves as it rises up and then falls to the ground.

- A stone is thrown vertically upwards from a $50\,m$ high cliff. The stone rises to a certain maximum height above the cliff, and then it falls to the ground below. The final displacement of the stone is $50\,m$ vertically downwards, but the distance moved by it is equal to twice the maximum height to which it rises above the cliff plus the height of the cliff which is $50\ m$

SI units of both displacement and distance are m.

2.1.2   Velocity and speed

Velocity is defined as the rate of change of displacement with time, and it is a vector quantity for which the direction must be specified along with the magnitude. If $\Delta\mathbf{r}$ is the displacement of a body in time interval Δt, then the velocity $\mathbf{v}$ is given as :

$$\mathbf{v} = \frac{\Delta\mathbf{r}}{\Delta t} \qquad (2.2)$$

The direction of velocity is along the direction of the displacement. Speed is the rate of change of distance with time, and it is a scalar quantity. If Δr is the distance moved by a body in time interval Δt, then the speed v is given as:

$$v = \frac{\Delta r}{\Delta t} \qquad (2.3)$$

The SI units of both velocity and speed are $m\,s^{-1}$.

Average and instantaneous velocities

Average velocity is the velocity calculated over an extended period of time, whereas instantaneous velocity is the velocity at a particular moment in time. For example, imagine you are driving from Gaborone to Francistown which is $450\,km$ away in the north-east direction, and you arrive there in 4.5 hours. Then your average velocity during the trip is $100\,km\,h^{-1}$ in the north-east direction. However, we know from experience that one does not maintain the velocity throughout the trip. The instantaneous velocity is the velocity that shows at the odometer of the car at a given instant. In order to determine the instantaneous velocity one must measure the displacement over an exceeding small interval of time, mathematically approaching to zero, and then calculate it from equation (2.2). Thus, the instantaneous velocity is defined as:

$$\mathbf{v} = \lim_{\Delta t \to 0} \frac{\Delta \mathbf{r}}{\Delta t} = \frac{d\mathbf{r}}{dt} \qquad (2.4)$$

Graphically, instantaneous velocity is the slope of the displacement-time graph. In the same manner one can define average and instantaneous speeds. The instantaneous speed and velocity are equal in magnitude, but the average speed and average velocity are not necessarily equal.

2.1.3 Acceleration

Acceleration is defined as the rate of change of velocity with time. If $\Delta \mathbf{v}$ is the change in velocity of a body in time interval Δt, then the acceleration $\mathbf{a}$ is given as:

$$\mathbf{a} = \frac{\Delta \mathbf{v}}{\Delta t} \qquad (2.5)$$

As the velocity of a moving object may either increase or decrease with time, correspondingly the acceleration takes positive or negative values. The negative acceleration is termed as deceleration. The SI unit of acceleration is $m\,s^{-2}$.

Acceleration is a vector quantity. The direction of acceleration is along the direction of the change in velocity, which is not necessarily the same as the direction of the velocity itself. In Chapter 3 on Newton's laws of motion, we shall learn that acceleration is caused by an external force, a vector quantity, applied to a moving object, and the direction of the acceleration is along the direction of the applied force. In the case of displacement and velocity vectors we defined their scalar counterparts distance and speed. But acceleration does not have a scalar counterpart, although some time one treats the magnitude of the acceleration as a scalar. This important feature of acceleration arises firstly because it is caused by a vector force, and secondly the velocity being a vector, a change in velocity can be affected in three different ways as follows:

(i) By changing the magnitude of the velocity while keeping its direction constant, for example when a car is accelerated or decelerated along a straight road. In this case the acceleration (deceleration) is parallel to the direction of the velocity. This chapter deals with this aspect of acceleration.

(ii) By changing the direction of the velocity while keeping the magnitude of the velocity constant, for example an object moving in a circular path with constant speed. The velocity of an object moving along a curved trajectory is tangential to the path, hence the change in the direction of the velocity. In this case the acceleration is perpendicular to the direction of the velocity. This aspect of acceleration is discussed in depth in Chapter 6 on circular motion.

(iii) By changing both the magnitude and the direction, for example motion of an abject along a general curved path with a changing speed. In this case the acceleration shall have a general direction which is neither parallel nor perpendicular to the velocity, but is along the direction of the force.

Lastly, one may define average and instantaneous acceleration in the same way as we discussed for the speed and velocity. The instantaneous acceleration following equation (2.4) is given as:

$$\mathbf{a} = \lim_{\Delta t \to 0} \frac{\Delta \mathbf{v}}{\Delta t} = \frac{d\mathbf{v}}{dt} \tag{2.6}$$

and it is the slope of the velocity-time graph.

2.2 One Dimensional ($1d$) Motion

2.2.1 Displacement, Velocity, Acceleration and the Kinematic Equations

In this section we shall consider *linear motion* which consists of $1d$ motion in a straight line. We shall derive the equations of motion under uniform acceleration. There are three important quantities in kinematics: displacement, velocity and acceleration, as defined below.

Displacement is a position vector which specifies the distance of a body relative to some selected point.

Consider motion in $1d$ such that a body has an initial position $x_i\mathbf{i}$ and a final position $x_f\mathbf{i}$. The distance Δx and displacement $\Delta\mathbf{x}$ are given by

$$\begin{aligned} \Delta x &= x_f - x_i \\ \Delta\mathbf{x} &= x_f\mathbf{i} - x_i\mathbf{i} \end{aligned}$$

where the *distance* is a *scalar quantity* and the *displacement* is a *vector quantity*.

Velocity is a vector that specifies how fast a body is moving in a given direction.

Consider a linear motion such that initially at time t_i a body is at some initial position $x_i\mathbf{i}$ moving with a velocity $\mathbf{u}$, and at time t_f later the body is at a position $x_f\mathbf{i}$ with a velocity $\mathbf{v}$. The average velocity $\bar{v}$ is defined as:

$$\bar{v} = \frac{u + v}{2} \tag{2.7}$$

or

$$\begin{aligned} \bar{v} &= \frac{\Delta x}{\Delta t} \\ &= \frac{(x_f - x_i)}{t_f - t_i} \\ &= \frac{s}{t} \end{aligned} \tag{2.8}$$

where $s = x_f - x_i$ is the distance moved, $\Delta t = t_f - t_i$ is the time interval, and if $t_i = 0$, $t_f = t$, then $t_f - t_i = t$.

From equation (2.7) and (2.8), the average velocity is given as

$$\bar{v} = \frac{s}{t} = \frac{u+v}{2} \tag{2.9}$$

from which we obtain

$$s = \frac{1}{2}(u+v)t \tag{2.10}$$

Acceleration is a vector that specifies how fast the velocity of a body is changing with time.

Consider a linear motion such that initially at time t_i a body has a velocity $\mathbf{v}_i$, and at time t_f the body has a velocity $\mathbf{v}_f$. The *average acceleration* $\mathbf{a}$ is defined as the ratio of change of velocity, $\Delta\mathbf{v} = \mathbf{v}_f - \mathbf{v}_i$ in the time interval $\Delta t = t_f - t_i$.

$$\begin{aligned}
\mathbf{a} &= \frac{\Delta\mathbf{v}}{\Delta t} \\
&= \frac{(\mathbf{v}_f - \mathbf{v}_i)}{t_f - t_i}
\end{aligned}$$

where if $\mathbf{v}_i = \mathbf{u}$, $\mathbf{v}_f = \mathbf{v}$, $t_i = 0$, $t_f = t$, then $t_f - t_i = t$, and hence

$$\mathbf{a} = \frac{\mathbf{v} - \mathbf{u}}{t}$$

from which we obtain

$$\mathbf{v} = \mathbf{u} + \mathbf{a}t \tag{2.11}$$

and in scalar notation:

$$v = u + at \tag{2.12}$$

where in equation (2.11) $\mathbf{u}, \mathbf{v}$ are *velocities* which are *vector quantities*, and in equation (2.12), u, v are *speeds* which are *scalar quantities*. a in equation (2. 12) is the magnitude of the acceleration $\mathbf{a}$ in equation (2.11). We are able to write equation (2.) as a scalar equation because in $1d$-motion, u, v and a are directed along the same straight line.

Inserting v from equation (2.12) in (2.10), we obtain

$$\begin{aligned}
s &= \frac{1}{2}(u + u + at)t \\
&= \frac{1}{2}(2u + at)t \\
s &= ut + \frac{1}{2}at^2 \tag{2.13}
\end{aligned}$$

If both sides of equation (2.12) are squared, we obtain

$$\begin{aligned}
v^2 &= (u + at)^2 \\
&= u^2 + a^2t^2 + 2uat \\
&= u^2 + 2a\left(ut + \frac{1}{2}at^2\right) \\
v^2 &= u^2 + 2as \tag{2.14}
\end{aligned}$$

where equation (2.13) has been used to obtain equation (2.14).

Equations (2.10), (2.12), (2.13) and (2.14) constitute what are known as *kinematic equations* for a uniformly accelerated motion, which are collected below in scalar form:

$$
\begin{aligned}
s &= \frac{1}{2}(u + v)t \\
v &= u + at \\
s &= ut + \frac{1}{2}at^2 \\
v^2 &= u^2 + 2as
\end{aligned}
$$

The kinematic equations can be applied to several cases of interest such as horizontal motion, vertical motion as well as projectile motion as will be done in the next sections.

Although we have expressed these equations as scalar equations, the vector nature of kinematical quantities involved in these equations must be born in mind. When we use these equations to study motion in $1d$, say in the vertical or horizontal direction, the vector nature of the various quantities shall be infused into the equations by following a sign convention. When we use these equations to study $2d$ projectile motion, we shall note that this is a special case of two simultaneous $1d$ motions in the horizontal and vertical directions. As we shall see in section (2.3), there is no acceleration in the horizontal direction but there is acceleration in the vertical direction. Details of the sign convention are discussed in the following sections.

2.2.2   Kinematics in $1d$ horizontal motion

In $1d$ horizontal motion, the kinematic equations discussed earlier can be applied to several cases of interest. There are several possibilities as discussed below.

Recipe 2.1: Identification of values

- Read the given problem carefully and list the known values

- Identify the unknown values

- When motion is from rest, $u = 0$

- When a particle stops, $v = 0$

- Express the unknown values in terms of known values and solve

Recipe 2.2: Uniformly accelerated motion along a horizontal line

- If the acceleration for the entire motion is not constant, divide the motion into separate segments such that the acceleration for each segment is constant.

- Solve for each segment of the motion with constant acceleration as follows.

- List all the given kinematical quantities for the segment of the motion under consideration, and convert them to SI units if given in any other units.
- Use a sign convention for the given acceleration, *i.e.*, positive for the acceleration and negative for deceleration.
- If acceleration is not given, and needs to be determined, then the sign of the calculated value obtained shall indicate whether it is an acceleration or deceleration.
- List all the quantities that need to be calculated.
- Identify appropriate kinematic equation(s) which must be used to determine the required parameters. Sometime one may need to use more than one equation in succession to determine one single parameter.
- Express your results with appropriate SI units and sign. If the results are required in any other units, convert the units.

- The final velocity for the previous segment of motion is the initial velocity for the next segment of the motion, and the initial time for each segment is taken to be zero.

- Total time and the total displacement for the entire motion is the sum of the times and displacements for individual segments of motion, from which other parameters may be calculated as may be required.

Example 2.1: A car starting from rest and moving with uniform acceleration picks a speed of 72 km/h in 20 s.
(a) What is the acceleration of the car?
(b) What distance does the car move in the 20s?

Solution
(a)

$$
\begin{aligned}
u &= 0 \\
v &= 72\text{km/h} = \frac{72 \times 1000}{60 \times 60}\text{m/s} = 20\text{m/s} \\
t &= 20\text{s} \\
a &= ?
\end{aligned}
$$

Using

$$
\begin{aligned}
v &= u + at \\
20 &= 0 + 20a \\
a &= 1 \ \text{m/s}^2
\end{aligned}
$$

(b)

$$
\begin{aligned}
s &= ut + \frac{1}{2}at^2 \\
&= 0 + \frac{1}{2} \times 1 \times 20 \times 20 \\
&= 200 \ \text{m}
\end{aligned}
$$

Exercise 2.1: A car starting from rest and moving with uniform acceleration picks a speed of 65 km/h in 30 s.
(a) What is the acceleration of the car?
(b) What distance does the car move in the 30s?

2.2.3   Kinematics in $1d$ vertical motion

In $1d$ vertical motion, the kinematic equations discussed earlier can be applied to several cases of interest, noting that under free motion the acceleration a is due to gravity. Furthermore, the acceleration during the motion is always constant, directed vertically downwards, whether the body is moving upwards or downwards. Therefore, for most applications the entire motion of the body including both ascent and descent can be treated as one single segment of the motion, without having to separate them into two segments for going up and coming down, a common time-consuming mistake made by students. However, to solve the problem in this manner one must be very careful about the sign convention and the vector nature of the kinematical quantities. One uses the following sign convention:

Sign convention:

All quantities directed vertically upwards are taken to be positive, and all quantities directed vertically downwards are taken to be negative.

The acceleration a in this case which is equal to the acceleration due to gravity g acts vertically downwards irrespective of the direction of the motion of the body, and hence is taken to be negative. The average value of g is $9.8\,m\,s^{-2}$, however, a value of $g = 10m\,s^{-2}$ is often adequate for simple calculations, unless stated otherwise in the given problem.

The velocity of a body moving upwards is positive, and negative when moving downwards.

The displacement s is positive if measured upwards, and is negative if measured vertically downwards. Equal displacements upwards and downwards add upto net zero displacement, for example an object thrown up from the ground shall have zero displacement on return to ground.

The problems of vertical motion under gravity can be solved following the recipe given below.

Recipe 2.3: Free motion under gravity.

- Remember that the proper use of sign convention and the vector nature of kinematical quantities is very crucial to such problems, and one need not split the up and down motions into two separate segments of motion. The effect of air resistance on the motion is ignored.

- List all the given quantities in SI units and with appropriate signs, keeping in mind that the acceleration $a = g = -10\,m\,s^{-2}$ or as given.

- at maximum height of the object the velocity is zero.

- List the quantities to be calculated.

- Identify the kinematical equation(s) that should be used to calculate the required quantities, and determine the unknown parameters in SI units.

Symmetry considerations of free motion under gravity

Free motion under gravity has the following symmetry aspects which can be very useful in solving problems, and in interpreting the results of calculations, when an answer has two roots.

- When an object passes through an imaginary horizontal plane on its round-trip motion under gravity, the magnitude of velocity going up is equal to the magnitude of velocity coming down, the directions of velocities, as obvious, being opposite to each other.

- When an object on its round-trip motion under gravity passes two imaginary, parallel horizontal planes separated by a vertical distance h, the times taken to pass the planes on upwards motion and on the downwards motion are equal. A direct consequence of this symmetry is that the time to reach the maximum height from a given point in a horizontal plane is equal to the time it takes to fall down from the maximum height to the same point.

Example 2.2: A ball is thrown vertically upwards from the ground with an initial velocity of 20 m/s.
(a) At what time does the ball reach a height of 15 m ?
(b) What is the velocity of the ball at the height of 15 m?
(c) What is the maximum height reached by the ball?
(d) How long does it take to reach the maximum height?
(e) What is the time of flight?

Solution
(a)

$$
\begin{aligned}
s &= ut + \frac{1}{2}at^2 \\
15 &= 20t + \frac{1}{2}(-10)t^2 \\
5t^2 - 20t + 15 &= 0 \\
t^2 - 4t + 3 &= 0 \\
(t-1)(t-3) &= \\
t &= 1 \text{ s and } 3 \text{ s}
\end{aligned}
$$

The ball reaches a height of 15 m when $t = 1$ s when going up and at $t = 3$ s when going down.

(b)

$$
\begin{aligned}
v^2 &= u^2 + 2as \\
&= 20^2 - 2 \times 10 \times 15 \\
&= 100 \\
v &= \pm 10 \text{ ms}^{-1}
\end{aligned}
$$

At a height of 15 m, the ball acquires a velocity of $+10$ ms^{-1} when going up and a velocity of -10 ms^{-1} when going down.

(c) Maximum height, h_{max}, occurs when the velocity $v = 0$, that is,

$$
\begin{aligned}
v^2 &= u^2 + 2ah_{max} \\
0 &= 20^2 - 2 \times 10 \times h_{max} \\
h_{max} &= \frac{400}{20} = 20 \text{ m}
\end{aligned}
$$

(d) At the maximum height, the velocity $v = 0$, hence:

$$
\begin{aligned}
v &= u + at \\
0 &= 20 - 10t \\
t &= \frac{20}{10} = 2 \text{ s}
\end{aligned}
$$

(e) By symmetry, the time of flight $T = 2t = 4$ s. Alternatively, the time of flight can be obtained by using the following equation, noting that the displacement is $s = 0$ when the motion has been completed.

$$
\begin{aligned}
s &= ut + \frac{1}{2}at^2 \\
0 &= 20t - \frac{1}{2}10t^2 \\
20t - 5t^2 &= 0 \\
t(20 - 5t) &= 0 \\
t &= 0 \text{ s and} 4 \text{ s}
\end{aligned}
$$

which gives the same answer for the time of flight as $t = 4$ s. Note that $t = 0$ simply signifies the initial time when the motion statrted with $s = 0$.

Exercise 2.2: A person standing at the edge of a $50\,m$ high cliff throws a ball vertically upwards with a velocity of $5\,m\,s^{-1}$. The ball rises to certain maximum height, passes the person on the cliff on its way down, and falls down to the ground below. Calculate the following:
(a) Time taken by the ball to hit the ground, and the velocity with which it hits the ground.
(b) Time and the velocity of the ball when it passes the person on the cliff on its way down.
(c) Time taken and the maximum height reached by the ball from the ground.
(d) Give an interpretation of the results for which two values of answers are obtained.

2.3 Two Dimensional ($2d$) Motion: Projectiles in a Vertical Plane

In this section we consider $2d$ motion in a vertical plane for a type of motion known as *projectile* motion. In projectile motion, there is constant acceleration (due to gravity) along the vertical direction and no acceleration along the horizontal direction. Hence the horizontal and vertical motion can be *decoupled*, that is, they can be treated separately.

2.3.1   Projectile motion in a vertical plane above a horizontal axis

Consider projectile motion in a vertical plane above the horizontal axis through the point of projection as illustrated in Figure (2.1). Let us study eight parameters of physical interest.

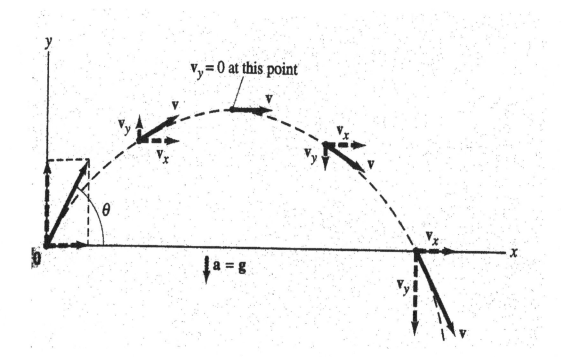

Figure 2.1: Projectile motion showing a projectile fired at an angle θ to the horizontal.

(i) *Angle of projection, θ*

The angle at which the projectile is fired is denoted by θ, measured relative to the horizontal direction.

(ii) *Velocities*

Suppose the initial velocity of the projectile is $\mathbf{u_0}$, given by

$$\mathbf{u_0} = u_{0x}\mathbf{i} + u_{0y}\mathbf{j} \qquad (2.15)$$

where u_{0x} and u_{0y} are the horizontal and vertical components of the initial velocity, given by

$$u_{0x} = u_0 \cos\theta \qquad (2.16)$$
$$u_{0y} = u_0 \sin\theta \qquad (2.17)$$

The horizontal component of the velocity, u_{0x}, remains constant throughout the motion since there is no acceleration in this direction. The vertical component changes because of the acceleration due to gravity, and the vertical component at any time t later is given by applying kinematic equations, and is given by

$$v_y = u_0 \sin \theta - gt \qquad (2.18)$$

(iii) *Acceleration,* **a**

The acceleration, **a**, can be written in the form

$$\mathbf{a} = a_x \mathbf{i} + a_y \mathbf{j} \qquad (2.19)$$

where

$$a_x = 0 \qquad (2.20)$$
$$a_y = -g \qquad (2.21)$$

since there is no acceleration along the horizontal direction, but there is acceleration in the vertical direction as a result of gravity.

(iv) *Time to reach the highest point,* t_B

The time to reach the highest point in the trajectory of the projectile is denoted by t_B, and this can be obtained as follows. The vertical component of velociy at any time is given by equation (2.18), and noting that the vertical component of velocity at the highest point is zero, with the corresponding time t_B.

$$v_y = u_0 \sin \theta - gt$$
$$0 = u_0 \sin \theta - gt_B$$

from which we obtain

$$t_B = \frac{u_0 \sin \theta}{g} \qquad (2.22)$$

(v) *Time of flight,* T

The time of flight, T, is the time taken by a projectile to reach a position which is on the same horizontal plane from which it was projected, as illustrated in figure (2.1). If air resistance is considered to be negligible, using symmetry considerations, the time of flight will be twice the time to reach the highest point.

$$T = 2t_B$$
$$T = \frac{2u_0 \sin \theta}{g} \qquad (2.23)$$

The time of flight can also be derived in another way. When the projectile returns to the horizontal plane of projection, the displacement along the vertical direction, s_y, is zero, and hence using one of the kinematic equations

$$s_y = u_{0y}t + \frac{1}{2}a_y t^2$$
$$0 = (u_0 \sin \theta) T - \frac{1}{2}gT^2$$

from which we obtain

$$T = \frac{2u_0 \sin \theta}{g}$$

which is the same as equation (2.23). The other possible solution is $T = 0$. What is the significance of this?

(vi) *Maximum height attained, H*

The maximum height attained by the projectile is H. Using one of the kinematic equations, say (2.14), and noting that the velocity at the highest point is zero, we have, along the y-direction,

$$\begin{aligned}
v_y^2 &= u_{0y}^2 + 2a_y s_y \\
0 &= u_{0y}^2 - 2gH \\
0 &= u_0^2 \sin^2 \theta - 2gH
\end{aligned}$$

from which we obtain

$$H = \frac{u_0^2 \sin^2 \theta}{2g} \tag{2.24}$$

(vii) *Range, R*

The Range attained by the projectile, denoted by R, is the displacement along the horizontal plane, and the corresponding time to cover the range is the time of flight, T. Using one of the kinematic equations, (2.13), along the x-direction, where there is no acceleration ($a_x = 0$), we obtain

$$\begin{aligned}
s_x &= u_{0x}t + \frac{1}{2}a_x t^2 \\
R &= (u_0 \cos \theta)\, T \\
&= \frac{(u_0 \cos \theta)(2u_0 \sin \theta)}{g} \\
&= \frac{u_0^2 2 \sin \theta \cos \theta}{g} \\
R &= \frac{u_0^2 \sin 2\theta}{g} \tag{2.25}
\end{aligned}$$

which gives an expression for the range. It can be noted that the maximum range, R_{max}, occurs when $\sin 2\theta$ is maximum, and this occurs when $2\theta = \pi/2$ or $\theta = \pi/4 = 45°$. The maximum range is

$$R_{max} = \frac{u_0^2}{g} \tag{2.26}$$

This interesting result of maximum range has important applications in the sporting events, which involve throwing of objects to achieve the largest possible range. It can be understood physically as follows.

The horizontal range of a projectile depends on the horizontal component of velocity, and the time of flight. The time of flight in turn depends on the vertical component of velocity. When the angle of projection is small, the horizontal components of velocity is large, but the vertical component is small resulting in a small time of flight. They both combine to give a smaller range. Imagine, if the projectile is thrown horizontally along the ground. It shall not rise to any height, and the range shall be zero. In contrast to this if the angle of projection is large, the vertical component of the velocity is large resulting in large time of flight, but the horizontal component being small, the combined effect of the two is a small range. In the extreme case when the object is thrown vertically upwards, its horizontal range is zero. It is only at the 45° angle of projection that the time of flight and the horizontal component of velocity reach such compatible values that their product result in the maximum range. A graph illustrating the relation between the angle and the range is shown in Figure 2.2

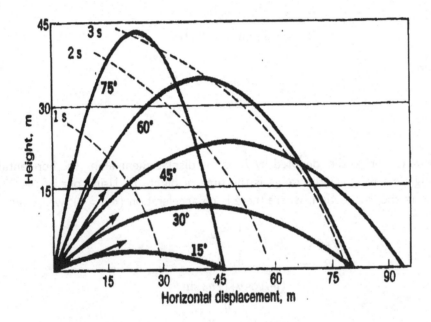

Figure 2.2: The relation between the angle of projection and the range for a projectile.

(viii) *Equation of the trajectory, $y = f(x)$*

The equation of the trajectory is given by $y = f(x)$. This can be obtained by considering horizontal motion along the $x - axis$, vertical motion along the $y - axis$ and relating the resulting equations through time t.

The horizontal displacement along the $x - axis$, where there is no acceleration ($a_x = 0$), is described by

$$s_x = u_{0x}t + \frac{1}{2}a_xt^2$$
$$x = u_0\cos\theta t$$

from which we obtain

$$t = \frac{x}{u_0 \cos \theta} \tag{2.27}$$

The vertical displacement along the $y - axis$, where there is acceleration due to gravity $(a_y = -g)$, is described below, and the expression for t from the horizontal motion in equation (2.27) will be used as well.

$$
\begin{aligned}
s_y &= u_{0_y} t + \frac{1}{2} a_y t^2 \\
y &= (u_0 \sin \theta)\, t - \frac{1}{2} g t^2 \\
&= u_0 \sin \theta \frac{x}{u_0 \cos \theta} - \frac{1}{2} g \frac{x^2}{u_0^2 \cos^2 \theta} \\
&= x \tan \theta - \frac{1}{2} g \frac{x^2}{u_0^2 \cos^2 \theta} \\
y &= bx - cx^2 \tag{2.28}
\end{aligned}
$$

is the equation of the trajectory, where $b = \tan \theta$ and $c = g/\left(2 u_0^2 \cos^2 \theta\right)$. This is an equation of a parabola.

In this section, we have considered the simplest case of a projectile motion, whereby the object is projected from the ground level, and at the end of its flights it returns to a point on the ground level in the same horizontal plane. The formulae derived for the time of flight, maximum height reached and the range derived earlier apply only to these types of projectile motions. However, there are innumerable possibilities of a projectile motion in a vertical plane, as many as one could imagine. Some such examples are:

- An object is projected from an elevated point above ground, and the angle of projection which could either be above horizontal, or below horizontal, or zero. At the end of the flight the object hits the ground.

- The object projected from the ground hits a target above ground, for example when a cricket ball batted from the ground hits a window of a building.

- An object projected from an elevated point from the ground reaches another elevated point from the ground which could either be higher or lower from the point of projection.

All such general cases of projectile motion can not be solved using the formulas of this section, and one must solve them from the first principles using the following recipe as illustrated by example (2.3).

Recipe 2.3: Projectile motion: A general case

- Resolve the velocity of projection into the horizontal and the vertical components.

- Motion of the projection along the horizontal is governed by the horizontal component of the velocity, and acceleration along this direction is zero.

- The motion along the vertical is governed by the vertical component of velocity of projection, and the acceleration is the acceleration due to gravity. In fact the vertical component of the motion is the same as the free motion under gravity discussed in an earlier subsection.

- Solve for horizontal and vertical motions independently using the kinematic equations of motion in the usual manner. The sign convention must carefully be adhered to.

- The time of flight for the two motions is the common parameter that connects the two motions. The effect of air resistance is ignored.

- Velocity of projectile at any point on its trajectory is obtained by combining the horizontal and vertical components of velocity at that point.

Example 2.3: A ball is projected with a velocity of $15\, m\, s^{-1}$ at 37° above horizontal from the top edge of a $50\, m$ high building. Calculate the following:
(a) Time it takes to reach the ground.
(b) Maximum height above the ground reached by the ball.
(c) Range of the ball on the ground.
(d) Velocity with which the ball hits the ground.

Solution: The horizontal and the vertical components of the velocity of projection are: $v_x = 15\cos 37 = 12\, m\, s^{-1}$ and $v_y = 15\sin 37 = 9\, m\, s^{-1}$

(a) For the time of flight we consider the vertical component of the motion for which $s = -50\, m$, $u = 9\, m\, s^{-1}$ and $a = g = -10\, m\, s^{-2}$. Calculate T using $s = ut + \frac{1}{2}at^2$.

$$-50 = 9T - \frac{1}{2} \times 10\, T^2 = 9T - 5T^2$$

This is a quadratic equation in T whose roots are $T_1 = 4.2\, s$ and $T_2 = -2.4\, s$. The required answer is the positive root, *i.e.*, the time of flight $T = 4.2\, s$. The significance of the negative root is discussed at the end of the example.

(b) The maximum height is also determined from the vertical component of the motion, using equation $v^2 = u^2 + 2as$ where $v = 0$ at the maximum height, and $s = h_{max}$ measured from the top of the building. This gives:
$$0 = 9^2 + 2 \times (-10)\, h_{max} \Rightarrow h_{max} = 4.05\, m$$

and the maximum height reached from the ground is: $H_{max} = 50 + 4.05 = 54.05\, m$

(c) The range is determined from the horizontal component of motion for which $u = 12\, m\, s^{-1}$, $a = 0$, $t = T = 4.2\, s$, and $s = R =?$. Using $s = ut + \frac{1}{2}at^2$, we get:

$$R = 12 \times 4.2 = 50.4\, m$$

(d) To determine the velocity $\mathbf{v_f} = (v_f, \theta)$ of the ball we must determine its $x-$ and $y-$components from the horizontal and the vertical motions, and then combine the components to determine the magnitude of the velocity and its direction. Since there is no acceleration along the horizontal direction, the horizontal component of velocity remains unchanged, and $v_{fx} = v_x = 12\,m\,s^{-1}$. Using principles of free motion under gravity, the vertical component is found to be $v_{fy} = \pm 32.9\,m\,s^{-1}$, which is left as an exercise for students. Since when reaching the ground the ball is traveling downwards, we take the negative root: $v_{fy} = -32.9\,m\,s^{-1}$. The interpretation of the positive root is discussed later. Combining the components v_{fx} and v_{fx} (left as an exercise for students), the velocity on hitting the ground is obtained as:

$$\mathbf{v_f} = (35\,m\,s^{-1},\, -70^o)$$

Interpretation of the negative root of time and the positive root of the v_{fy} component of the velocity: During the course of solving the problem we obtained $T_2 = -2.4\,s$ and $v_{fy} = 32.9\,m\,s^{-1}$ which we rejected on physical grounds, and chose the other roots. However, since these roots are also mathematically correct, they are not wrong, and must have physical significance. Combining the positive root $v_{fy} = 32.9\,m\,s^{-1}$ to the horizontal component of the velocity $v_x = 12\,m\,s^{-1}$, we obtain a value of velocity $\mathbf{u} = (35\,m\,s^{-1},\,70^o)$. These values point to the possible *previous history* of the projection of the ball. If the ball were thrown from the ground at $T_2 = -2.4\,s$ with velocity $\mathbf{u} = (35\,m\,s^{-1},\,70^o)$, the ball will follow exactly the same motion as the ball thrown from the top of the building at $t = 0\,s$.

Exercise 2.3: A projectile projected from the ground has a range of $30\ m$ and reaches a maximum height of $20\ m$. Calculate the velocity and the angle of projection. For the same velocity of projection what is the maximum range that can be achieved by the projectile, and what should be the angle of projection?

Symmetry considerations of projectile motion

The projectile motion, like the free motion under gravity along the vertical has certain symmetry features that can be beneficially exploited in solving problems. They are:

- Consider a projectile passing an imaginary horizontal plane during the course of its flight. If its velocity while going up is $\mathbf{v_1} = (v_1, \theta)$ then its velocity on the way down is $\mathbf{v_2} = (v_1, -\theta)$

- Consider a projectile passing two imaginary parallel horizontal planes separated by a vertical distance h. Then the time taken by the projectile to pass the planes on the way up is equal to the time taken to pass the same planes on the way down. A direct consequence of this is that the time taken to reach the highest point of trajectory from a point in a horizontal plane is equal to the time taken to return to a point in the same horizontal plane.

- It has already been proved that the range of a projectile is maximum for the angle of projection equal to 45°. For any other angle of projection the range is always smaller such that the range of projectiles projected at angles $(45 - \phi)^o$ and $(45 + \phi)^o$ are equal where $\phi < 45^o$. However the maximum height reached, and the time of flights in two cases are different. Discuss qualitatively, without solving as to in which case the maximum height and the time of flight are larger, and why?

2.3.2   Projectile motion over an inclined plane

Consider projectile motion over an *inclined plane* at an angle α above the horizontal, as illustrated in Figure (2.3). Let us study five parameters of physical interest.

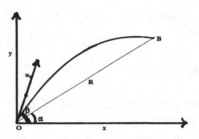

Figure 2.3: Projectile fired at an inclined plane.

(i) *Angle of projection, β*

The angle at which the projectile is fired is denoted by β, measured relative to the horizontal direction.

(ii) *Equation of the inclined plane*

The inclined plane is inclined at an angle α relative to the horizontal. Hence the equation of the inclined plane is

$$y = x \tan \alpha \qquad (2.29)$$

(iii) *Velocities*

Suppose the initial velocity of the projetile is $\mathbf{u}_0$, given by

$$\mathbf{u}_0 = u_{0x}\mathbf{i} + u_{0y}\mathbf{j} \qquad (2.30)$$

where u_{0x} and u_{0y} are the horizontal and vertical components of the initial velocity, given by

$$u_{0x} = u_0 \cos \beta \qquad (2.31)$$
$$u_{0y} = u_0 \sin \beta \qquad (2.32)$$

The horizontal component of the velocity, u_{0x}, remains a constant throughout the motion since there is no acceleration in this direction. The vertical component changes because of the acceleration due to gravity, and the vertical component at any time t later is given by applying kinematic equations, and is given by

$$v_y = u_0 \sin \beta - gt \qquad (2.33)$$

(iv) *Displacements and time of flight*

The horizontal and vertical components of displacement at any time are x and y respectively, given by

$$x = u_0(\cos\beta)t \tag{2.34}$$

$$y = u_0(\sin\beta)t - \frac{1}{2}gt^2 \tag{2.35}$$

From equation (2.29), (2.34) and (2.35), the projectile's path and the inclined plane intersect when

$$u_0(\sin\beta)t - \frac{1}{2}gt^2 = u_0 t \cos\beta \tan\alpha \tag{2.36}$$

which has two solutions

$$t = 0 \text{ at the origin O, and} \tag{2.37}$$

$$t = \frac{2u_0\sin(\beta-\alpha)}{g\cos\alpha} \text{ at B} \tag{2.38}$$

(v) *Range on the inclined plane, R*

The range R on the inclined plane is given by

$$
\begin{aligned}
R &= x\sec\alpha \\
&= u_0\cos\beta\frac{2u_0\sin(\beta-\alpha)}{g\cos\alpha}\sec\alpha \\
&= \frac{2u_0^2\sin(\beta-\alpha)\cos\beta}{g\cos^2\alpha}
\end{aligned} \tag{2.39}
$$

from which the maximum range is obtained as

$$R = \frac{u_0^2}{g(1+\sin\alpha)} \tag{2.40}$$

The proof of equation (2.40) is left as an exercise for the students.

2.4   Problems

2.1 A car on a road where the speed limit is 120 km/h went out of control and made skids 0.3 km long. Given that the deceleration was $0.4g$,
(a) Calculate the speed (in m/s and km/h) when the car just went out of control , and state whether the driver was maintaining the speed limit or not.
(b) How long did it take for the car to come to stop from the moment it went out of control?

(c) What was the speed at a point halfway the skid?

2.2 A plane starts from rest and has an acceleration of 3.6 ms^{-2} for 5 s. Full power is then turned on and an acceleration of 5.1 ms^{-2} is achieved and the plane acquires a speed of 300 km/h just before take off.
(a) How much time elapses between starting and taking off?
(b) Calculate the total distance of the runway the plane covers before take off .

2.3. A car starts from rest and accelerates uniformly at $5\,m\,s^{-2}$ for $10\,s$. Then it continues to drive at constant speed for $2\,minutes$ when the driver sees a police road block $100\,m$ ahead. The driver immediately starts to decelerate uniformly, and comes to a smooth stop at the road block. Calculate the following:
(a) Maximum velocity of the car during the entire journey.
(b) Distance moved in order to reach the maximum velocity.
(c) Time taken to stop the car, and the acceleration during stopping.
(d) Average velocity of the car from the start to stop.

2.4. A person driving at $90\,km\,h^{-1}$ sees a cattle in the road ahead. The reaction time of the driver is $0.75\,s$, but (s)he is able to stop the car smoothly just close to the cattle. If the deceleration of the car from the breaking force was $8\,m\,s^{-2}$, calculate the distance covered by the car, and the time taken to stop the car from the moment when the cattle was spotted by the driver.

2.5. A driver in a $120\,km\,h^{-1}$ speed zone approaches the $80\,km\,h^{-1}$ speed zone. He decelerates the car uniformly from $120\,km\,h^{-1}$ to $80\,km\,h^{-1}$ in $30\,s$ just next to the lower speed sign. Neglecting the reaction time of the driver, how far was (s)he from the $80\,km\,h^{-1}$ speed sign when (s)he first saw the sign, and what was the deceleration of the car?

2.6. A particle moving along the $+x-$axis passes the origin with a velocity of $20\,m\,s^{-1}$ at $t = 0\,s$. A uniform deceleration of $-4.5\,m\,s^{-2}$ by some external means (for example a spring or an elastic band pulling on it) is applied such that at certain point along the axis it momentarily comes to stop and its direction of motion is reversed. Calculate the following:
(a) Distance from the origin, and time when the direction of motion of the particle is reversed.
(b) At what times does the particle passes the $15\,m$ point on the axis, and what are the corresponding velocities?
(c) What is the velocity of the particle at $x = \pm 25\,m$, and the times when it passes these points on the axis.
(d) What is the velocity of the particle at $x = 55\,m$? Discuss the significance of your answer.

2.7. An object moving with a velocity of $5\,m\,s^{-1}$ slides up an inclined smooth plane, and stops after moving a distance of $2.5\,m$. Calculate the deceleration of the object, and hence the angle of inclination of the plane.

2.8. Two trains start from rest from stations A and B $150\,km$ apart traveling towards each other (on separate parallel tracks of course). Train from station A accelerates for $10\,s$ at a rate of $10\,m\,s^{-2}$,

and train from station B accelerates for $20\,s$ at a rate of $10\,m\,s^{-2}$ after which they maintain their velocities.
(a) Calculate the distance from station A when the engines of the two trains pass each other.
(b) After the engines pass each other, both trains start to decelerate uniformly, and come to stop at their respective stations. Calculate the time taken by each train to complete the journey.

2.9 A ball is thrown vertically upwards from the ground with an initial velocity of 25 m/s.
(a) After what times does the ball reach a height of 20m?
(b) What is the velocity of the ball at the height of 20m?
(c) What is the maximum height reached by the ball?
(d) How long does the ball take to reach the maximum height?

2.10 A ball is thrown vertically upwards from the ground with an initial velocity of 10 m/s. One second later, a stone is also thrown vertically upwards from the same position with an initial velocity of 25 m/s. Calculate
(a) the time when the ball and the stone are at the same level.
(b) the velocity of the ball and the stone when they are at the same level, and state the direction of motion of each of the particles.
(c) the total time of flight for each of the particles.

2.11. A player kicks a football which just clears the $3.5\,m$ high goal post at a $30\,m$ horizontal distance from the player. Calculate the velocity with which the ball was kicked, if it was kicked at 45° angle. How far behind the goal post shall the ball hit the ground?

2.12. A dart player throws a dart horizontally aiming at the bull's eye from a distance of $3.0\,m$. The dart hits the board at $2.0\,cm$ below the bulls eye. Calculate the velocity with which the dart was thrown.

2.13. An open train is moving at $40\,m\,s^{-1}$ due north. A boy on the train kicks a ball towards south with a velocity of $10\,m\,s^{-1}$ at 30° above horizontal, The train is $1\,m$ high from the ground. Calculate the time taken by the ball to hit the ground, and its range on the ground. Qualitatively sketch the trajectory of the ball as seen by an observer on the ground, labeling the north and south directions on the sketch.

2.14. A bottle is just released from a hot air balloon and the bottle hits the ground $20\,s$ later. Calculate the height of the balloon from the ground if: *(i)* it was stationary in the air, *(ii)* it was ascending vertically with a speed of $15\,m\,s^{-1}$, and *(iii)* it was descending vertically with a speed of $15\,m\,s^{-1}$. What is the separation between the bottle and the balloon at $t = 10\,s$ in each case?

2.15 A soccer ball is kicked off the ground with a speed of 12 m/s at an angle of $50°$ above the horizontal. Calculate
(a) the time the ball is in the air.
(b) the maximum height reached by the ball.
(c) the horizontal displacement between where the ball was kicked off and where it lands.

(d) the velocity of the ball with which it hits the ground.

2.16 A cliff 52 m high is overlooking an ocean. A stone is thrown off at the edge of the cliff with a velocity of 48 m/s at an angle of $36°$ above the horizontal. Calculate
(a) the time the stone takes to hit the ocean.
(b) the horizontal distance from the foot of the cliff to the point the stone hits the ocean.
(c) the maximum height above the ocean reached by the stone.
(d) the velocity of the stone with which it hits the ocean.

2.17 A child kicks a ball off the ground at a velocity of 14 m/s and at an angle of $36°$ above the horizontal. After 2s, the ball strikes a window in a nearby building. Calculate
(a) the position above the ground where the window is struck.
(b) the horizontal distance the ball has travelled.
(c) the velocity of the ball with which it strikes the window.

2.18 A projectile is thrown horizontally with a velocity of 20 m/s from the top of a building 50m high. Calculate
(a) the time the projectile is in the air.
(b) the range on the ground.
(c) the velocity of the projectile with which it hits the ground.

2.19 A ball is thrown from the top of a building 50m high with a velocity of 12 m/s at an angle of $15°$ below the horizontal. Calculate
(a) the time the ball is in the air.
(b) the horizontal distance from the foot of the building to the point the ball hits the ground.
(c) the velocity of the ball with which it hits the ground.

2.20. A ball is thrown with a velocity of $12\,m\,s^{-1}$ at $30°$ above horizontal. Calculate the velocities of the ball and times when it is half the maximum height from the ground.

2.21. A cricket player hits a ball with a velocity of $30m\,s^{-1}$ at $37°$ above horizontal. After $1.5\,s$ the ball hits a window of a opposite building. How far is the building from the batter, and what is the height of the window from the ground?

2.22. A boy throws a stone with a velocity of $10\,m\,s^{-1}$ at an angle of $53°$ above horizontal at a snake advancing towards him with a constant velocity of $5\,ms^{-1}$. If the stone hits the snake, calculate the following:
(a) How far was the snake when the boy threw the stone?
(b) If the boy had intended to hit a bird flying above, what is the maximum height of the bird, and its horizontal distance from the boy that can be hit by the stone.

2.23. Two projectiles A and B are thrown with the same velocity v_o at angles of projection $\theta_A = (45 + \alpha)°$ and $\theta_B = (45 - \alpha)°$ respectively, where $\alpha < 45°$. Prove that:
(a) Both the projectile have the same range.

(b) The difference between their times of flight is:

$$(t_A - t_B) = \frac{4\,v_o}{g}\,\sin\alpha\,\cos 45^o$$

(c) The difference between the maximum height reached by them is:

$$(h_A - h_B) = \frac{v_o^2}{2\,g}\,\sin(2\,\alpha)$$

2.24. A cricketer bats a ball, and at the same time a fielder at $50\,m$ distance starts to run to take the catch. The fielder catches the ball after $5\,s$ at a distance of $70\,m$ at $2\,m$ above the ground. Calculate:
(a) The velocity and the angle of projection of the ball.
(b) velocity of the fielder.
(c) velocity of the ball when it was caught.

2.25. A canon is fired with a velocity of $95\,m\,s^{-1}$ at 50° above horizontal, and after $5\,s$ it hits the top of a high rise building. Calculate the:
(a) Height of the building.
(b) Horizontal distance of the building.
(c) Velocity with which the canon hits the building.

2.26. A ball is thrown with a velocity of $20\,m\,s^{-1}$ at 40° above horizontal from the top of a tall building. The ball strikes the wall of another tall building $50\,m$ away. How far above or below the level of the building from which the ball was thrown, and with what velocity does the ball strike the other building?

2.27. A projectile projected with velocity v from ground is aimed at a target on the ground. When the angle of projection is α the projectile falls $a\,cm$ short of the target, and when the angle of projection is β the projectile falls $b\,cm$ beyond the target. Prove that the correct angle of projection θ to hit the target is given by:

$$\frac{1}{2}\,sin^{-1}\left(\frac{b\,sin(2\alpha) + a\,sin(2\beta)}{(a+b)}\right)$$

2.28. The muzzle velocity of a bullet from a gun is $200\,m\,s^{-1}$. Calculate the possible angles of firing when for the range of the bullet to be 80% of the maximum range.

2.29. A bomber flying $3\,km$ above sea with a speed of $300\,m\,s^{-1}$ releases a bomb which hits an enemy ship at the sea. If the ship was moving away at a constant speed of $36\,km\,h^{-1}$, calculate the location of the ship relative to the bomber when the bomb was released. At the same time when the bomb was released, the ship also fires a missile at the bomber at an angle of 53° above horizontal. The missile hits the bomber horizontally. Calculate the muzzle velocity of the missile. Which is hit first, and what is the time difference between the times when the two are hit?

2.30. Two hunters fire simultaneously at a bird on a tree, and both hit the bird at the same time. The first hunter shoots with a speed of $100\,m\,s^{-1}$ at an angle of 30°. The second hunter who is $50\,m$ ahead of the first hunter shoots at $80\,m\,s^{-1}$ at an angle θ. Calculate:
(a) The angle θ.
(b) Distance of the first hunter from the tree.
(c) Height of the bird from the ground.

2.31. Two stones A and B are projected from the same point on the ground at $t = 0$. The horizontal components of their velocities are u_a and u_b and the vertical components are v_a and v_b respectively. Both the stones pass through a common points on their trajectory at times t_a and t_b respectively. Prove the following:

$$(t_a - t_b) = 2\,\frac{v_a\,u_b - v_b\,u_a}{g\,(u_a + u_b)}$$

Chapter 3

Newton's Laws of Motion

In the Chapter on Kinematics we defined displacement, distance, speed, velocity, and acceleration, established how these parameters are related to each other, and how the motion of a body progressed with time. Amongst the examples, we discussed the motion of bodies moving with uniform velocity and with uniform acceleration. However, some questions there remained unanswered, such as: What determines whether a body shall have a uniform motion or an accelerated motion? What causes acceleration? How the 'cause and effect' in the context of acceleration are related? These questions and more are answered by the three *Newton's Laws of Motion*. Newton's laws of motion were first put forward by Sir Isaac Newton in his book *Philosophiae Naturalis Principia Mathematica* published in Latin in 1687. These are amongst the oldest known laws of nature, and more than three centuries later they still continue to be the most discussed laws in Physics. The laws are introduced to students at the Junior school level, if not earlier, and the discussion of the laws goes on right upto the tertiary level in the basic physics and mechanics courses. As for their applications, Newton's laws are also the most widely applied laws to various areas of classical and mechanical sciences. It is for this reason that we devote one full chapter to Newton's Laws and their applications, a good understanding of which is crucial not only to the study of physics but also to other disciplines for example mechanical engineering. We shall first state and discuss the three Newton's laws of motion. This shall be followed by their applications to simple physical systems.

3.1   Newton's laws of motion

Since Newton's laws are principally concerned with acceleration or the absence of it, it is appropriate to revisit the definition of acceleration. The force which causes acceleration is also defined.

Definition 3.1: Acceleration *is defined as the rate of change of velocity with time.*

If $\mathbf{v_i}$ and $\mathbf{v_f}$ are the initial and final velocities at times t_1 and t_2 respectively, then the average acceleration $\mathbf{a}$ over the time interval under consideration is given by:

$$\mathbf{a} = \frac{\mathbf{v_f} - \mathbf{v_i}}{t_2 - t_1} = \frac{\Delta \mathbf{v}}{\Delta t} \qquad (3.1)$$

In the limit of the time interval being infinitesimal, approaching zero, the instantaneous acceleration is given as:

$$\mathbf{a} = Lim_{\Delta t \to 0} \left(\frac{\Delta \mathbf{v}}{\Delta t} \right) = \frac{d\mathbf{v}}{dt} \tag{3.2}$$

Acceleration could result either from the change in the magnitude of velocity while its direction remained unchanged, or it could result from the change in the direction of the velocity while its magnitude remained constant, or it could result from the change in both, the magnitude and the direction. It is a vector quantity which does not necessarily lie along the same direction as the velocity. In a 3-D space it can be expressed in terms of three components in the usual manner. The SI unit of acceleration is $m\,s^{-2}$.

Definition 3.2: Force *is an external push or pull which when applied to a body may produce any one or more of the following effects:*

- *It may cause a change in the state of rest or motion of the body, i.e., a body at rest may gain acceleration and start moving, or a body moving with a uniform velocity may experience a change either in the magnitude or the direction of its velocity or both.*

- *It may produce an elastic deformation in the body, for example the compression or extension of a spring or a rubber band.*

- *It may produce permanent, non-reversible deformation of the body, for example stretching a spring or a wire beyond its elastic limit, or the deformation of vehicles involved in an accident.*

- *It may cause two objects which exert mutual forces on each other to bind together as one system, for example the binding of neuclons in a nucleus.*

- *Mutual forces between two objects may cause one of the objects to go around another object in an orbital motion, for example the planetary orbital motion or the orbital motion of an electron in an atom.*

These are just few of the many effects that may be produced by an external force. Students are assigned to list other possible effects produced by forces. Although any force can be viewed as a push or a pull, but in terms of the physical nature of forces there are many different types of forces. The effects produced by a force depend on the nature of force. Different types of forces, its SI unit, and a discussion of the vector nature of the force is deferred to a later section. At this stage this basic understanding of force shall suffice to indulge in the discussion of the Newton's laws of motion. In this context, it is the change in the state of rest or motion of a body produced by a force that is of relevance.

3.1.1 Newton's first law of motion

Statement of the Law: *A body in a state of rest or in a state of uniform motion continues to be in that state unless it is changed by an external force.*

From the above statement of the first law it is interesting to note that one of the definitions of force stated in the previous section has now been stated as the Newton's first law of motion with a rearrangement of the words. This arises from the fact that force as such does not have an independent formal definition. One understands force only in terms of the effects it can produce, the change in the state of rest and motion being one of them.

There are a few key words which are important in the above statement of the first law, without which either the statement of the law is incomplete or it loses its meaning. These key words are: the state of 'rest', 'uniform motion' and the 'external' force. Both the natural states of equilibrium of a body namely the state of 'rest' and 'uniform motion', known as the static and dynamic states of equilibrium respectively, should be considered. The uniform motion means a constant velocity in both magnitude and direction, *i.e.*, the acceleration is zero. The force which can change these states of a body is an external force as against an internal force such as the forces of interaction between atoms and molecules in an object. Such internal forces can not bring about a change in the static or dynamical equilibrium state of a body.

Newton's first law displays an important property of a body, its resistance to a change in its state of equilibrium. Every object offers a resistance to a change it its state of rest or uniform motion which is overcome by an external force, resulting in the acceleration of the body. This resistance of the body is known as the *inertia* of the body, and is given by the mass of the body. The larger the mass of the body the larger is the inertia. It is for this reason that the rest mass of a body is also referred to as the *inertial mass*. Newton's first law is thus also known as the law of inertia. One can compare this to electrical resistance, whereby a potential difference applied across the resistance results in the flow of current. One can see an equivalence between the Ohm's law ($V = RI$) and the Newton's second law ($\mathbf{F} = m\mathbf{a}, discussed in section 3.1.2$) as: $Force \rightarrow Voltage; mass \rightarrow Electrical resistance$, and $acceleration \rightarrow electrical current$

Some well known examples that illustrate Newton's first law are:

- When a car suddenly starts from rest, passengers are pushed backwards, and when a car suddenly stops passengers are thrown forward. Explain, why?

- If a running horse suddenly stops (due to some obstacle), the rider is thrown off the horse in the forward direction.

Some students, particularly at an early stage of learning, find it difficult to comprehend how an object once set in motion could continue to move forever without a change in its velocity unless it was continually pushed by a force. Contrary to the statement of the law, general observation and experience tells them that all moving objects must eventually stop once the external push applied to it is withdrawn. They fail to visualize the presence of a force like the force of friction which acting opposite to the motion is responsible for stopping of a moving object. This difficulty may be overcome by experimental demonstration using linear air-track or the air-table equipment. Such equipment are generally not readily available in schools in the remote rural areas of developing countries due to many constraints, and the problem remains status quo. Here we suggest a low cost alternative demonstration of the first law for which the equipment can be easily improvised even in the remotest poor schools. The experiment is that of the terminal velocity of a small steel ball (may be obtained from some old ball bearings) dropped through a column of viscous fluid. The forces acting on the ball are: *(i)* the constant force of gravity acting downwards which is responsible for its downwards acceleration, and *(ii)* the velocity dependent force of fluid friction acting opposite to the motion of the ball, which increases with the increase in the velocity of the ball. At some stage the two forces acting in opposite directions become equal, rendering the total external force on the ball equal to zero. From here onwards the ball falls through the column of the liquid with a constant velocity, and even if the length of the liquid column was made very large, the ball shall continue to fall with constant velocity. We refer students to a self-study project to learn more about the terminal velocity experiment from appropriate books in the library, and answer the question: What forces constitute the net downward force on the ball?

3.1.2   Newton's second law of motion

Newton's first law of motion is a qualitative law that does not provide quantitative relationship between the cause (force) and the effect (acceleration). Such a relationship is provided by Newton's second law. Let us consider a mass m to which an external force $\mathbf{F}$ is applied, and the mass undergoes an acceleration $\mathbf{a}$. From simple experiments, by varying the force and the mass of the object one can make following observations.

- Keeping the force constant, if the mass of the object is changed, the acceleration of the object changes inversely to the mass, *i.e.*, heavier is the object smaller is the acceleration from the same external force.

- Keeping the mass constant, if the force is varied, the acceleration changes in proportion to the applied force, *i.e.*, larger is the force larger is the acceleration when applied to the same mass.

Both of these observations can be expressed mathematically by the expression:

$$\mathbf{F} = k\,m\,\mathbf{a} \qquad\qquad (3.3)$$

where k is the constant of proportionality. Equation (3.3) provides two important observations about the force: *(i)* Since the mass m is scalar, $\mathbf{a}$ is a vector, and k is a scalar constant, the force is therefore a vector quantity. Both acceleration and the force act along the same direction. *(ii)* The relation defines the secondary (derived) units of the force. In the SI system of units, the unit of force is selected in such a way that the constant k becomes dimensionless one ($k = 1$), and equation (3.3) becomes:

$$\mathbf{F} = m\,\mathbf{a} \qquad\qquad (3.4)$$

and the SI unit of force is $kg\,m\,s^{-2}$ which is given a name Newton (N), *i.e.*, $1N = 1\,kg\,m\,s^{-2}$. From equation (3.4) we can now state the Newton's second law of motion as follows.

Statement of the second law: *When an external force* $\mathbf{F}$ *acts on a body of mass m, it experiences an acceleration parallel to the direction of the applied force given by the relation:* $\mathbf{F} = m\mathbf{a}$.

Equation (3.4) is a vector equation, and it can be expressed in terms of the components of the force and acceleration. In Cartesian coordinates in 3-D space the three independent component equations are:

$$
\begin{aligned}
F_x &= m\,a_x \\
F_y &= m\,a_y \\
F_z &= m\,a_z
\end{aligned}
\tag{3.5}
$$

If there are more forces than a single force acting on a body, it is important to note that the acceleration is proportional to *net forces*, and thus Newton's second law can also be written as

$$
m\mathbf{a} = \sum \mathbf{F}
$$

where $\sum \mathbf{F}$ denotes the net forces. This form will be used frequently in the later chapters.

Alternate forms of the second law:

Equations (3.4) and (3.5) are the most commonly used forms for the Newton's second law of motion. There are yet other forms of expressing Newton's second law, for example, in terms of momentum and impulse of a force. These forms are very useful for certain calculations pertaining to the change in momentum. The momentum $\mathbf{P}$ of a mass m, moving with velocity $\mathbf{v}$ is given as $\mathbf{P} = m\,\mathbf{v}$, and if a force $\mathbf{F}$ is applied for a Δt duration, then the impulse ($\mathbf{I}$) of the force is given as: $\mathbf{I} = \mathbf{F}\,\Delta t$. It is left as an exercise for students to figure out the nature and SI units of momentum and the impulse. In case of any difficulties, a detailed discussion of both is presented in a later chapter on the conservation of momentum.

By using the definition of acceleration (equation 3.1), Newton's second law (equation 3.4) can be expressed as:

$$
\mathbf{F} = m\,\frac{\Delta \mathbf{v}}{\Delta t} = \frac{m\,\Delta \mathbf{v}}{\Delta t} = \frac{\Delta(m\,\mathbf{v})}{\Delta t} = \frac{\Delta \mathbf{P}}{\Delta t} \Rightarrow \frac{d\mathbf{P}}{dt}
\tag{3.6}
$$

where ΔP is the change in the momentum of the body during the time interval Δt. Thus from equation (3.6) *the rate of change of momentum of the body is equal to the external force.* This is one of the alternate statements of the second law which leads to the law of conservation of momentum discussed in a later chapter.

We can also express equation (3.6) as:

$$
\mathbf{F}\,\Delta t = \mathbf{I} = \Delta \mathbf{P}
\tag{3.7}
$$

This is yet another form of the second law according to which *the change in the momentum of a body is equal to the impulse of the applied force.* Both equations (3.6) and (3.7) can also be expressed in terms of the components in the usual way. Students are advised to write the component equations in Cartesian coordinates in 3-D space.

A special case of the second law and its consequences:

(i) If the external force is zero, *i.e.*, $\mathbf{F} = 0$, then from equation (3.4) the acceleration $\mathbf{a} = 0$, and a body at rest continues to remain at rest and a body in motion continues in the state of uniform motion. This is basically the same as the Newton's first law of motion. However, when stated in this manner, this does not emphasize the inertial property of a mass, and therefore, in esence it is not the same as the first law. This is further discussed in detail in a later section on the law of translational equilibrium.

(ii) Again if the external force is zero, *i.e.*, $\mathbf{F} = 0$, then from equation (3.6), $\frac{d\mathbf{P}}{dt} = 0$ and from equation (3.7) $\Delta\mathbf{P} = 0$. Thus the rate of change of momentum of the body is zero, and its momentum remains constant. This is the law of conservation of momentum discussed with applications in Chapter 4. This conclusion again mathematically points to the first law of motion, but the concept of inertia, the most important feature of the first law, is lost.

3.1.3   Newton's third law of motion

When two objects interact with one another in such a way that one of the objects exerts a force on the other, then the other object responds by exerting a force on the first object. The third law of motion relates these two forces which states:

" To every action there is an equal and opposite reaction".

In the above statement, action is the force exerted by the first object on the second object, and reaction is the force generated in response by the second object and exerted on the first one. The reaction force acts normally to the common surface of contact between the two objects, and is also referred to as the *normal reaction force*. The reaction force comes into play only when an action force is applied, *i.e.*, in the absence of the action force there is no reaction force whereas the action force may exist on its own. This distinguishes them from each other. It is important to note that the action and reaction forces act on two different objects that are involved in interaction, and never on the same object. These features of action-reaction pair of forces will become clear from the examples given below.

- Referring to Figure (3.1), when an object of mass m and weight $m\,\mathbf{g}$ is placed on a plane surface, it exerts a downward force equal to its weight on the surface, where $\mathbf{g}$ is the acceleration due to gravity. This is the action force. The weight of the object is an ever present force due to gravity whether it is placed on a surface or not. The moment the object is placed on the surface, the reaction force of the surface comes into play, which was not there before the object was placed on it. The reaction force acts on the object, and according to third law it is equal and opposite to the action force, *i.e.*, $\mathbf{R} = (-)m\,\mathbf{g}$.

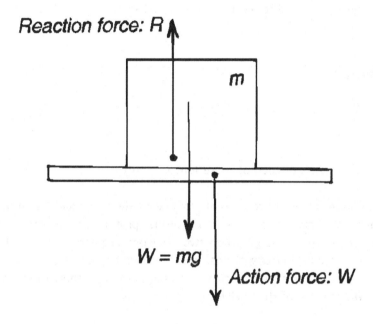

Figure 3.1: An example showing the pair of action and reaction forces

In this particular example the reaction force is found to be equal to $(-)m\,\mathbf{g}$. But this is not always the case. A common mistake made by students is to take $\mathbf{R} = (-)m\,\mathbf{g}$ most of the time. The reaction force of the surface on which the object rests in fact depends on whether the surface is horizontal or inclined, and what other forces are acting on the object. This shall be illustrated later by examples.

- When a bullet is fired from a gun, the force produced from the combustion of the gun powder with which the bullet is propelled forward is the action force. The gun recoils back due to the force of reaction which is equal and opposite to the action force.

- When a person steps ashore from a boat, (s)he pushes (her)himself forward with the action force, and the boat is pushed back into the water with an equal and opposite reaction force. It is for this reason that the boat must always be anchored to the shore before passengers from it disembark.

- When two objects collide, for example the collision of two balls in the game of billiards (pool) or the collision of two vehicles on a highway, the colliding objects exert equal and opposite forces on each other that constitute the pair of action-reaction forces. In this case a distinct identification of action and reaction forces depends on which of the two objects is the projectile and which is the target that absorbs the collision. The force exerted by the projectile is the action force, and the force exerted by the target is the reaction force.

3.2   Forces

3.2.1   Fundamental forces of nature

Although there are many observable forces in nature, they can, however, be classied into four fundamental forces, conventionally referred to as the fundamental interactions. They are:

- Strong interaction

- Electromagnetic interaction

- Weak interaction, and

- Gravitational interaction

Let us discuss the nature and scope of each of these forces in turn.

(i) Strong nuclear force: These forces are responsible for the binding of nucleons (protons and neutrons) in the nucleus, and acts between protons and protons, protons and neutrons, neutrons and neutrons. They are attractive forces, except when two nucleons are very close ($\sim 10^{-16}\, m$) they repel each other. They are extremely strong forces, in fact the strongest of all forces, $\sim 10^2$ times stronger than the electromagnetic force. The nuclear forces are also very short range, acting only up to a distance of $\sim 10^{-15}\, m$, the size of the nucleus.

(ii) Electromagnetic force: This force consists of two parts, the electrostatic part and magnetostatic part,

$$\mathbf{F}_{em} = q(\mathbf{E} + \mathbf{v} \wedge \mathbf{B}) \tag{3.8}$$

where $q\mathbf{E}$ is the electrostatic force, due to a charge q in the electric field $\mathbf{E}$, and $q\mathbf{v} \wedge \mathbf{B}$ is the magnetic force due the charge moving with a velocity v in a magnetic field $\mathbf{B}$. Further explanation of this force follows below.

Electrostatic force: Two point charges q_1 and q_2 separated by a distance r exert an electrostatic force on each other, the magnitude of which is given by the Coulomb's law:

$$F = \frac{1}{4\,\pi\,\epsilon_o}\, \frac{q_1\, q_2}{r^2} = k\, \frac{q_1\, q_2}{r^2} \tag{3.9}$$

where $\epsilon_o = 8.854 \times 10^{-12}\, C^2\, N^{-1}\, m^{-2}$ is the permittivity of free space and $k = \frac{1}{4\,\pi\,\epsilon_o} = 9 \times 10^9\, N\, m^2\, C^{-2}$ is the Coulomb's (electrostatic) force constant for free space. The force is repulsive if the two charges are similar, and attractive if they are dissimilar. It is a central force acting along the radius vector joining the two charges, and it is also a long range force like the force of gravity. It is this force which is responsible for the orbital motion of electrons about the nucleus in an atom. The Coulomb force is a much stronger force compared to the gravitational force. For example, the gravitational force between two electrons separated by $1\, m$ is $F_G = 5.5 \times 10^{-71}\, N$ whereas the Coulomb force between the same two electrons is $F_C = 2.3 \times 10^{-28}\, N$.

Magnetostatic force: When two magnetic poles lie in the field of each other, they exert a force on each other which is repulsive between the similar poles, and attractive between dissimilar poles. A magnet also exerts force on a magnetic material in its vicinity due to temporary or permanent magnetization of the material by the magnet. A current carrying conductor produces a magnetic field so that the conductor acts like a magnet and exerts force on other current carrying conductors, permanent magnets, and the magnetic materials in the vicinity. In magnetostatics, there are no mono poles like the free charges in electrostatics. A dipole is the smallest unit. Therefore, the magnetostatic forces are more complicated, and can not be expressed by a simple expression of the type used for the Coulomb's force. The force between two current carrying conductors can be arrived at by using the Biot-Savart law which is beyond the scope of this discussion.

Arising from the relationships between the electric current and the magnetic fields, the magneto-static and the electrostatic interactions are unified as electromagnetic interaction (equation 3.8). The binding of atoms and molecules in solids, the elastic forces, tension, friction, are all electrostatic interactions. It is responsible for all chemical and biological processes, DNA and the cellular structure, the heart beat, the activity of the brain and the nerves etc. The gravity keeps us bound to the earth, earth bound to the solar system, solar system to the galaxy and so on. The strong interaction is responsible for the binding of the nucleus of every single atom in the universe. Thus, whereas the strong interaction and the gravity permeate the entire universe, it is perhaps the electromagnetic interaction that governs our life and the matter around us most closely.

(iii) Weak interaction: These forces are responsible for processes such as β−decay in the nucleus. They are short range ($\sim 10^{-15}\,m$), acting within the nucleus, and are weaker than the electromagnetic force but stronger than the gravitational force.

(iv) Gravitational force: Two objects of mass M and m separated by a distance r attract each other with a force of magnitude given by:

$$F_G = G\,\frac{M\,m}{r^2} \tag{3.10}$$

where G is the universal gravitational constant, and has a value: $G = 6.673 \times 10^{-11}\,N\,m^2\,kg^2$. The law given by equation (3.10) is known as the universal law of gravitation. F_G is a central force, *i.e.*, the force acts along the radius vector r joining the two objects, and it is a long range force, approaching zero only when $r \to \infty$.

The gravitational force acts between any pair of two objects whatsoever, but because the value of the constant G is very small, it is a very weak force. It assumes significant magnitude only in the case of large planetary bodies even though they are separated by large distances. It is the force of gravity which is responsible for the orbital motion of planets, and produces the acceleration due to gravity.

The strong and the weak interactions are very short range, and the other two are long range, extending practically to infinity. There is also a vast variation in their strengths. The strong interaction is the strongest of all, and gravity is the weakest. On a scale of zero to one, if we assign 1 to the strength of the strong interaction, then the electromagnetic interaction shall be 10^{-2}, the weak interaction shall be 10^{-9} and the gravity shall be 10^{-39}. It has been the quest of Physicists to discover a theory (or model) unifying all the four fundamental interactions into one single force ever since it was perceived by Einstein for the first time in the early 1900's. Since then only a partial success has been achieved. The electromagnetic and the weak interactions were unified into electroweak interaction in 1970's by Glashow, Weinberg and Salam for which they were awarded the 1979 Physics Nobel prize. Their theory had predicted the existence of new particles, called the W and Z bosons, which were discovered in 1983. The unification of all four interactions into one grand interaction still remains a golden dream of Physicists the world over.

3.2.2 Examples manifesting the fundamental forces

Technically, all the forces are manifestations of the four fundamental interactions discussed above. However, irrespective of the fundamental nature and source, we can treat any force as either a 'push' force like pushing an object along the surface on which it is resting, or a 'pull' like pulling a box placed on the floor with a string attached to it. We can also call these forces as the 'repulsive' force acting away from the source of the force, or the 'attractive' force acting towards the source of the force respectively. Nevertheless, despite this general categorization, forces from various sources differ significantly from each other. In this section we define and discuss some examples of forces which are manifestations of the four fundamental forces of nature. Can students visualize how each of these forces is related to the fundamental forces?

(i) Push and pull forces: These are the forces resulting from the simple acts of pushing and pulling. The attractive and repulsive forces can also be visualized as push and pull forces.

(ii) Weight: A mass m held in the gravitational field of the earth experiences a force mg known as weight, where g is the acceleration due to gravity towards the center of the earth. When the mass is allowed to fall freely it approaches the earth with this acceleration, and when the object is thrown upwards, it experiences the same magnitude of deceleration in its velocity while moving away from the earth. From the Newton's second law we can write the earths gravitational force on the mass in terms of the acceleration as:

$$F = mg \qquad\qquad (3.11)$$

Comparing equations (3.10) and (3.11) we get:

$$F = mg = G\frac{Mm}{r^2}$$

or

$$g = G\frac{M}{r^2} \qquad\qquad (3.12)$$

where in equation (3.12) M is the mass of the earth, $r = (R+h)$ is the distance of mass m from the center of the earth, R is the average radius of the earth, and h is the vertical height of the object from the earth's surface. For small height h, $R \gg h$ and one can take $r \approx R$. The equation (3.12) then becomes:

$$g \approx G\,\frac{M}{R^2} \tag{3.13}$$

Thus the gravitational acceleration due to a planetary body depends on its mass and size. Equation (3.13) is a very rough approximation for the acceleration due to gravity g, because as we know the earth is not a perfect sphere, and there is inhomogeneity in the density and its mass distribution. Average value of the acceleration due to earth's gravity for routine applications is taken to be: $g = 9.8\,m\,s^{-2}$, but it varies with the latitude because of the shape of the earth which is not a perfect sphere, and with altitude, particularly at heights comparable to the radius of earth. Table (3.1) gives average values of g at some select latitudes and altitudes to illustrate its variation.

Table 3.1: Variation of the acceleration due to earth's gravity with altitude at 45° latitude, and with latitude at the seal level

Altitude (m)	$g\ (m\,s^{-2})$	Latitude	$g\ (m\,s^{-2})$
0 (Sea level)	9.806	0° (Equator)	9.780
100,000	9.598	20°	9.786
1000,000	7.410	40°	9.802
12,740,000 (2 × Radius of earth)	2.452	60°	9.819
380,000,000 (Moon's surface)	0.0027	$90^o(Poles)$	9.832

Mass and the weight: The distinction

As an everyday common practice, one uses mass and weight interchangeably. In fact a common person does not know the distinction between these two terms. More often than not one uses the term weight to express the mass through statements such as: 'The weight of this object is 80 kg'. This is an accepted practice in a general conversation, and there is no confusion as to what is being said. However, in Physics mass and weight are two distinctly different physical quantities, representing entirely different properties of the object, and accordingly in a scientific context one must use these terms carefully .

Mass (M) of an object is its matter content. It is a scalar quantity, and its SI unit is kg. The mass of an object is measured by comparing it to standard masses using equipments such as a beam balance. Mass of an object is universally constant no matter where it is measured, whether on the earth or on the moon.

Weight (W) of an object is the force of gravity on the object. For an object of mass m it is given as: $W = mg$, where g is the acceleration due to gravity. Being a force, it is a vector quantity, acting vertically downwards, and its SI unit is N. Because of the variation of g with location and altitude, the weight varies accordingly. It also varies if measured at different planetary bodies, for example the weight of an object on the moon's surface is about one-sixth of its weight on the earth's surface because of the different values of 'g' on moon and the earth. It is for this reason, that the astronauts find it difficult to walk on the moon's surface where their weight is about one-sixth of their weight on the earth. Weight is measured using a spring balance. A weight suspended freely with a spring produces an extention of the spring. If the force constant of the spring is known, the weight can be determined from the extention of the spring. A spring balance can also be calibrated in terms of the mass in kg, and they are commonly used in supermarkets to measure produce. However, it must be noted that due to the variation of g with location, a spring balance calibrated at one location say in London is not valid for use elsewhere, say in Botswana. While a small variation in the calibration resulting from a changed value of g is not important in measuring low cost commodities such as fruits and vegetables, the spring balances are never used to measure costly commodities such as gold, precious stones etc.

(iii) Elastic force: This force comes into play due to the elastic properties of the source of the force. For example if we extend or compress a spring by applying an external pull or push, the spring exerts an equal and opposite restoring force. The force is generated from the elastic property of the spring which tends to bring it to its natural state when extended or compressed. Same force acts when a rubber band is stretched, a wire is pulled by loading it with a weight within its elastic limit, and a solid is compressed or sheared. In an ideal system, the elastic or the spring force is proportional to the extention (compression) given by the following relation:

$$\mathbf{F} = -k\,\mathbf{x} \qquad\qquad (3.14)$$

where k is the force constant in $N\,m^{-1}$, and x is the extension (compression). The negative sign signifies the direction of the force being opposite to the direction of the change in length (a restoring force). The force law given by equation (3.14) is known as the *Hooke's law*, and applies to ideal springs, and elastic materials deformed within the elastic limit.

(iv) Tension in a string: When a weight is suspended freely with an unstreachable string, or when the string is pulled by some other means, the reaction force of the string is known as the tension in the string. It is the tension in the string which is responsible for various harmonics of vibration of the string, and one tunes a string musical instrument by varying the tension in the string.

(v) Normal reaction force: As per the Newton's third law of motion, this is the reaction force of the surface on which an object is placed. It acts normal to the common surface of contact. It plays an important role in determining the force of friction. We shall illustrate its calculation with examples later in the chapter.

(vi) Force of friction: When one object moves in contact with another object, solid or fluid (liquid or air), a force of friction acts on the object opposite to the direction of motion and causes the motion to decelerate. If no other external forces are present, the moving object eventually comes to rest losing all its mechanical energy to do work against the force of friction. An important point to remember here is that the force of friction comes into play only when the object is in motion, or when the object at rest has a tendency to move on being pushed or pulled. Depending on whether motion is translational or rotational, the friction force encountered is the translational friction or the rolling friction respectively. In this chapter we consider only the translational motion, and the friction arising from it.

When an object moves through a fluid, fluid friction, besides other parameters, as a first order approximation is proportional to the velocity of the object. This can be experienced when one tries to walk through a swimming pool. Faster one tries to run, more difficult it gets due to the increased fluid friction. It is the fluid friction of the atmosphere which causes most of the meteorites from outer space to be burnt-out in the upper atmosphere before reaching the earth. These burning meteorites are in fact the popularly known shooting stars. Imagine if this was not the case, the earth shall be under constant shower of space debris consisting of stones, rocks, and large to very large planetary fragments, and the life on earth would not be sustained. Likewise space ships on entering the earth's atmosphere at high speed become very hot due to the large atmospheric friction. In this chapter we shall be concerned only with the solid friction, *i.e.* the friction force that comes into play when a solid object translates in contact with a solid surface.

The (translational) solid friction is proportional to the normal reaction force (R), and is given by:

$$F_f = \mu R \qquad\qquad (3.15)$$

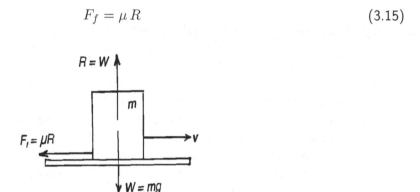

Figure 3.2: Solid (translational) friction is proportional to the normal reaction force and acts opposite to the direction of motion.

where μ is a dimensionless constant, known as the coefficient of friction. Its value depends on the smoothness (or roughness) of the contact surfaces of the object and the plane along which it moves. It is a common experience, that if one tries to move a heavy object along the floor, one requires a large push or pull force to start the motion. But once the motion has started, the force required to maintain the motion is less than what was required to start the motion. Accordingly, there are two types of translational friction, the static friction and the kinetic friction.

Let us consider an object of mass m at rest on a horizontal surface. We start with a small push (force) F_1 which is not enough to start the motion, but the object shall have a tendency to move in the direction of the applied force. As a result of this tendency the force of friction F_f comes into play. Since the object continues to be at rest, the force of friction is just equal to the external push-force and acts opposite to it, *i.e.*, $F_f = -F_1$. As we increase the push force, the friction force also increases, and so long as there is no motion and there is just a tendency to move the two forces remain equal and opposite to each other. On further increasing the push a stage is eventually reached when the object is just about to start moving (not moving yet!). An infinitesimal increase in the push shall set the object in motion. The maximum friction force at this stage which is equal to the maximum push is the static friction $(F_{f,s})$, and is given by:

$$F_{f,s} = \mu_s R \tag{3.16}$$

where μ_s is the coefficient of static friction.

Once the motion has started, the force of kinetic friction comes into play which is given by:

$$F_{f,k} = \mu_k R \tag{3.17}$$

where μ_k is the coefficient of kinetic friction. Since $F_{f,s} > F_{f,k}$ as discussed, the respective coefficient of friction are related as: $\mu_{f,s} > \mu_{f,k}$.

3.3 Applications of Newton's laws of motion and examples

The examples of the applications of Newton's laws of motion involve the motion of masses along smooth and rough horizontal and inclined planes, motion of masses attached with strings passing over pulleys, configuration of masses held in equilibrium, masses moving with acceleration, and any possible combination of these. To solve these problems, one begins by identifying all the forces acting on each of the components of the system, namely the masses, surfaces, strings etc., and then considering one component at a time one applies the relevant laws to them. In doing so the two steps, where applicable, are rather important: *(i)* To draw freebody diagrams for each component of the system, and *(ii)* to determine the normal reaction and friction forces. We first discuss both these aspects before we go on to solving the problems.

3.3.1 Freebody diagrams

A freebody diagram is a diagram of a mass in the system, or of a point of intersection (junction) where lines of action of different forces meet. It shows all the forces acting on the individual mass or the point under consideration. The free body diagram is drawn independent of the rest of the system, and as many free body diagrams are drawn as may be necessary to solve a given problem. The following recipe may be useful in drawing the free body diagrams.

Recipe 3.1: Drawing the free body diagrams.

- Draw a neat and clear sketch of the system described by the problem.

- Identify all the forces acting on each of the component of the system, and mark them on the diagram of the system.

- Also show all the relevant angles between the lines of action of various forces on the diagram.

- Consider each mass and junction in the system one at a time, and draw their freebody diagrams showing all the forces acting on them and the angles between the forces.

Here it is important to mention that whereas free body diagrams are very useful in solving Newton's laws of motion problems, but they are not absolutely necessary. The diagrams simply minimize or eliminate careless mistakes that may creep in during the working out of the problem. If one is careful enough, and knows the work involved well, one may work directly with the main diagram of the system. Students' success in solving problems shall depend on their ability to identify all the forces on the system, choice of appropriate perpendicular axes, and calculation of normal reaction and friction forces

Example 3.1: A $50\,kg$ mass is suspended with two cables as shown in Figure (3.3(a)). Assuming the cables to be massless, draw freebody diagrams for the mass and the junction where the cables meet.

 Solution: The freebody diagram for the mass and the junction are shown in Figures (3.3(b)) and

(3.3(c)) respectively.

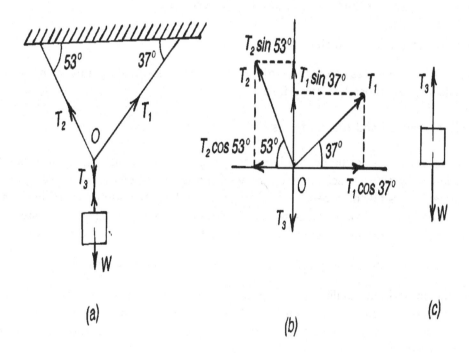

(a)   (b)   (c)

Figure 3.3: Freebody diagrams of a mass suspended with cables.

Note that a free body diagram is an important step in solving mechanics problems since it assists in the visualization of the forces acting on a particular object. From these diagrams, one can resolve the forces into components and solve the resulting equations.

Exercise 3.1: Consider the masses supported with strings as shown in Figure (3.4). A force of friction also acts on the mass placed on the horizontal surface, and the system is in equilibrium. Assuming the strings to be massless, draw freebody diagrams for both the masses and the junction where the strings meet.

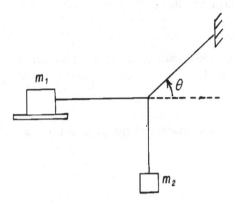

Figure 3.4: Figure for Exercise (3.1)

3.3.2   Normal reaction and friction forces

Normal reaction and friction forces can be determined from the freebody diagrams of the masses on which they act following the procedure given below.

Recipe 3.2: Normal reaction and friction forces.

- Choose an appropriate set of perpendicular axes. Generally one of the axis is taken parallel to the plane (direction) along which the mass moves, and the force of friction is along the same axis. The other axis is taken perpendicular to the plane, along the normal reaction force. The choice of axes is not necessarily the same for all the masses and junctions in a system. More often than not each mass and junction has its own set of axes different from the others.

- Resolve all the forces into perpendicular components along the chosen set of axes.

- Invariably, the mass is in equilibrium along the perpendicular axis along which the normal reaction force acts, *i.e.*, the acceleration along this axis is zero. Apply Newton's second law of motion to the net, resultant force along this direction (net force $= 0$, because the acceleration is zero), and determine the normal reaction force R.

- Next the friction force which is along the other axis is calculated from $F_f = \mu R$.

Example 3.2: A mass m placed on a rough horizontal plane is pulled along the plane with a force **F**, applied at an angle θ above horizontal. The coefficient of friction is μ. Find the normal reaction and the friction forces.

Solution: In this example and the ones that follow, we shall not draw the freebody diagrams explicitly. All forces, and their components are shown in one single diagram of the system. Students who wish to take the freebody diagram route, must try so for practice.

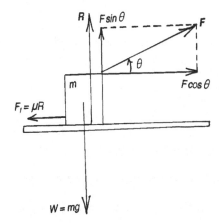

Figure 3.5: Forces and their components for the system of Example 3.2.

Figure (3.5) shows the forces acting on the mass, which are:
Weight of the mass: $W = m\,g$ acting vertically down.
Normal reaction force: R acting vertically up.
Pulling force F applied at angle θ above horizontal.
Force of friction: F_f horizontally along the plane, opposite to the direction of motion.

We choose the $x-$axis parallel to the horizontal plane and the $y-$axis perpendicular to it along the vertical. Forces W, R, and F_f are aligned parallel to the chosen axes as shown in the Figure. We only need to resolve force F into components. These components are:

$F_x = F\,cos\,\theta$ along the $(+)x-$axis, and $F_y = F\,sin\,\theta$ along the $(+)y-$axis.

Applying Newton's second law to the net force along the $y-$axis, we get: $(R + F\,sin\,\theta - m\,g) = 0$ because the acceleration along $y-$axis is zero. This gives:

$$R = (m\,g - F\,sin\,\theta)$$

We note that the normal reaction force in this case is less than the weight of the mass. This is so because a component of the pulling force is partially lifting (supporting) the mass upwards, and the mass does not press against the plane with its full weight.

Next, the friction force is obtained from the normal reaction force as:

$$F_f = \mu\,R = \mu\,(m\,g - F\,sin\,\theta)$$

Note that the pulling of a mass with a force at an angle results in reduced friction, which makes pulling of objects easier.

Exercise 3.2: A mass m placed on a rough horizontal plane is pushed along the plane with a force **F**, applied at an angle θ above horizontal. The coefficient of friction is μ. Find the normal reaction and friction forces. Is it better to push or to pull in order to move a heavy object along a surface? Explain why.

3.3.3 Motion along horizontal and inclined planes without friction

Recipe 3.3: Motion along planes without friction.

- Identify all the forces acting on the mass.

- Choose an appropriate set of perpendicular axes. Generally one of the axis is taken along the plane of motion and the other axis is perpendicular to it.

- Resolve all forces into perpendicular components along the chosen set of axes.

- Find net resultant forces along each of the axes from the algebraic sum of the corresponding components of the forces.

- Since there is no friction force, the net force perpendicular to the plane in this case is of no interest. However, if desired one may calculate the normal reaction force by equating the net perpendicular force to zero (why?).

- Apply Newton's second law of motion to the net force parallel to the plane of motion, and determine the acceleration of the mass along the plane, and (or) other quantities that may be required.

- Repeat the procedure for each mass in the system to determine its motion. Remember that each mass has its own set of perpendicular axes.

Example 3.3: A mass m freely slides down a smooth plane inclined at an angle θ above horizontal. Find the acceleration of the mass down the plane.

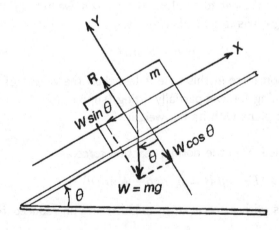

Figure 3.6: Forces and their components for the system of Example 3.3.

Solution: We take the x−axis parallel to the plane directed up, and the y−axis is perpendicular to the plane also directed up. The mass slides down the plane along the $(-)x$−axis. Figure 3.6 shows the forces acting on the mass, and their components along the perpendicular axes.

The forces acting on the mass are:
Weight of the mass: $W = m\,g$ acting vertically down.
Normal reaction force R perpendicular to the plane along the $(+)y$−axis.

We only need to resolve the weight W of the mass into components along the axes, which are:
The component along the x−axis: $W_x = (-)\,m\,g\,sin\,\theta$ acting down the plane.
The component along the y−axis: $W_y = (-)\,m\,g\,cos\,\theta$ acting opposite to the normal reaction force.

Applying Newton's second law to the net forces along the perpendicular axes gives:
Normal reaction force $R = m\,g\,cos\,\theta$. (This in fact is not required for this example).
From the forces along the plane: $W_x = (-)\,m\,g\,sin\,\theta = m\,a_x$. This gives $a_x = (-)\,g\,sin\,\theta$

Thus the mass slides down the plane with an acceleration $(-)\,g\,sin\,\theta$

Exercise 3.3: A mass m placed on a smooth plane inclined at an angle θ above horizontal is pulled up the plane with a force F applied at an angle α above the plane. Find the acceleration of the mass up the plane.

3.3.4 Motion along horizontal and inclined planes with friction

These problems are similar to the problems of motion along planes without friction described in subsection 3.3.3, except that in this case a force of friction acting opposite to the direction of the motion is present. Therefore, recipe 3.3 with modifications as follows applies to this class of problems and applications.

Recipe 3.4: Motion along planes with friction.

- Following recipe 3.3 first determine the normal reaction force from the net force perpendicular to the plane of motion.

- From the normal reaction force calculate the force of friction which acts along the plane of motion opposite to the direction of motion.

- Next proceed with the net force parallel to the plane of motion to determine the acceleration of the mass.

- Repeat the procedure for each mass.

Example 3.4: A mass m placed on a rough plane inclined at an angle θ above horizontal is pulled up the plane with a force F applied at an angle α above the plane. The coefficient of friction for the mass on the plane is μ. Find the acceleration of the mass up the plane.

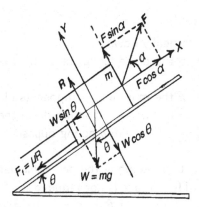

Figure 3.7: Forces and their components for the system of Example 3.4.

Solution: We choose the perpendicular axes as in Example 3.3. Figure (3.7) shows the forces and their components acting on the mass.

The forces acting on the mass are:
Weight of the mass: $W = mg$ acting vertically down.
Normal reaction force R perpendicular to the plane along the $(+)y$−axis.
The pulling force F at angle α above the inclined plane.
Force of friction: $F_f = \mu R$ down the plane, opposite to the direction in which the mass is pulled.

We only need to resolve weight W and force F into components along the axes. These components are:
Component of W along the x−axis: $W_x = (-)mg\sin\theta$ acting down the plane.
Component of W along the y−axis: $W_y = (-)mg\cos\theta$ acting opposite to the normal reaction force.
Component of F along the x− axis: $F_x = F\cos\alpha$ along the $(+)x$−axis.
Component of F along the y− axis: $F_y = F\sin\alpha$ along the $(+)y$−axis.

From the y− components of the forces we get the normal reaction force as: $R = (mg\cos\theta - F\sin\alpha)$, which gives the friction force as:

$$F_f = \mu R = \mu(mg\cos\theta - F\sin\alpha)$$

Applying Newton's second law to the x−components of forces, we get the acceleration of the mass up the plane as:

$$a_x = \frac{F\cos\alpha - mg\sin\theta - \mu(mg\cos\theta - F\sin\alpha)}{m}$$

The details are left for the students to workout.

Exercise 3.4: A mass m freely slides down a rough plane inclined at an angle θ above horizontal. The coefficient of friction for the plane is μ. Find the acceleration of the mass down the plane.

3.3.5   Weightlessness: The elevator problem

Those who had an opportunity to use an elevator (lift) to go up or down in a multi-storey building must have experienced the following:

- When the elevator starts from rest to go upwards, one finds oneself pressing against the elevator floor with an increased force as if one's weight had increased. Same sensation of increased weight is experienced when the elevator traveling downwards stops at a lower floor. In both these cases the elevator accelerates upwards.

- When the elevator traveling from a lower floor stops at a higher floor, a sensation of decreased reaction force against the elevator floor is experienced, as if one's weight was decreased. Same sensation of decreased weight is experienced when the elevator at an upper floor starts from rest to go down. In these cases the elevator accelerates downwards.

- No sensation of changed weight is experienced when the elevator is moving in between the floors at constant speed. In this case the acceleration of the elevator is zero.

Similar sensations of changed weight are experienced when one rides a giant-wheel rotating in a vertical plane. When the wheel goes up, an increase in weight is felt, and when it goes down, the weight appears to have decreased from the normal. These experiences are not just sensations, but in reality if one stood on a spring balance during these motions, the scale will in fact register changes in weight, *i.e.*, it shall show increased or decreased weight in accordance with the respective sensation. This happens because of the acceleration during these motions which translates into a force, resulting in the changed weight. When the elevator and the giant wheel accelerate upwards the weight increases, and when they accelerate downwards the weight decreases. In between the floors the elevator moves with a constant speed, and no change in weight is experienced.

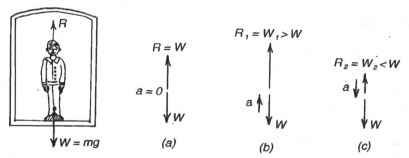

Figure 3.8: Forces on a person in an elevator: (a) at rest or moving with constant speed, (b) accelarating upwards, and (c) accelerating downwards.

The forces acting on a person of mass m standing on a (elevator) floor are his weight $W = mg$ acting vertically downwards, and the normal reaction R of the floor acting vertically upwards (Figure

3.8 (a)).

When the elevator is at rest or is moving with a constant speed, the acceleration $a = 0$, so that the net force on the person: $(R - mg) = 0$. This gives: $R = mg$. If the person stood on a spring balance instead of the floor, the balance shall record the normal reaction force R of the spring . Hence, the normal reaction force R is the weight of the person, which in this case is equal to the true weight W, *i.e.*, $R = mg = W$. Now consider that the elevator is accelerating upwards with an acceleration a (Figure 3.8 (b)), and let R_1 be the normal reaction force, which is the apparent weight W_1 recorded by the spring balance. In order to accelerate upwards, there must be a net upwards force on the person, *i.e.*, $R_1 > mg$, and from Newton's second law:

$$
\begin{aligned}
(R_1 - mg) &= ma \\
\text{or} \qquad R_1 &= (mg + ma) = W_1
\end{aligned}
\tag{3.18}
$$

Thus in this case the apparent weight $W_1 = R_1 = (mg + ma) = (W + ma) > W$.

When the elevator accelerates downwards with acceleration a (Figure 3.8 (c)), let the normal reaction force be R_2. The apparent weight W_2 is given by:

$$
R_2 = (mg - ma) = W_2
\tag{3.19}
$$

The detailed explanation and working out of equation (3.19) is left as an exercise for the students.

From equation (3.19), we note that $W_2 = R_2 = (mg - ma) = (W - ma) < W$. The decreased weight is the sensation of partial weightlessness. If the elevator falls freely under gravity $(a = g)$, then the apparent weight $W_2 = 0$ and one shall experience a complete weightlessness. It is this weightlessness that the astranauts experience because in their orbital motion they are virtually in a state of free fall. What shall be the weight recorded by a spring balance, if one tied it to one's feet and jumped out from a high window? (Do not try to experiment this to find the answer!).

3.3.6   Mass-pulley systems

These category of problems involve two or more masses connected by string(s), and the strings pass over the pulleys. The mass(es) may be hanging freely, or may be placed on horizontal or inclined planes, with or without friction, or any combination of these possibilities. We consider an ideal system with the following simplifying assumptions.

- String is massless and unstreachable.

- The pulley is massless and frictionless.

The consequences of these assumptions are:

- Both masses on either side of the string have the same magnitudes of acceleration and velocity.

- The tension in the string on both sides of the pulley is the same.

To solve this category of problems, one follows the following procedure.

Recipe 3.5: Mass-pulley problems.

- Identify all the forces including tension(s) in the string(s) acting on each mass, whether hanging freely or placed on a plane. For mass(es) placed on a plane this shall also involve the normal reaction and friction forces.

- Choose an appropriate set of perpendicular axes for each mass, one of the axis being along the direction of motion of the mass. Resolve all forces on the mass along the chosen set of axes.

- For the masses placed on planes, determine the normal reaction and the friction forces.

- Apply Newton's second law of motion to the components of forces along the direction of motion for each mass, and determine their acceleration, or calculate whatever may be required.

Example 3.5: Two masses m_1 and m_2 are attached with a string which passes over a pulley. Mass m_1 rests on a smooth horizontal plane, whereas mass m_2 hangs freely along the vertical from over the pulley fixed to the edge of the plane. Determine the acceleration of the masses and tension in the string. State the assumptions made to solve the problem.

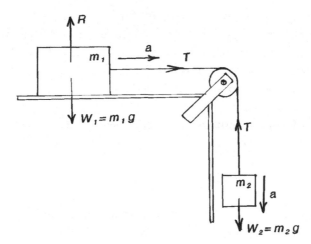

Figure 3.9: Forces on the system of Example 3.5.

Solution: The assumptions made are: *(i)* the pulley is frictionless and massless, and *(ii)* the string is massless and unstretchable.

Let T be the tension in the string and a be the common acceleration of the masses as shown in Figure (3.9). The choice of axes for both the masses are also shown in the figure. Applying Newton's second law to the net forces on each mass along the direction of motion, we get:

$$T = m_1 a$$
$$T - m_2 g = -m_2 a$$

Solving these equations for a and T we get:

$$a = \frac{m_2 g}{(m_1 + m_2)}$$
$$T = \frac{m_1 m_2 g}{(m_1 + m_2)}$$

Details are left as an exercise for the students to workout.

Exercise 3.5: Repeat example (3.5) if the coefficient of friction for the horizontal plane is μ.

3.3.7   Condition of equilibrium

There are two states of equilibrium for a body, the state of translational equilibrium and the state of rotational equilibrium. These two states are described by two laws (conditions) of equilibrium, namely the 'first law of equilibrium' and the 'second law of equilibrium' respectively. Here we deal only with the first law, the second law is discussed in the chapter on rotational motion.

A body is said to be in a state of translational equilibrium if it is at rest, or is moving with a uniform velocity, *i.e.*, with a velocity which is constant in both magnitude and direction. In both the cases of rest and uniform motion, the acceleration of the body is zero, and hence from Newton's second law of motion, the net external force on the body must be zero. This defines the *First law (condition) of equilibrium*, also known as the *Law of translational equilibrium*.

Statement of the First law of equilibrium: *A body is in a state of translational equilibrium if the net external force acting on it is zero, i.e.,*

$$\mathbf{F_{ext}} = \sum_i \mathbf{F_i} = 0 \tag{3.20}$$

where $\mathbf{F_1}$, $\mathbf{F_2}$ $\mathbf{F_3}$, ...$\mathbf{F_n}$ ... are the various external forces acting on the body. In a 3-D space, the law applies to each component of the net external force, independent of the other components. For example if the net x-component of the external force is zero, then the object is in the state of translational equilibrium along the x-axis, whereas along the other axes it may or may not be so depending on the net forces along those axes.

Comparing this to the Newton's first law of motion, one notes some similarity. There also it was stated that a body shall continue to be in the state of rest or uniform motion (both are states of equilibrium) unless the state was changed by applying an external force. However, the two laws can not be treated as equal, the reason being that the condition of equilibrium is just that, and does not imply the inertial feature of the body which is an important aspect of the first law. Likewise, implicitly the first law of equilibrium does point to the conservation of linear momentum as well, but does not state it explicitly, and hence it can not be used to substitute for the law of conservation of linear momentum.

As far as dealing with the problems involving equilibrium is concerned, one does not need a specific recipe or examples for this. One simply need to determine the net external force on a body in a particular direction, and from that conclude whether the body is in a state of translational equilibrium or not along that direction or determine the unknown force(s) which shall lead to the state of equilibrium.

3.4 Problems

3.1 An electron moving with a velocity of $4 \times 10^3 \, m \, s^{-1}$ passes through a uniform electric field, and after $2 \times 10^{-7} \, s$ emerges with a velocity of $5 \times 10^5 \, m \, s^{-1}$ along the original direction of motion. Calculate the average electrostatic force on the electron.

3.2 A $2 \, kg$ mass moving with a velocity of $\mathbf{u} = (10 \, \mathbf{i} + 8 \, \mathbf{j} - 12 \, \mathbf{k})$ is subjected to a constant external force. After $30 \, s$ the velocity of the object is found to be $\mathbf{v} = (8 \, \mathbf{i} + 10 \, \mathbf{j} + 12 \, \mathbf{k})$. Determine the force applied to the mass.

3.3 A tow-truck is pulling a car of mass $1,500 \, kg$ up an inclined plane of 37° slope with an average coefficient of kinetic friction 0.3 (Note that in this case, friction is in fact the rolling friction which we are representing by an average kinetic friction). The tow-rope is parallel to the incline-surface. Calculate the tension in the rope if: *(i)* they are moving up the incline plane with a constatnt speed of $60 \, km \, h^{-1}$. *(ii)* They accelerate up the plane at a rate of $2 \, m \, s^{-2}$. *(iii)* If the rope suddenly breaks when the car was moving at a constant speed of $60 \, km \, h^{-1}$ what shall happen to the car? Describe the different stages of the motion of the car briefly and clearly, and calculate values of related acceleration(s).

3.4 A $10 \, kg$ mass placed on a plane inclined at an angle of 37° above horizontal is acted upon by three forces: $\mathbf{F_1} = 100 \, N$ applied horizontally pushes the mass up the plane, $\mathbf{F_2} = 50 \, N$ applied perpendicular to the plane presses the mass against the plane, and $\mathbf{F_3} = 700 \, N$ applied parallel to the plane pulls the mass up the plane. Calculate the acceleration of the mass up the plane if: *(i)* the plane is smooth, and *(ii)* the coefficient of kinetic friction for the plane is 0.1.

3.5 Directions of all the three forces $\mathbf{F_1}$, $\mathbf{F_2}$ and $\mathbf{F_3}$ in problem (3.4) are reversed. Calculate the acceleration of the mass down the plane for both cases, *i.e.*, when the plane is smooth, and when the coefficient of the kinetic friction is 0.1.

3.6 The force needed to impart an acceleration of $1.5 \, m \, s^{-2}$ to a locomotive of mass $20,000 \, kg$ on a level track is $36 \, kN$. What is the average force of friction, and the average coefficient of friction (assuming it to be the sliding friction)? What shall be the force required if the locomotive was to be pulled up the tracks inclined at an angle of 15° above horizontal with the force applied parallel to the tracks assuming the same coefficient of friction?

3.7. In Example (3.2), section 3.3.2, find the acceleration of the mass along the horizontal surface. If the coefficient of static friction for the mass-surface is μ_s, what is the maximum pull force that can be applied to the mass so that it shall remain at rest?

3.8. In Exercise (3.2), section 3.3.2, find the acceleration of the mass along the plane. If the co-efficient of static friction for the mass-surface is μ_s, what is the maximum push force that can be applied to the mass so that it shall remain at rest?

3.9 A mass m placed on a smooth plane inclined at an angle θ above horizontal is pushed up the plane with a force F applied at an angle α above the plane. Find the acceleration of the mass up the plane.

3.10. In Example (3.4), section 3.3.4, if the coefficient of static friction for the mass-surface is μ_s, what is the maximum pull force that can be applied to the mass so that it shall remain at rest?

3.11 A mass m placed on a rough plane inclined at an angle θ above horizontal is pushed up the plane with a force F applied at an angle α above the plane. The coefficient of friction for the plane is μ. Find the acceleration of the mass up the plane. If the coefficient of static friction for the mass-surface is μ_s, what is the maximum push force that can be applied to the mass so that it shall remain at rest.

3.12 A mass m placed on a rough plane inclined at an angle θ above horizontal is pushed down the plane with a force F applied at an angle α above the plane. The coefficient of friction for the plane is μ. Find the acceleration of the mass down the plane.

3.13 A mass m placed on a rough plane inclined at an angle θ above horizontal is pulled down the plane with a force F applied at an angle α above the plane. The coefficient of friction for the plane is μ. Find the acceleration of the mass down the plane.

3.14 A person of weight $750\,N$ stands on a spring balance inside an elevator. Calculate the acceleration of the elevator if the successive readings on the spring balance shows the weight of the person to be *(i)* $1000\,N$, *(ii)* $750\,N$ *(iii)* $500\,N$, and *(iv)* $0\,N$. Describe the motion of the elevator for these sequence of weight readings.

3.15 A mass m is suspended from the ceiling of an elevator with a spring balance. When the elevator is moving upwards with an acceleration of $3\,m\,s^{-2}$, reading on the spring balance shows the weight of the mass to be $25\,N$. What shall be the weight registered if the elevator was moving down with the same acceleration?

3.16. Two masses m_1 and m_2 attached to the two end of a string passing over a pulley hang freely. When released, the masses move freely under gravity. Find the acceleration of the masses and tension in the string if: *(i)* $m_1 > m_2$, and *(ii)* $m_1 < m_2$.

3.17. Two masses m_1 and m_2 are attached to the two ends of a string that passes over a pulley.

Mass m_1 hangs freely, and mass m_2 is placed on a shooth plane inclined at an angle θ above horizontal. Find the acceleration of the masses and tension in the string if: *(i)* mass m_2 slides up the plane, and *(ii)* mass m_2 slides down the plane.

3.18. Two masses m_1 and m_2 are attached to the two ends of a string that passes over a pulley. Mass m_1 hangs freely, and mass m_2 is placed on a rough plane inclined at an angle θ above horizontal. The coefficient of friction for the mass-plane is μ. Find the acceleration of the masses and tension in the string if: *(i)* mass m_2 slides up the plane, and *(ii)* mass m_2 slides down the plane. *(iii)* What are the maximum and minimum values of mass m_1 in terms the other given parameters which shall maintain mass m_2 at rest on the inclined plane?

3.19. Two masses m_1 and m_2 are attached to the two ends of a string that passes over a pulley. Mass m_1 is placed on a plane inclined at angle α, and mass m_2 is placed on another plane inclined at angle β above the horizontal, and the pulley is fixed at the common apex of the planes. Both the planes are smooth. Calculate the acceleration of the masses and tension in the string if: *(i)* mass m_1 slides up the plane, *(ii)* mass m_1 slides down the plane, and *(iii)* Find the ratio of the masses if they are in a state of equilibrium.

3.20. Two masses m_1 and m_2 are attached to the two ends of a string that passes over a pulley. Mass m_1 is placed on a rough plane inclined at angle α, and mass m_2 is placed on another rough plane inclined at angle β above the horizontal, and the pulley is fixed at the common apex of the planes. The coefficients of friction of the two planes are μ_1 and μ_2 respectively. Calculate the acceleration of the masses and tension in the string if: *(i)* mass m_1 slides up the plane, and *(ii)* mass m_1 slides down the plane. *(iii)* What are the maximum and minimum values of mass m_1 in terms the other given parameters which shall maintain mass m_2 at rest on the plane?

3.21. Two strings are attached to the two opposite sides of a mass m placed on a smooth horizontal plane. The strings are pulled in the opposite directions, one along the $(+)x-$ axis and the other along the $(-)x-$axis, and pass over the pulleys fixed to the opposite edges of the plane. Other ends of the strings are attached to two masses m_1 and m_2 which hang freely. Find the acceleration of mass m and tensions in the strings if: *(i)* mass m_1 ascends, and *(ii)* mass m_1 descends.

3.22. Two strings are attached to the two opposite sides of a mass m placed on a rough horizontal plane with coefficient of friction μ. The strings are pulled in the opposite directions, one along the $(+)x-$ axis and the other along the $(-)x-$axis, and pass over the pulleys fixed to the opposite edges of the plane. Other ends of the strings are attached to two masses m_1 and m_2 which hang freely. Find the acceleration of mass m and tensions in the strings if: *(i)* mass m_1 ascends, and *(ii)* mass m_1 descends. *(iii)* What are the maximum and minimum values of mass m_2 in terms of the other given parameters which shall maintain mass m in equilibrium?

3.23 A $5\,kg$ mass is pulled at a constant speed up a plane inclined at an angle θ above horizontal by a $2\,kg$ mass attached to it by a string that passes over the pulley, and hangs freely. Calculate the angle θ, and tension in the string if: *(i)* the plane is smooth, and *(ii)* the coefficient of kinetic friction is 0.1.

3.24 A $10\,kg$ mass placed on a rough ($\mu = 0.1$) plane inclined at an angle of 37° above horizontal is attached to a $7\,kg$ mass with a string that passes over a pulley and hangs freely. *(i)* Will the $10\,kg$ mass slide up the plane or down the plane? *(ii)* Determine its acceleration and tension in the string. *(iii)* If the coefficient of static friction is 0.15, what are the minimum and maximum mass that can be supported on the plane in equilibrium by the freely hanging $7\,kg$ mass?

3.25 Two masses m_1 and m_2 are connected with a string that passes over a pulley. Mass m_2 hangs freely, whereas mass m_1 is placed on a smooth horizontal plane, and is pulled away from the pulley by a horizontal force F. Find the acceleration of the masses and tension in the string if mass m_1: *(i)* ascends, *(ii)* descends. *(iii)* For what value and direction of force F the tension in the string is zero?

3.26 A $20\,kg$ mass suspended with a sting is pulled to the side with a horizontal force F till the string makes an angle of 37° from the vertical. Calculate the tension in the string and the force applied.

3.27 In Example (3.1), section 3.3.1, determine the tension in the three cables.

3.28. In Exercise (3.1), section 3.3.1, if the coefficient of friction between mass m_1 and the horizontal surface is μ, determine the tension in each of the strings.

3.29 A $5\,kg$ mass placed on a rough plane inclined at 30° above horizontal is just about to slide down. *(i)* Calculate the coefficient of static friction between the plane and the mass. *(ii)* If the coefficient of kinetic friction for the mass-plane is 0.4, calculate the acceleration of the mass down the plane. *(iii)* Calculate the force that should be applied to the mass parallel to the plane so that the mass moves with a constant velocity: *(a)* up the plane, *(b)* down the plane.

3.30 A mass $m_1 = 10\,kg$ is placed on a rough horizontal plane, and is attached to a string which passes over a pulley. The other end of the string is attached to a mass $m_2 = 5\,kg$ and hangs by the side of the plane. A horizontal force $\mathbf{F}$ is applied at the mid-point of the string with which m_2 hangs pulling it away so that the upper half of the segment makes an angle of 37° from the vertical. The mass m_1 is just about to slide. *(i)* Calculate tensions in each segment of the string and the force F. *(ii)* Determine the coefficient of static friction between the mass and the surface.

Chapter 4

Conservation of Linear Momentum

Two most important properties of physical systems are the energy and the momentum. There are two types of momenta, one associated with the translational motion, called *linear momentum* or simply *momentum* and the other associated with the rotational motion, known as *angular momentum* or *rotational momentum*. Both energy and momentum follow conservation principles under certain conditions, which have wide ranging applications not only in mechanics, but in almost all branches of science and technology. These conservation principles have led to the discoveries of new laws, principles and properties on one hand, and to the development of new machines and systems on the other. For example, Einstein's Nobel Prize winning discovery of the "photoelectric effect equation" is nothing but the statement of the law of conservation of energy. Likewise, the law of conservation of momentum led to the discovery of the fundamental particle "neutrino" in beta decay, and has application in rocket propulsion and design of guns and heavy artillery. It is for these reasons that both energy and momentum and laws governing them are given far greater attention in mechanics than any other principle, and their clear understanding is crucial to the understanding of Physics as well as other branches of science and technology. In this chapter we shall devote ourselves to the principles governing the linear momentum, and their applications. The energy and angular momentum will be dealt with in other chapters. At the end of this chapter, it is expected that students shall have acquired skills of applying the principle of conservation of linear momentum.

4.1 Linear momentum and its conservation

Definition 4.1: *The (linear)momentum* **P** *of an object of mass m, moving with velocity* **v** *is defined as the product of the mass and the velocity as:*

$$\mathbf{P} = m\mathbf{v} \tag{4.1}$$

Its SI unit is $kg\ m\ s^{-1}$ which can also be expressed as $N\ s$ (students, verify this!)

The velocity being a vector quantity and mass a scalar quantity, the momentum is thus a vector quantity directed along the direction of the velocity with magnitude equal to the magnitude of the velocity scaled by a factor equal to the mass of the object (Figure 4.1).

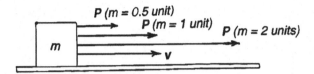

Figure 4.1: Linear momentum of a mass m moving with velocity **v**.

In 3-D space, momentum can be expressed in terms of three components, for example in Cartesian coordinates the components are:

$$\mathbf{P} = P_x\mathbf{i} + P_y\mathbf{j} + P_z\mathbf{k} = (P_x, P_y, P_z)$$

where

$$P_x = m\,v_x, P_y = m\,v_y, \quad \text{and} \quad P_z = m\,v_z, \tag{4.2}$$

4.1.1   Conservation of momentum

The Newton's second law of motion for an object of constant mass m acted upon by an external force **F** can be expresses in terms of the momentum **P** of the object as follows:

$$
\begin{aligned}
\mathbf{F} &= m\mathbf{a} = m \times \frac{\mathbf{v_f} - \mathbf{v_i}}{t_2 - t_1} = \frac{m\mathbf{v_f} - m\mathbf{v_i}}{t_2 - t_1} \\
&= \frac{\mathbf{P_f} - \mathbf{P_i}}{t_2 - t_1} = \frac{\Delta \mathbf{P}}{\Delta t} = \frac{d\,\mathbf{P}}{dt}
\end{aligned}
\tag{4.3}
$$

where $\mathbf{v_i}$ and $\mathbf{v_f}$ are the initial and final velocities of the mass at times t_1 and t_2 respectively, and other symbols have their usual meanings. Thus, from equation (4.3), the rate of change of momentum of the object is equal to the external force acting upon it. If the external force is equal to zero, then the rate of change of the momentum is zero, *i.e.* the momentum of the body remains constant. This is the law of conservation of momentum.

Definition 4.2 Law of conservation of momentum: *If the net external force acting on a body is zero, its (linear) momentum is conserved.*

In a 3-D space, equation (4.3) can be expressed in terms of the components of the force and the momentum as three equations. For example, in Cartesian coordinates the three equations are:

$$F_x = \frac{d\,P_x}{dt}, \ F_y = \frac{d\,P_y}{dt}, \quad \text{and} \quad F_z = \frac{d\,P_z}{dt}, \tag{4.4}$$

The law of conservation of momentum applies to each of the components independently. For example, if $F_x = 0$, then the $x-$component of the momentum (P_x) is conserved irrespective of the value of the other components of the force and momentum.

4.1.2 Impulse of force and momentum

Definition 4.3: *If a force* **F** *acts on a body for a duration* Δt, *the impulse* **I** *of the force is defined as:* $\mathbf{I} = \mathbf{F}\,\Delta t$.

Its SI units are Ns. It is a vector quantity directed along the direction of the force. It is left as an exercise for students to express the three components of impulse in a 3-D space in terms of the three components of the force.

From equation (4.3), impulse of the force is related to the change in the momentum of the body as:

$$\mathbf{I} = \mathbf{F}\,\Delta t = \Delta \mathbf{P}$$

or in terms of the components:

$$I_x = \Delta P_x,\ I_y = \Delta P_y,\ \text{ and }\ I_z = \Delta P_z, \tag{4.5}$$

From equation (4.5), if the total impulse or any of its component is zero, the change in the total momentum or the change in the corresponding component of the momentum respectively is zero. This is the law of conservation of momentum in terms of the impulse of the force. It is particularly useful when a force acts for an infinitesimal small duration as in the case of collisions and explosions, and the force can not be measured conveniently but changes in momentum can be measured.

Recipe 4.1: Application of conservation of momentum.

- Identify all external forces, and express them in terms of their components.

- Express all given velocities in terms of their components, and calculate the corresponding components of momentum.

- Find the rate of change for each component of momentum.

- By equating the components of forces to the corresponding components of the rate of change of momentum, calculate the unknown quantities, or draw conclusions about which components of momentum are conserved and which components of the total external force must be zero.

- A similar approach is applied if the impulse of force is used. By equating the changes in the various components of momentum to the corresponding components of impulse, the duration of force or other unknown parameters can be calculated.

Example 4.1: A force $\mathbf{F} = (10N, 37^o)$ acts on a 500 g mass for 20 s moving with velocity $\mathbf{u} = (3\mathbf{i} + 2\mathbf{j} - \mathbf{k})\ ms^{-1}$ at $t = 0$. Calculate: *(i)* the rate of change of momentum of the mass, and *(ii)* its velocity at $t = 20$ s.

Solution: The components of the external force are: $F_x = 10\cos 37^o = 8N$, $F_y = 10\sin 37^o = 6N$, and $F_z = 0$

(i) Since $F_z = 0$ the z−component of momentum is conserved. The rate of change of the components of momentum are:

$$\frac{d\,P_x}{dt} = 8\,N$$
$$\frac{d\,P_y}{dt} = 6\,N$$
$$\frac{d\,P_z}{dt} = 0$$

and $\quad \dfrac{d\,\mathbf{P}}{dt} = (\mathbf{i}\,8 + \mathbf{j}\,6)\,N \qquad\qquad (4.6)$

(ii) The changes in the three components of momentum after 20 s are:

$$
\begin{aligned}
\Delta P_x &= F_x \times \Delta t = 8 \times 20 = 160\,Ns = 0.5 \times (v_x - 3) \\
\Delta P_y &= F_y \times \Delta t = 6 \times 20 = 120\,Ns = 0.5 \times (v_y - 2) \\
\Delta P_z &= F_z \times \Delta t = 0 = 0.5 \times (v_z + 1)
\end{aligned}
\qquad (4.7)
$$

Solving for v_x, v_y and v_z, we get the final velocity of the mass as: $\mathbf{v} = (\mathbf{i}\,323 + \mathbf{j}\,242 - \mathbf{k})ms^{-1}$

Exercise 4.1: The velocity of a 4 kg mass at $t = 0$ is $\mathbf{u} = (\mathbf{i}\,5 + \mathbf{k}\,4)ms^{-1}$ and at $t = 30\,s$ is $\mathbf{v} = (\mathbf{i}\,25 + \mathbf{j}\,10 - \mathbf{k}\,20)ms^{-1}$. Calculate the external force on the mass. What additional external force $\mathbf{F_2}$ is required so that the y-component of the momentum would remain conserved?

4.2   Collision and the conservation of momentum

When two or more objects approaching each other (some may be at rest) collide, on coming in contact they exert forces on each other. As a result of this the momentum of each object involved in the collision changes. The change in the momentum of each objects is such that the total momentum in the collision process is conserved, i.e., the total momentum of the objects just before the collision is equal to the total momentum just after the collision. The problem of the conservation of momentum on collision can be analyzed in the following three different ways leading to the same conclusions. Here, for the sake of simplicity, we shall consider collision between two objects only, but the discussion applies to any number of colliding objects.

(i)Analysis in terms of the internal and external forces: Let us visualize the two colliding objects constituting a closed physical system. Then the forces of collision are internal to the system. If there are no other external forces present, from the principle of the conservation of momentum, the total momentum of the system which is equal to the sum of the momentum of individual objects must be conserved despite the collision. Let m_1 and m_2 be the two the masses, $\mathbf{u_1}$ and $\mathbf{u_2}$ be their velocities just before the collision, and $\mathbf{v_1}$ and $\mathbf{v_2}$ are the velocities immediately after the collision. Then:

$$\mathbf{P_i} = m_1\mathbf{u_1} + m_2\mathbf{u_2} = m_1\mathbf{v_1} + m_2\mathbf{v_2} = \mathbf{P_f} \qquad\qquad (4.8)$$

Equation (4.8) is a vector equation. In a 3-D space this gives the following three independent equations in Cartesian coordinates.

$$P_{ix} = m_1 u_{1x} + m_2 u_{2x} = m_1 v_{1x} + m_2 v_{2x} = P_{fx}$$
$$P_{iy} = m_1 u_{1y} + m_2 u_{2y} = m_1 v_{1y} + m_2 v_{2y} = P_{fy}$$
$$P_{iz} = m_1 u_{1z} + m_2 u_{2z} = m_1 v_{1z} + m_2 v_{2z} = P_{fz} \qquad (4.9)$$

However, in the case of the collision of two particles, three component-equations given by (4.9) is an over specification of the problem. For two particles approaching each other with velocities $\mathbf{u_1}$ and $\mathbf{u_2}$, one can always define a 2-D plane containing the velocity vectors $\mathbf{u_1}$ and $\mathbf{u_2}$. Let this plane be the $x - y$ plane. On collision, the forces of collision exerted by the particles on each other lie within the same plane, and have no components perpendicular to the plane of motion. As there is no other mechanism available to impart momentum to the particles in the direction perpendicular to the plane of motion, the particles continue to move within the same plane after collision. Thus the problem of collision of two particles is a 2-D problem, and one requires only the first two $x-$ and $y-$component equations out of the three (4.9) equations. The problem is discussed in further detail in section (4.5.2).

(ii) Analysis in terms of the forces of collision: Let $\mathbf{F_1}$ and $\mathbf{F_2}$ be the average forces exerted on the two objects on collision. The two forces constitute the pair of action and reaction forces, and from the Newton's third law of motion they must be equal and opposite, *i.e.*, $\mathbf{F_1} = (-)\mathbf{F_2}$. From the Newton's second law (equation 4.3), the rate of change of momentum of the two objects on collision is:

$$\mathbf{F_1} = \frac{d\,\mathbf{P_1}}{dt} \qquad (4.10)$$

$$\mathbf{F_2} = \frac{d\,\mathbf{P_2}}{dt} \qquad (4.11)$$

Adding equations (4.10) and (4.11) one gets:

$$\mathbf{F_1} + \mathbf{F_2} = 0 = \frac{d\,\mathbf{P_1}}{dt} + \frac{d\,\mathbf{P_2}}{dt} = \frac{d\,(\mathbf{P_1} + \mathbf{P_2})}{dt} \qquad (4.12)$$

Thus, the rate of change of total momentum of the system during collision is zero and thus the momentum of the system remains constant, *i.e.*, $\mathbf{P_1} + \mathbf{P_2} = $ constant. This in turn means that the total momentum in a collision event is conserved leading to equations (4.8) and (4.9).

(iii) Analysis in terms of the impulse of the forces of collision: Let Δt be the duration for which the objects are in contact during collision. Then the impulse of the forces of collision on each object are $\mathbf{F_1}\,\Delta t$ and $\mathbf{F_2}\,\Delta t$, and from the principle of action and reaction forces $\mathbf{F_1}\,\Delta t = (-)\mathbf{F_2}\,\Delta t$. The change in momentum of each objects, in terms of the impulse of forces is given as:

$$\mathbf{F_1}\,\Delta t = \Delta\mathbf{P_1} \qquad (4.13)$$
$$\mathbf{F_2}\,\Delta t = \Delta\mathbf{P_2} \qquad (4.14)$$

Adding equations (4.13) and (4.14):

$$\mathbf{F_1}\,\Delta t + \mathbf{F_2}\,\Delta t = 0 = \Delta \mathbf{P_1} + \Delta \mathbf{P_2} \qquad (4.15)$$

Once again the change in momentum in the collision process is zero, *i.e.*, the momentum is conserved, and equations (4.8) and (4.9) apply.

4.3   Types of collision

Based on the relative motion of the two colliding objects after collision as compared to their relative motion before collision, collisions are divided into three categories: *(i) (perfectly) elastic collision, (ii) partially elastic or partially inelastic collision,* and *(iii) (perfectly) inelastic collision.* Let $\mathbf{u_1}$ and $\mathbf{u_2}$ be the velocities of two masses m_1 and m_2 respectively just before the collision, and $\mathbf{v_1}$ and $\mathbf{v_2}$ are the velocities immediately after the collision. To distinguish between the three types of collisions we define the following quantities.

Definition 4.4: The velocity of approach $(\mathbf{u_a})$ *is the velocity with which the two objects approach each other before collision. It is given as:* $\mathbf{u_a} = (\mathbf{u_1} - \mathbf{u_2})$

Definition 4.5: The velocity of separation $(\mathbf{v_s})$ *is the velocity with which the two objects separate from each other after collision. It is given as:* $\mathbf{v_s} = (\mathbf{v_2} - \mathbf{v_1})$.

Note the order in which the difference of the velocities of the first and the second object is taken to obtain $\mathbf{u_a}$ and $\mathbf{v_s}$. Explain why?

Definition 4.6: The coefficient of restitution (e) *for a collision process is defined as the ratio of the velocity of separation to the velocity of approach, i.e.,*

$$e = \frac{\mathbf{v_s}}{\mathbf{u_a}} = \frac{(\mathbf{v_2} - \mathbf{v_1})}{(\mathbf{u_1} - \mathbf{u_2})} \qquad (4.16)$$

The coefficient of restitution can take values from 0 to 1, and the three types of collisions are distinguished from each other in terms of the value of e as follows.

- If $e = 1$ then $(\mathbf{v_2} - \mathbf{v_1}) = (\mathbf{u_1} - \mathbf{u_2})$, *i.e.*, the velocity of separation of the objects after collision is equal to their velocity of approach before collision. This is called the *(perfectly) elastic collision.*

- If $e = 0$ then $(\mathbf{v_2} - \mathbf{v_1}) = 0$, *i.e.* the velocity of separation of the objects after collision is zero. This means that the object after collision are stuck together and move as one composite body of mass $m = (m_1 + m_2)$. This is called the *(perfectly) inelastic collision.*

- If $0 < e < 1$, then $(\mathbf{v_2} - \mathbf{v_1}) < (\mathbf{u_1} - \mathbf{u_2})$ and $(\mathbf{v_2} - \mathbf{v_1}) \neq 0$, *i.e.*, the two objects after collision move separately and their velocity of separation is less than their velocity of approach. This is called the *partially-elastic* or *partially-inelastic collision.*

In the discussion on the conservation of momentum in a collision process, we did not make any assumptions about the type of collision directly or implicitly. Hence, the conservation of momentum applies to all types of collisions; elastic and the non-elastic collisions. There is one simplifying aspect of momentum conservation in the case of inelastic collision, given by the following equation, obtained from equation (4.8):

$$m_1 \mathbf{u_1} + m_2 \mathbf{u_2} = (m_1 + m_2)\mathbf{v} = m\mathbf{v} \tag{4.17}$$

where $\mathbf{v}$ is the velocity of the composite mass $m = (m_1 + m_2)$ after the collision. Equation (4.17) can be expressed as two independent equations in terms of the components of the velocities as:

$$m_1 u_{1x} + m_2 u_{2x} = m v_x$$
$$m_1 u_{1y} + m_2 u_{2y} = m v_y \tag{4.18}$$

It is the elastic and the inelastic collisions that are of much interest for physical applications.

4.4 Energy considerations in collision

Although the work, energy and the principles that govern them are dealt with in a separate chapter where they are discussed in depth, here we consider the energy aspect of the collision process only for completeness sake, so that these aspects along with the conservation of momentum could be applied to physical systems involving collision. Let us first consider the elastic collision for which the coefficient of restitution $e = 1$. This, from equation (4.16), gives the following relation between the velocities of the colliding masses.

$$(\mathbf{v_2} - \mathbf{v_1}) = (\mathbf{u_1} - \mathbf{u_2})$$
or
$$(\mathbf{u_1} + \mathbf{v_1}) = (\mathbf{u_2} + \mathbf{v_2}) \tag{4.19}$$

Rearranging equation (4.8)for the conservation of momentum, we get:

$$m_1 (\mathbf{u_1} - \mathbf{v_1}) = m_2 (\mathbf{v_2} - \mathbf{u_2}) \tag{4.20}$$

Take the dot product of vector equations (4.19) and (4.20), multiply across with $\frac{1}{2}$, and simplify to obtain the following equation (this is left as an exercise for students):

$$\frac{1}{2} m_1 u_1^2 + \frac{1}{2} m_2 u_2^2 = \frac{1}{2} m_1 v_1^2 + \frac{1}{2} m_2 v_2^2 \tag{4.21}$$

The left side of equation (4.21) is the total kinetic energy of the objects before collision (KE_i), and the right side is the total kinetic energy of the objects after collision (KE_f) According to this equation $KE_i = KE_f$. Thus for elastic collision the kinetic energy of the colliding objects is conserved. This is a consequence of the fact that for elastic collision the velocity of separation is equal to the velocity of approach. In this context equations (4.19) and (4.21) are equivalent, not indepen-dent, and either of the two equations can be used to express this additional feature of elastic collision.

Equation (4.21) is a scalar equation, and it cannot be split into three independent component-equations like equations (4.9). However, it can still be expressed in terms of the components of the velocities by using expressions of the form $u_1^2 = (u_{1x}^2 + u_{1y}^2 + u_{1z}^2)$. (*Reminder:* Note that the collision of two particles can be reduced to a 2-D problem, and one uses expressions of the form: $u_1^2 = (u_{1x}^2 + u_{1y}^2)$.

For all non-elastic collisions, *i.e.* the inelastic and partially-elastic collisions, the kinetic energy is not conserved. In fact, in such collisions the kinetic energy of the system after collision decreases. In such cases, one only has the equations of conservation of momentum at ones disposal to solve problems.

Here the conservation and the non-conservation of kinetic energy for elastic and non-elastic collisions respectively should be clearly distinguished from the universal law of the conservation of total energy, and the mass-energy conservation which applies to all physical systems without exception. In non-elastic collisions the kinetic energy lost is converted to the internal energy of the colliding objects in the form of heat energy or is used up for the deformation of objects as in the case of the collisions of cars. Collision of particles and nuclei, and the disintegration of nuclei are common processes in nuclear physics that are used to study nuclear reactions. Such process can also be studied using the principles of collision (and explosion discussed in a later section), except that in nuclear collisions and explosions the masses of the particles may also change. The total mass of the particles and nuclie after collision (explosion) compared to the total mass before collision (explosion) may either decrease or increase or remain the same, and correspondingly the change in mass would either be positive or negative or zero. Such a change in mass (Δm) is known as the mass defect of the nuclear process. From the mass-energy conservation principle: $E = \Delta mc^2$, there is a corresponding change in the total energy of the system. Further details of this are discussed along with the relevant examples.

4.5   Application of the principles of collisions

In the following subsections we discuss in depth the examples and applications of the principles of collisions to 1- and 2-D spaces. Collision of more than two particles in 3-D space and its implications are discussed qualitatively.

4.5.1   Collision in 1-D

When two objects approaching each other along a straight line collide head-on, the objects after the collision continue to move along the original straight line of motion. Such a collision with all velocities directed along a straight line, let us say the $x-$ axis, is treated as a 1-D problem in conjunction with a sign convention for the direction of velocities. The velocities along one direction of the straight line are taken to be positive and in the opposite direction are taken to be negative. Consider the head-on collision of two masses m_1 and m_2 traveling with velocities u_1 and u_2 respectively along the

x−axis. We shall take all velocities along the $+x$-axis. The signs of the unknown velocities obtained on solving the equations of collision shall determine the direction of those velocities. We deal with elastic and inelastic collisions separately.

(i) Elastic collision: Let v_1 and v_2 respectively be the velocities of the two masses after collision, then from conservation of momentum:

$$m_1\, u_1 + m_2\, u_2 = m_1\, v_1 + m_2\, v_2 \qquad (4.22)$$

and from the conservation of kinetic energy:

$$\frac{1}{2}\, m_1\, u_1^2 + \frac{1}{2}\, m_2\, u_2^2 = \frac{1}{2}\, m_1\, v_1^2 + \frac{1}{2}\, m_2\, v_2^2 \qquad (4.23)$$

We have two independent equations, (4.22) and (4.23) which can be solved to determine any two unknown parameters provided the rest of the parameters are known. These equations can be further simplified as follows:

From equation (4.22):

$$m_1\,(u_1 - v_1) = m_2\,(v_2 - u_2) \qquad (4.24)$$

and from equation (4.23):

$$m_1\,(u_1^2 - v_1^2) \;=\; m_2\,(v_2^2 - u_2^2)$$

or

$$m_1\,(u_1 - v_1)\,(u_1 + v_1) \;=\; m_2\,(v_2 - u_2)\,(v_2 + u_2) \qquad (4.25)$$

Dividing equation (4.25) with equation (4.24) one obtains:

$$(u_1 + v_1) = (v_2 + u_2) \qquad (4.26)$$

Equations (4.24) and (4.26) can now be easily solved for the unknown parameters.

(ii) Inelastic collision and the ballistic pendulum: The two objects approaching each other along a straight after headon collision are stuck together, and move as one composite body of mass $m = (m_1 + m_2)$ with velocity v. Only momentum is conserved, and not the kinetic energy. Hence:

$$m_1\, u_1 + m_2\, u_2 = m\, v \qquad (4.27)$$

Here, we have only one equation from which only one unknown parameter can be determined provided all other parameters are known. From the vector considerations it is clear from equation (4.27) that velocity v of the composite mass m is along the same straight line as the velocities of the masses before collision. The inelastic collision has an important application as ballistic pendulum which is used to determine the velocities of projectiles such as the velocity of a bullet fired from a gun.

In a ballistic pendulum experiment, a bullet of mass m_1 moving with velocity u_1 is fired into a heavy block of mass m_2, $(m_2 \gg m_1)$ at rest, and gets imbeded into it. The composite mass $m = (m_1 + m_2)$ starts off with a velocity v after collision. The equation of conservation of momentum is:

$$m_1 u_1 = (m_1 + m_2) v = m v \qquad (4.28)$$

The velocity v of the composite mass is determined from its subsequent motion. Some possible motions of the composite mass are given as exercises at the end of the chapter. Pendulum like motion is one of possibilities from which this class of experiments derive the name *ballistic pendulum*. Here, we take up the case of one possible motion which is not the pendulum-motion.

If the bullet is fired along the vertical from below, the composite mass after collision rises freely to a height h in the gravitational field. The velocity v is determined by applying the law of conservation of mechanical energy to the post-collision motion in the conservative gravitional force field. The relevant equation is:

$$\frac{1}{2} m v^2 = \frac{1}{2} (m_1 + m_2) v^2 = (m_1 + m_2) g h \qquad (4.29)$$

where g is the acceleration due to gravity. Here, it is important to note that the conservation of energy in this case is not a part of the collision process, rather it is an independent after-event governed by free motion under gravity. Once the velocity v of the composite mass is determined from equation (4.29), the velocity of the bullet can be found from equation (4.28).

Let us give further consideration to the kinetic energy in inelastic collision which is not conserved. From the above example, the kinetic energy of the system before collision (KE_i) is:

$$KE_i = \frac{1}{2} m_1 u_1^2 \qquad (4.30)$$

and the kinetic energy of the composite mass after collision (KE_f) is:

$$KE_f = \frac{1}{2} (m_1 + m_2) v^2 \qquad (4.31)$$

From equations (4.28), (4.30) and (4.31), we obtain the ratio of kinetic energies as:

$$\frac{KE_f}{KE_i} = \frac{m_1}{m_1 + m_2} \qquad (4.32)$$

A simple proof of equation (4.32) is left as an exercise for students. From this equation one notes that in inelastic collision there is always a loss of kinetic energy. Although we have proved this for the case of ballistic pendulum, but the conclusion applies to all inelastic collisions, perfect as well as partial. The energy lost during collision appears as the internal energy of the colliding particles resulting in the rise in temperature of the composite mass or used up for deformation if any.

Recipe 4.2: The Recipe applies to all elastic, inelastic, and nuclear collisions problems in 1-D and 2-D some of which are discussed in the sections that follow. 3-D problems are not fully solvable unless most of the parameters except a very few are provided by experimental measurements.

- Identify suitable set of perpendicular axes in the plane of collision. Generally one of the axis is along the direction of motion of the projectile. In 1-D only one axis and a sign convention are used.

- Write down the components of the initial and the final momenta of the system along the chosen axes. Apply the law of conservation of momentum to each component of momentum. In 2-D it will give two equations, and in 1-D only one equation is obtained.

- If the collision is elastic, write down the equation of conservation of kinetic energy. This does not apply to non-elastic collisions.

- If there is an after-event following the collision as in the case of ballistic pendulum, write down the appropriate conservation equation(s) or the equation of motion as may be appropriate.

- In case of nuclear collision events, calculate the mass defect of the event and apply mass-energy conservation to the kinetic energies of the particles.

Example 4.2: A 10 g bullet is fired vertically upwards, from below into a 2 kg wooden block resting on a horizontal support with a central hole. The bullet, passing through the hole gets embedded into the wooden block, which then rises to a height of 1 m before dropping down. Calculate the speed of the bullet.

Solution: It is a 1-D inelastic collision problem along the vertical with subsequent free motion under gravity. Let v be velocity of the bullet and V be the velocity of the bullet and the wood block immediately after inelastic collision, both directed vertically upwards. From the conservation of momentum on collision:

$$0.01 \times v + 2 \times 0 = (0.01 + 2) \times V$$
$$\text{or} \quad 0.01\, v = 2.01\, V$$

When the block rises vertically, there is a conservation of mechanical energy. The initial kinetic energy of the bullet-block is equal to its gravitational potential energy at the highest point. Using $g = 9.8\, m\, s^{-2}$ we have:

$$\frac{1}{2} \times 2.01 \times V^2 = 2.01 \times 9.8 \times 1$$
$$\text{or} \quad V = 4.43\, m\, s^{-1}$$

Substituting this value of V in the equation of momentum conservation and solving, we get: $v = 889.9 ms^{-1}$.

Exercise 4.2: A neutron moving with a velocity u collides head-on with a nucleus of atomic mass number A at rest. Show that *(i)* the velocity of the neutron (v_n) and that of the nucleus (v_a) after collision are:

$$v_n = \frac{1-A}{1+A}u$$

$$v_a = \frac{2}{1+A}u$$

and *(ii)* fractional loss in the kinetic energy of the neutron is: $\frac{4A}{(1+A)^2}$. Explain why heavy water (containing deuterium atoms: ^2_1D) is used to slow down neutrons in nuclear reactor?

4.5.2   Collision in 2-D

As discussed earlier, collision of two particles is in general a 2-D problem for which conservation of momentum gives two independent equations. If the collision is elastic third equation is obtained from the conservation of kinetic energy . From these equations one can determine three unknown parameters for the elastic collision and two for the inelastic collision, while all the other parameters must be known. Figure 4.2 shows a general elastic collision of two particles. From the figure the following equations for the conservation of momentum and the kinetic energy (for elastic collision) can be written.

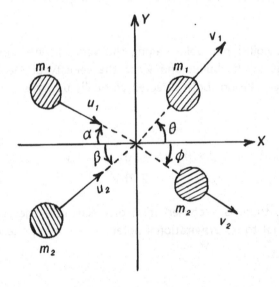

Figure 4.2: 2-D elastic collision of two particles.

$$\mathbf{P_{ix}} = m_1\,u_1\cos\alpha + m_2\,u_2\cos\beta \;=\; m_1\,v_1\cos\theta + m_2\,v_2\cos\phi = \mathbf{P_{fx}}$$
$$\mathbf{P_{iy}} = -m_1\,u_1\sin\alpha + m_2\,u_2\sin\beta \;=\; m_1\,v_1\sin\theta - m_2\,v_2\sin\phi = \mathbf{P_{fy}} \qquad (4.33)$$

and

$$KE_i = \frac{1}{2}m_1\,u_1^2 + \frac{1}{2}m_2\,u_2^2 = \frac{1}{2}m_1\,v_1^2 + \frac{1}{2}m_2\,v_2^2 = KE_f \qquad (4.34)$$

Assuming that the velocities before collision are known, there are four unknown parameters for the velocities after collision; their magnitudes v_1 and v_2 and directions θ and ϕ out of which one can determine only three of the parameters using equations (4.33) and (4.34). We must, therefore, have other means at our disposal, say some experimental observations, to help us solve the problem of elastic collision between two particles. We consider the following special case to further develop this problem.

Let mass m_1 be moving along the $x-$ axis before collision and mass m_2 is at rest, *i.e.*, $\alpha = 0$ and $u_2 = 0$. Then equations (4.33) become:

$$
\begin{aligned}
m_1 u_1 &= m_1 v_1 \cos\theta + m_2 v_2 \cos\phi \\
0 &= m_1 v_1 \sin\theta - m_2 v_2 \sin\phi
\end{aligned}
\tag{4.35}
$$

We still have the same four unknown parameters, which still can not be determined from the available equations. We further simplify the problem by considering two possible collision scenarios.

(i) Glancing collision: After the collision the particles move in such a way that $\theta = 0$ and $\phi = 90^o$, *i.e.*, the incident particle continues undeflected along the $x-$ axis and the target particle is deflected at 90° to the direction of the incident particle. Then from equations (4.35):

$$
\begin{aligned}
m_1 u_1 &= m_1 v_1 \\
0 &= m_2 v_2
\end{aligned}
\tag{4.36}
$$

These equations give $v_1 = u_1$, and $v_2 = 0$. Thus after glancing collision the incident particle continues to move undeflected without losing its kinetic energy, whereas the target particle continues to be at rest. This is an idealized scenario of glancing collision. In actual glancing collision $\theta \approx 0$ and $\phi \approx 90^o$, *i.e.* the angle θ is very small and the incident particle carries almost all the kinetic energy, whereas the target particle is deflected at 90° to the incident particle and carries extremely small fraction of the total kinetic energy. The total kinetic energy is conserved.

(ii) Head-on collision: The target particle, originally at rest, after collision moves in the direction of the incident particle, *i.e.*, $\phi = 0$. In this case equations (4.35) reduce to:

$$
\begin{aligned}
m_1 u_1 &= m_1 v_1 \cos\theta + m_2 v_2 \\
0 &= m_1 v_1 \sin\theta
\end{aligned}
\tag{4.37}
$$

There are two implications of equations (4.37): *(i)*Either $v_1 = 0$, *i.e.* the incident particle comes to a stop after collision and the target particle takes off in the direction of the incident particle before collision, or *(ii)* $\theta = 0$ and both the incident and the target particles move along the direction of the incident velocity with their new acquired velocities after collision. These conclusions are in agreement with the discussion on collision in 1-D in an earlier section.

Inelastic collision of two particles: From the conservation of momentum, we have only two equations at our disposal:

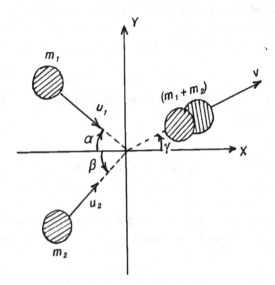

Figure 4.3: 2-D inelastic collision of two particles.

$$m_1\, u_1\, cos\,\alpha + m_2\, u_2\, cos\,\beta \;=\; (m_1 + m_2)\, v\, cos\,\gamma$$
$$-m_1\, u_1\, sin\,\alpha + m_2\, u_2\, sin\,\beta \;=\; (m_1 + m_2)\, v\, sin\,\gamma \qquad\qquad (4.38)$$

where (v, γ) is the velocity vector of the composite mass after collision which can be determined from equations (4.38). The problem is fully solvable provided precollision velocities are known.

Example 4.3: A 100 g mass moving with a velocity of 4 ms^{-1} along the $+x-$axis collides with a 200g mass moving with a velocity of 2 $m\,s^{-1}$ along the $+y-$axis in a horizontal plane. After the collision the 100 g mass comes to rest. *(i)* Determine the velocity of the 200 g mass after collision. *(ii)* Is the collision elastic or non-elastic? If non-elastic, calculate the fractional loss in kinetic energy due to collision. *(iii)* Now consider the (perfectly) inelastic collision of the two masses. Calculate the fractional loss in the kinetic energy.

Solution: Using the symbols defined in the text, given: $m_1 = 0.1\,kg$, $m_2 = 0.2\,kg$, $\mathbf{u_1} = u_{1x} = 4\,m\,s^{-1}$, $\mathbf{u_2} = u_{2y} = 2\,m\,s^{-1}$, and $\mathbf{v_1} = 0$. Let the velocity of m_2 after collision be $\mathbf{v_2} = (v_{2x},\, v_{2y}) = (v_2,\, \theta)$.

(i) From the conservation of the $x-$component of momentum:

$$0.1 \times 4 = 0.2 \times v_{2x},$$

and from the conservation of the $y-$component of momentum:

$$0.2 \times 2 = 0.2 \times v_{2y}$$

.

Solving for v_{2x} and v_{2y} we get: $v_{2x} = 2\,m\,s^{-1}$, and $v_{2y} = 2\,m\,s^{-1}$. This gives the magnitude of the velocity of m_2 after collision: $v_2 = 2.83 ms^{-1}$ and direction; $\theta = 45^o$.

(ii) Kinetic energy of the system before collision:

$$KE_i = \frac{1}{2}\left(0.1 \times 4^2 + 0.2 \times 2^2\right) = 1.2 J$$

Kinetic energy of the system after collision:

$$KE_f = \frac{1}{2} \times 0.2 \times 2.83^2 = 0.8 J$$

From the above calculations we note that $KE_i \neq KE_f$, *i.e.*, the kinetic energy is not conserved. Hence the collision is non-elastic. The fractional loss in KE is:

$$\Delta KE\% = \frac{(1.2 - 0.8)}{1.2} \times 100 = 33.3\%$$

(iii) Let $\mathbf{v} = (v_x, v_y) = (v, \phi)$ be the velocity of the composite mass of 0.3 kg after collision. Again from the conservation of the $x-$ and $y-$components of momentum:

Conservation of P_x:

$$0.1 \times 4 = 0.3 \times v_x$$

and from the conservation of P_y:

$$0.2 \times 2 = 0.3 \times v_y$$

Solving these expressions we get the components of the velocity of the composite mass: $v_x = 1.33 ms^{-1}$, and $v_y = 1.33 ms^{-1}$. From the components one obtains the magnitude of the velocity after collision: $v = 1.89 ms^{-1}$ and direction; $\phi = 45^o$.

From this one can calculate the fractional loss in KE as: $\Delta KE\% = 55.6\%$, the details of which are left for the students to work out.

This example demonstrates that the loss in kinetic energy in partial elastic collision is less than the loss in perfectly inelastic collision. In fact the loss of kinetic energy is maximum when the collision is perfectly inelastic.

Exercise 4.3: A neutron (mass $= 1u = 1.67 \times 10^{-27}\,kg$) moving with a velocity of $10^4\,m\,s^{-1}$ makes an elastic collision with a carbon atom (mass 12 u) at rest, and is deflected at 90° from its original direction of motion. Calculate the recoil velocity of the carbon atom and the kinetic energies of the neutron and the carbon atom.

4.5.3   Collision in 3-D

Collision in 3-D is of relevance only for the collision of three or more particles. For these particles to collide at the same precise instant, their velocities must be highly synchronized, an event which is highly improbable in nature and in a real physical system. Even if such an event were to take place, for collision between $n(> 2)$ number of particles there are $3\,n$ unknown components of velocities after collision (9 components of velocities for $n = 3$ particles). For the elastic collision we have three equations from the conservation of momentum, and one equation from the conservation of kinetic energy at our disposal. Hence the problem of elastic collision between $n(> 2)$ particles is not solvable and one can not draw any useful conclusions. The problem is, therefore, neither of any practical application nor of any theoretical interest.

However, if an inelastic collision is to take place between $n(> 2)$ number of particles in which all particles are stuck together after the collision, the three equations of the conservation of momentum as given below are sufficient to solve the problem completely, $i.e.$, to determine the components of the velocity of the composite mass provided the velocities of all the particles before collision are known.

$$\sum_{i=1}^{n}(m_i u_{ix}) = \left(\sum_{i}^{n} m_i\right) v_x$$
$$\sum_{i=1}^{n}(m_i u_{iy}) = \left(\sum_{i}^{n} m_i\right) v_y$$
$$\sum_{i=1}^{n}(m_i u_{iz}) = \left(\sum_{i}^{n} m_i\right) v_z \tag{4.39}$$

Alternately, collision between $n > 2$ number of particles may take place as a series of collisions one after the other. In that case, one solves the problem for one collision at a time starting from the very first collision, and works through to the last collision. Some assumptions must be made about the velocities of particles between two successive collisions, $i.e.$, either the velocity of a particle after previous collision to the next collision remains constant, or if it does not then the mechanisms leading to the change in velocity between two successive collisions must be well understood.

4.6   Explosion and the conservation of momentum

When an object spontaneously explodes into two or more fragments, the forces of explosion are internal to the system. There being no external forces involved in spontaneous explosion, the total momentum during the process is conserved, $i.e.$, the total momentum of all the fragments immediately after collision is equal to the momentum of the object before collision. If M is the mass and $\mathbf{u}$ is the velocity of the object before collision, and m_i are the masses and $\mathbf{v_i}$ are the velocities of the n fragments after collision then from the conservation of momentum:

$$M\,\mathbf{u} = \sum_{i=1}^{n}(m_i\,\mathbf{v_i}) \qquad (4.40)$$

where $\sum_{i=1}^{n} m_i = M$. Equation (4.40) can be expressed in terms of the component of velocities in the usual manner.

Kinetic energy is not conserved in explosion. In fact the kinetic energy of the system increases due to the energy released during explosion, for example from the combustion of the gun powder in the case of a bullet being fired or in the explosion of a bomb.

In 3-D space, equation (4.40) gives three independent equations from which only three unknown parameters can be determined. Assuming that the mass of the object and its velocity before explosion, and masses of all the fragments after explosion are known, then equation (4.40) is not adequate to determine components of velocities of all the fragments after collision except when there are only two fragments of explosion. However, from general observations of explosions, some interesting features can be inferred:

- If an object at rest spontaneously explodes into two fragments, the two fragments must move in diagonally opposite directions to conserve momentum, *i.e.*, the velocities of the fragments are along a straight line, opposite to each other and the problem can be treated as a 1-D problem.

- If the two fragments are of unequal masses, one being substantially smaller than the other ($m_1 \ll m_2$), then the smaller fragment has much larger velocity than the larger fragment ($v_2 \ll v_1$). Consequently, the smaller fragment on explosion is projected to a larger distance than the larger fragment. This applies even when there are many fragments, and it is not surprising that in the explosion of bombs, small sharpnels can fly long distances to injure people at large distances from the site of explosion. Likewise when a glass shatters, small pieces of glass can be found in the remotest corners of the room.

Explosion can be visualized as an inelastic collision in reverse. This is seen by comparing equation (4.40) to equations (4.39). Equation (4.40) can be obtained by interchanging the right and the left sides of equations (4.39). In the following subsections we discuss some specific examples of explosion.

4.6.1 Recoiling gun and bullet

Let a gun of mass M loaded with a bullet of mass m be at rest. On firing the gun, the bullet leaves the gun with a velocity $\mathbf{v}$, and the gun recoils with a velocity $\mathbf{V}$. From the conservation of momentum:

$$\mathbf{P_i} \;=\; 0 = M\mathbf{V} + m\,\mathbf{v} = \mathbf{P_f}$$

or

$$M\mathbf{V} \;=\; -m\,\mathbf{v} \qquad (4.41)$$

From equation (4.41) we draw two conclusions:

- The velocities $\mathbf{v}$ and $\mathbf{V}$ are directed opposite to each other, and the problem is a 1-D problem. The bullet (projectile) carries momentum in the forward direction, whereas the gun recoils back and carries an equal momentum in the backward direction. If one of the two velocities is known, the other can be calculated.

- Since $M \gg m$, $V \ll v$, i.e., the lighter bullet moves with a much higher velocity compared to the recoil velocity of the heavier gun.

Both these conclusions were also reached previously from general considerations of explosions. The second conclusion has important application in the design of guns. A gun must be designed in such a way that not only the barrel of the gun should be strong enough to withstand the pressure of the exploding gun powder - an obvious requirement, the gun itself should be much heavier (as heavy as possible) compared to the bullet so that it has a very small recoil velocity. This is why large powered guns can not be designed for hand-held firing, and cannons for firing heavy artillery and missiles are fixed to the ground or to a gun carriage to minimize their recoil.

4.6.2   Alpha decay

In Alpha decay, an unstable, radioactive nucleus at rest spontaneously decays (or disintegrates or explodes) by emitting an $\alpha-$particle. An $\alpha-$particle is a bare helium atom ($_2^4 He^{2+}$) with atomic mass number four and atomic number two. The remainder of the nucleus after decay, referred to as the daughter nucleus, is a new element whose atomic mass number is less by four and the atomic number is less by two compared to those of the original (parent) nucleus before the decay. The $\alpha-$decay of a nucleus can be expressed by the following equation:

$$_Z^A X \rightarrow {}_{(Z-2)}^{(A-4)}Y + {}_2^4 He \tag{4.42}$$

In equations representing nuclear processes like the one above, one uses an arrow rather than the equal sign unless the equation is energy balanced, showing the energy released or consumed during the process. The $\alpha-$decay is similar to the gun and the bullet problem discussed in the previous section with daughter nucleus akin to the empty gun after firing and $\alpha-$particle akin to the bullet. The $\alpha-$particle is emitted, let us say, in the forward direction with a large velocity and the daughter nucleus recoils in the opposite direction with a small velocity carrying a momentum equal and opposite to that carried by the $\alpha-$particle. Thus, the momentum is conserved, but not the kinetic energy. However, the total energy in $\alpha-$decay is conserved as per the universal law of mass-energy conservation which applies to all physical and natural processes.

In $\alpha-$decay, mass of the parent nucleus is always larger than the combined mass of the daughter nucleus and the $\alpha-$particle, i.e., the mass defect Δm is positive. From the mass-energy relationship: $E = \Delta mc^2$, the energy of disintegration or the $Q-$ value of $\alpha-$decay is positive. This energy, in accordance with the energy conservation principle appears as the kinetic energy of the daughter nucleus and the $\alpha-$ particle.

Students are assigned to write a short discussion on the conservation of energy in non-elastic collisions and explosion although the kinetic energy in these processes is not conserved.

4.6.3 Rocket propulsion

In rocket propulsion, hot gases from the combustion of fuel escape at a high velocity from a nozzle at the rear end of the rocket, and the rocket is propelled in the forward direction against the earth's gravitational force. The problem is a 1-D explosion problem like the firing of a bullet from a gun, but it is a relatively much more complex because of the changing mass of the system from to the combustion of the fuel on board, the external force on the rocket due to earth's gravity, and the relative velocity considerations for the combustion gases ejected from a moving rocket.

Let M_o be the mass of the empty rocket, and m_o the mass of the total fuel on board. At $t = 0$ the rocket loaded with fuel is at rest. Let at $t = t$ the mass of the rocket and fuel be m which is moving up against the force of gravity with velocity v. At this moment, in a short time interval Δt a small amount of fuel Δq is combusted and hot gases are ejected from the rocket with velocity $-v_e$ relative to the moving rocket. The velocity of the rocket increases to $(v + \Delta v)$ while its mass decreases to $(m - \Delta q)$. The velocity of the combustion gases relative to an observer on the ground is $(v - v_e)$. Initial momentum of the rocket and fuel at $t = t$ is:

$$P_i = m\,v \tag{4.43}$$

and the final momentum of the rocket and the ejected combustion gases at $t + \Delta t$ is:

$$
\begin{aligned}
P_f &= (m - \Delta q)(v + \Delta v) + \Delta q(v - v_e) \\
&= m\,v + m\,\Delta v - v\,\Delta q - \Delta q\,\Delta v + v\,\Delta q - v_e\,\Delta q \\
&= m\,v + m\,\Delta v - v_e\,\Delta q \tag{4.44}
\end{aligned}
$$

where we have neglected the small term $\Delta q\,\Delta v$. The change in momentum in time interval Δt is:

$$\Delta P = (P_f - P_i) = m\,\Delta v - v_e\,\Delta q \tag{4.45}$$

and the rate of change of momentum of the system is given by:

$$\frac{\Delta P}{\Delta t} = \frac{d\,P}{dt} = m\frac{d\,v}{dt} - v_e\frac{d\,q}{dt} \tag{4.46}$$

The rate of fuel combustion $\frac{d\,q}{dt}$ is just equal to the rate of the decrease in the mass of the rocket $-\frac{d\,m}{dt}$, i.e., $\frac{d\,q}{dt} = -\frac{d\,m}{dt}$. Substituting this in equation (4.46) we get:

$$\frac{d\,P}{dt} = m\frac{d\,v}{dt} + v_e\frac{d\,m}{dt} \tag{4.47}$$

From the Newton's second law of motion, the rate of change of momentum of the rocket-fuel system is equal to the external force $-m\,g$ due to gravity against which the rocket ascends. Hence:

$$\frac{d\,P}{dt} = -m\,g = m\,\frac{d\,v}{dt} - v_e\,\frac{d\,m}{dt} \tag{4.48}$$

We rearrange equation (4.48) and assuming the acceleration due to gravity g to be constant with altitude, we integrate to obtain velocity of the rocket at $t = t$:

$$\int_{v=0}^{v} dv = -g\int_{t=0}^{t} dt + v_e\int_{M_o+m_o}^{m} \frac{1}{m}\,dm$$

or

$$v = -g\,t + v_e\,ln\left(\frac{M_o + m_o}{m}\right) \tag{4.49}$$

The final velocity V reached by the rocket at $t = T$ after the entire fuel has been consumed is obtained by substituting $m = M_o$ in equation (4.49):

$$V = -g\,T + v_e\,ln\left(1 + \frac{m_o}{M_o}\right) \tag{4.50}$$

Equations (4.49) and (4.50) are idealized expressions for the velocity of the rocket because of the assumptions made, namely the constancy of g which is not a true scenario. We have also used a constant rate of fuel consumption and constant velocity of the ejection of the combustion gases which may not necessarily be so. Even then from equation (4.50) we can reach to the following important conclusions about the final speed of the rocket.

- The speed of the rocket depends directly on the speed of ejection of the combustion gases and not on the rate of combustion of the fuel. The larger the value of v_e, the larger is the speed of the rocket. This is achieved by maintaining very high temperatures inside the fuel combustion chamber.

- To achieve a final large speed of the rocket, the ratio of the mass of the fuel to the mass of the empty rocket, $\frac{m_o}{M_o}$, should be large.

4.6.4   Beta decay and neutrino emission

Beta (β) decay like the α−decay is a radioactive disintegration process in which an unstable nucleus decays by emitting an electron (e^-). The electron emitted is not the orbital electron, rather it is emitted from the nucleus, and one neutron in the nucleus is converted to a proton as per the charge conservation principle. The daughter nucleus has, therefore, the same atomic mass number as the parent nucleus but its atomic number is increased by one. The decay can be expressed by an equation of the form (which we shall soon see is not the correct representation):

$$^A_Z X \; \longrightarrow \; ^A_{(Z+1)} Y + e^- \tag{4.51}$$

Some radioactive elements decay by emitting a positron (e^+) as:

$$^A_Z X \; \longrightarrow \; ^A_{(Z-1)} Y + e^+ \tag{4.52}$$

Both electron and the positron are referred to as the β particles, and the disintegration process is termed as the $\beta-$ decay. We shall center our discussion around equation (4.51) which shall apply to equation (4.52) as well.

In early experiments on $\beta-$ decay only the $\beta-$particle and the daughter nucleus were detected, and no other product of decay was detected. Therefore, equation (4.51) was considered to be the only possible decay scheme. If equation (4.51) were a correct representation of $\beta-$ decay, then it is no different from the $\alpha-$ decay. The momentum must be conserved with the daughter nucleus and the $\beta-$particle going in diagonally opposite directions, and the energy must also be conserved. However, all possible experiments on $\beta-$ decay gave results contrary to equation (4.51) which are summarized below:

- The daughter nucleus and the $\beta-$particle after disintegration do not move in opposite directions, rather the daughter nucleus recoils at various angles to the direction of the emitted $\beta-$particle. Therefore, the momentum in $\beta-$decay as given by equation (4.51) is not conserved.

- In a two particle decay process the kinetic energy of all $\beta-$particles from the same parent nucleus must be equal, whereas experimentally the $\beta-$particles were found to have a range of kinetic energies.

- The total energy is not conserved, *i.e.*, the total kinetic energy of the $\beta-$ particle and the daughter nucleus is not equal to the $Q-$value of the decay process.

- Another important conservation law of nature is the conservation of angular momentum. For $\beta-$decay as expressed by equation (4.51) the angular momentum is also not conserved.

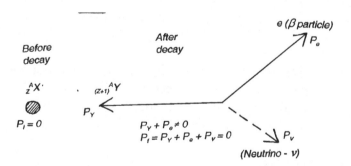

Figure 4.4: Beta decay and neutrino emission.

Equation (4.51) is, therefore, not the correct representation of the $\beta-$decay, and the failure of conservation laws baffled scientists for decades. The problem was resolved by Pauli in 1930 who proposed that a third particle must be emitted during $\beta-$ decay to account for the missing energy and momentum. The particle had to be charge neutral because of the charge conservation, and was give a name *neutrino* (the little neutral one or the little neutron in Italian). The neutrino (denoted by the Greek symbol nu: ν) was assigned the following properties:

- It is charge neutral.

- Its mass is either zero, in which case it travels with the speed of light like a photon, or the mass must be exceedingly small.

- It has spin $\frac{1}{2}$ to account for the conservation of angular momentum.

- It interacts with matter extremely weakly as a result of which, and together with its zero charge and exceedingly small (or zero) mass it escapes undetected.

Although neutrino could not be detected experimentally till 1950, its presence in β−decay was universally accepted because with it all the conservation principle in β−decay could be satisfied. Recent experiments confirm that the mass of neutrino is not zero and it has a small mass. In fact there are two types of neutrinos called the neutrino and the anti-neutrino ($\bar{\nu}$) which are emitted in positron and electron decays respectively. Thus the correct forms of equations (4.51) and (4.52) for β−decay are:

$$\begin{aligned}
{}_{Z}^{A}X &\rightarrow {}_{(Z+1)}^{A}Y + e^- + \bar{\nu} \\
{}_{Z}^{A}X &\rightarrow {}_{(Z-1)}^{A}Y + e^+ + \nu
\end{aligned} \qquad (4.53)$$

Let the parent nucleus be at rest before the decay. After the decay, considering only the daughter nucleus and the β−particle, one can define a plane containing the motion of both of them. In order to conserve momentum, the neutrino can not have a component of momentum perpendicular to the plane containing the daughter nucleus and the β−particle, and must also lie within the same plane. Hence, β−decay is a coplanar process in which all the three products of decay lie.

Another important aspect of β−decay is the very large kinetic energy of the β−particles, in the range of $1\ MeV$ compared to the rest mass of an electron of $0.511\ MeV$. Hence relativistic expressions are used for energy of the β−particle. Other nuclear process in which neutrino participates is the electron capture.

This example illustrates the importance of the conservation laws. The conservation laws are universally accepted laws that must apply to all physical processes. It is this faith of scientists in these laws that led to the postulation of the presence, and eventual detection of an elusive fundamental particle, neutrino.

Recipe 4.4: Explosion and nuclear decay.

- These are non-elastic like events. Only conservation of momentum, and the mass-energy conservation in case of nuclear decay events applies. Conservation of kinetic energy does not apply.

- Write down the conservation of momentum equations for each component of momentum.

- Where applicable, calculate the mass defect Δm and the Q-value for the decay process ($Q = \Delta m\, c^2$), and apply energy conservation to the process.

- Use available equations to calculate the desired quantities.

Example 4.4: An unstable nucleus of atomic mass 150 u at rest undergoes $\beta-$decay by emitting the electron and the neutrino at right angle to each other. The momenta of the electron and the neutrino are measured to be $2 \times 10^{-22} kg\ m\ s^{-1}$ and $10^{-22} kg\ m\ s^{-1}$ respectively. Find the recoil velocity of the daughter nucleus.

Solution: As an approximation, we ignore the mass defect of disintegration, and assume the mass of the daughter nuclei (m_d) to be the same as that of the parent nucleus, *i.e.*, $m_d = 150 \times 1.67 \times 10^{-27}\ kg = 2.505 \times 10^{-25}\ kg$. Let the electron and the neutrino be ejected along $x-$ and $y-$axes respectively, and let $\mathbf{v_d} = (v_{dx}, v_{dy})$ be the velocity of the daughter nucleus. Then from the conservation of momentum:

$$P_x \Rightarrow 2.505 \times 10^{-25} \times v_{dx} = -2 \times 10^{-22}, \text{ and } P_y \Rightarrow 2.505 \times 10^{-25} \times v_{dy} = -1 \times 10^{-22}.$$

Solving for the components of velocities we get: $v_{dx} = -798.4m\ s^{-1}$ and $v_{dy} = -3999.2m\ s^{-1}$.

From the components, magnitude of the velocity and its direction respectively can be obtained as: $v = 892.6\ m\ s^{-1}$, and $\theta = 26.6^o$ or 206.6^o. 26.6^o is the direction of the sum of the momenta of the electron and the neutrino which are along the $x-$ and the $y-$ axis respectively. Hence, the momentum of the recoil nucleus must lie diagonally opposite to it in the third quadrant at 206.6^o. Thus $\mathbf{v_d} = (892.6\ m\ s^{-1}, 206.6^o)$.

NOTE: If we calculate the velocity of the electron from the non-relativistic relation: $m_e v_e = 2 \times 10^{-22}\ kg\ m\ s^{-1}$, we get $v_e = 2.2 \times 10^8\ m\ s^{-1}$. This velocity is very close to the speed of light ($\approx 0.73\ c$), which supports the earlier statement that for $\beta-$decay one must use relativistic momentum and energy relationships. In the present example we did not have to worry about it because we were given the momenta of the electron and the neutrino and not their velocities. We also note that the recoil velocity of the daughter nucleus is very small, and its momentum and energy are treated non-relativistically.

Exercise 4.4: An object is projected with a velocity of $10\ m\ s^{-1}$ at 37° above horizontal. At its highest point the object explodes into two equal parts. Following the explosion, one part assumes a motion along the vertical and rises up another $2\ m$ before falling to ground. Calculate the components of velocity of the other part, and describe its motion. Ignore the effect of friction due to air.

4.7 Simultaneous collision and explosion

Other important category of nuclear reactions that can be dealt with by using the principles of collision and explosion, and the conservation of energy are the ones in which two particles of masses m_1 and m_2 enter the reaction, and two different particles of masses m_3 and m_4 emerge from the reaction. Principally the reaction can be visualized as a two step process. In the first step the masses m_1 and m_2 collide inelastically, forming a single particle of $\approx (m_1 + m_2)$ mass. Then the composite particle explodes to yield particles m_3 and m_4. Since both collision and explosions are inelastic processes, the momentum is conserved, and not the kinetic energy. The theoretically conceived intermediate composite particle is so short lived, and the explosion is almost so intantaneous that one only detects the final products of the reaction as follows:

$$(m_1 + m_2) \rightarrow (m_3 + m_4) \tag{4.54}$$

The Q–value of the reaction is: $Q = \Delta m \, c^2 = \{(m_3+m_4) - (m_1+m_2)\} \, c^2$ which could be positive or negative. The conservation of energy i.e., $\Delta mc^2 = \Delta KE = (KE_f - KE_i)$ gives the kinetic energy of the products of collision. If the Q–value is positive the reaction is exoergic (or exothermic) and the kinetic energy increases, and if it is negative the reaction is endoergic (or endothermic) and the kinetic energy decreases. In case the Q–value is zero, the kinetic energy is conserved, and the reaction can be treated as an elastic collision even though the products of collision are different from the incident particles. Examples of this type of reactions are fusion of deuterium (^2_1H) and tritium (^3_1H) isotopes of hydrogen as follows:

$$
\begin{aligned}
^2_1H + {}^2_1H &= {}^3_2He + {}^1_0n + 3.2 MeV \\
^2_1H + {}^2_1H &= {}^3_1H + {}^1_1H + 4.0 MeV \\
^2_1H + {}^3_1H &= {}^4_2He + {}^1_0n + 17.6 MeV
\end{aligned}
\tag{4.55}
$$

The energies in MeV are the Q–values of the reactions.

4.8   Problems

Note: Some of the problems may require work and energy conservation principles discussed in Chapter 5.

4.1: An external force $\mathbf{F} = (\mathbf{i}\,12 + \mathbf{k}\,6)\ N$ is applied to an object moving with momentum $\mathbf{P} = (\mathbf{i}\,4 - \mathbf{j}\,8 + \mathbf{k}\,12)\ kg\ m\ s^{-1}$ at $t = 0\ s$. *(i)* Which component of the momentum is conserved, and what is its value? *(ii)* What is the rate of change of momentum of the object? *(iii)* If the duration of the force is $20\ s$, what is the momentum of the object at $t = 20\ s$?

4.2: A mass of $4\ kg$ is moving with a velocity $(\mathbf{i}\,12 + \mathbf{j}\,6 - \mathbf{k}\,10)\ m\ s^{-1}$, and another mass of $8\ kg$ is moving with a velocity of $(\mathbf{i}\,4 - \mathbf{j}\,6 + \mathbf{k}\,2)\ m\ s^{-1}$. *(i)* Which of the two objects has a larger momentum? *(ii)* What is the impulse of the force that shall render the momenta of both objects equal, when applied to the object with (a) smaller momentum, (b) larger momentum?

4.3: Momentum of an object at $t = 0\ s$ is $(\mathbf{i}\,3 + \mathbf{j}\,5 - \mathbf{k}\,4)\ kg\ m\ s^{-1}$ and at $t = 15\ s$ the momentum is $(\mathbf{i}\,9 - \mathbf{j}\,5 - \mathbf{k}\,4)\ kg\ m\ s^{-1}$. *(i)* Which component of the force acting on the object is zero. *(ii)* What is the impulse of the force acting on the body? *(iii)* Calculate the average external force acting on the body.

4.4: A squash player smashes a ball of $100\ g$ horizontally against the wall, and the ball rebounds in the same direction with the same speed. Calculate: *(i)* The impulse of the force exerted by the wall on the ball. *(ii)* If the ball was in contact with the wall for $1\ ms$, what is the average force exerted by the wall on the ball?

4.5: At the Marupule power station, coal is brought to the furnace by a conveyor belt. Coal is deposited on the belt at a rate of $20\ kg\ s^{-1}$ which is moving at a rate of $1\ m\ s^{-1}$. Ignoring the velocity with which the coal is dropped on the belt, calculate the force exerted by the belt on the coal.

4.6: A mass m_1 moving to the right with a velocity u_1 undergoes elastic collision with mass m_2 also moving to the right with a velocity u_2. If v_1 and v_2 respectively are the velocities of the two masses after collision, show that:

$$v_1 = \frac{m_1 - m_2}{m_1 + m_2}\ u_1 + \frac{2\ m_2}{m_1 + m_2}\ u_2$$
$$v_2 = \frac{2\ m_1}{m_1 + m_2}\ u_1 + \frac{m_2 - m_1}{m_1 + m_2}\ u_2$$

Calculate the coefficient of restitution for the collision. Describe the motion of the masses after collision if their masses were equal.

4.7: A proton of mass $1.66 \times 10^{-27}\ kg$ collides head-on with a helium atom at rest. Mass of the helium atom is $6.64 \times 10^{-27}\ kg$, and it recoils with a speed of $5 \times 10^5\ m\ s^{-1}$. If the collision is perfectly elastic, calculate: *(i)* Initial and the final velocities of the proton. *(ii)* Percentage of the initial kinetic energy of the proton that is transferred to the helium atom.

4.8: A proton (mass $1\ u$) moving with a velocity of $2 \times 10^6\ m\ s^{-1}$ undergoes head-on elastic collision with *(i)* an oxygen (mass $16\ u$) atom, *(ii)* a neutron (mass $16\ u$), and *(iii)* an electron (mass $\frac{1}{1836}\ u$), all of them initially at rest. Calculate the velocities of the proton and the target particle in each case.

4.9: Two protons (mass $1\ u$ each) moving along the $x-$axis with velocities $+u$ and $-u$ respectively undergo an elastic collision. After collision one of the protons shoots off at 37° from the $+x-$axis. In which direction does the other proton go, and calculate their speeds after collision?

4.10: A $0.4\ kg$ ball moving with a velocity of $3\ m\ s^{-1}$ along the $x-$axis has a head-on elastic collision with a $0.6\ kg$ ball at rest. Calculate the velocities of both the balls after collision. What shall be the velocity of the composite mass if the collision was inelastic?

4.11: A car traveling at certain speed along a straight road collides from behind with an identical car stopped at the traffic light. After the collision the two cars are dragged together along the road making $20\ m$ long skid marks. If the average coefficient of friction during the skid is assumed to be 0.6, what was the speed of the car that hit the stationary car? Solve the problem using the same basic data as given above for the cases if:*(a)* the moving car was twice as heavy as the stationary car, *(b)* the stationary car was twice as heavy as the moving car.

4.12: Two vehicles, one twice as heavy as the other traveling along a straight highway towards each other collide head-on, and both are dragged together along the road. *(i)* If the average coefficient of friction is 0.7, what shall be the length of the skid mark if they both were traveling at $60\ km\ h^{-1}$? *(ii)* If after collision the collided wrack of the cars comes to an instant stop without skidding, what conclusions can you draw about the speeds of the two cars before collision?

4.13: A $10\ g$ bullet is fired vertically upwards from below into a $3\ kg$ wooden block resting on a horizontal support with an opening for the bullet to penetrate the block. On collision, the bullet is embedded into the block and it rises vertically up to a height of $2\ m$ before falling back. Calculate the speed of the bullet.

4.14: A $30\ g$ bullet is fired horizontally into a $3\ kg$ wooden block placed on a smooth horizontal surface. The wooden block is fastened to a fixed spring of force constant $500\ N\ m^{-1}$. The spring can be freely compressed or extended horizontally. On collision, the bullet gets embedded into the wooden block and the spring is compressed by $2\ cm$. Calculate the speed of the bullet. (Hint: Energy of a compressed spring is $= \frac{1}{2}kx^2$, where $k =$ Force constant of the spring and $x =$ compression of the spring.)

4.15: A $25\ g$ bullet is fired horizontally into a $5\ kg$ wooden block suspended freely with a light inextensible string. On collision the bullet gets embedded into the block, and the block swings up by 30° from its equilibrium position. Calculate the speed of the bullet.

4.16: A $15\ g$ bullet is fired horizontally into a $4\ kg$ wooden block resting on a rough horizontal surface for which the coefficient of friction is 0.4. On collision the bullet gets embedded into the block, and it slides $10\ cm$ along the surface before coming to rest. Calculate: *(i)* the speed of the bullet, and *(ii)* the fractional loss in the kinetic energy of the bullet on impact.

4.17: A $15\ g$ bullet is fired horizontally into a $2.85\ kg$ wooden block placed at the edge of a $1.25\ m$ high table, and the bullet gets embedded into the block. The block then falls to the ground and lands at a horizontal distance of $1.5m$ from its initial position. Calculate the speed of the bullet.

4.18: A $4\ kg$ block is resting on an inclined plane, raised by 37° above horizontal. A $10\ g$ bullet traveling at $100m\ s^{-1}$ is fired into the block parallel to the plane directed upwards. The bullet gets embedded into the block. Upto what vertical height will the block rise along the plane if *(i)* the plane is smooth, *(ii)* the coefficient of friction is 0.3?

4.19: Two Wooden blocks, A of mass $m_1 = 2\ kg$ and B of mass $m_2 = 2.98\ kg$ are placed parallel to each other on a smooth horizontal surface separated by a distance. A $20\ g$ bullet is fired horizontally into the blocks, it passes through the block A, and gets embedded into the block B. At the end of the collision, both the blocks are found to be moving at the same speed, parallel to each other. Calculate the percentage loss in the speed of the bullet after passing through the first block. Ignore the loss in its mass when the bullet passes through it.

4.20: A 2 kg mass moving with a velocity of 0.5 m s^{-1} collides with a 4 kg mass moving in the opposite direction. After the collision the two masses stick together and come to rest. Calculate the velocity of the second mass before collision.

4.21. A mass m_1 with kinetic energy K_1 undergoes perfectly inelastic collision with a mass m_2 at rest. Calculate the kinetic energy of the composite mass if *(i)* $m_1 = m_2$, *(ii)* $m_1 = 0.01 \times m_2$, *(iii)* $m_1 = 0.1 \times m_2$, *(iv)* $m_1 = 10 \times m_2$, and *(v)* $m_1 = 100 \times m_2$. What general conclusion can you draw from these calculations about inelastic collision between unequal masses and the loss in kinetic energy?

4.22. Two identical balls are suspended freely with light, inextensible strings side by side touching each other. One of the ball is lifted to 45° position from the vertical, and let go. On collision both the balls swing together to the opposite side. Calculate the angle from the vertical to which they shall rise.

4.23: A 1 kg mass moving with a velocity of 3 m s^{-1} along the $x-$axis and a 3 kg mass moving along the $y-$axis at 1 m s^{-1} collide simultaneously with a 3 kg mass at rest at the origin. The collision is perfectly inelastic. Calculate: *(i)* the velocity of the composite mass, *(ii)* the loss in kinetic energy of the system as a result of collision.

4.24: Consider the first stage of nuclear fission in which a nucleus of fissile material of mass m_f moving with velocity u_f absorbs a slow neutron of mass m_n moving with velocity v_n. Let the directions of motion of the nucleus and the neutron be along $x-$ and the $y-$axes respectively. Find the magnitude and direction of the nucleus after the capture of the neutron. Ignore any mass defect resulting from the capture.

4.25: Nuclear fusion: A proton of mass 1.00759 u moving with a velocity of 2500 m s^{-1} along the $x-$axis collides with a neutron of mass 1.00898 u moving with a velocity of 2000 m s^{-1} along the $y-$axis, and the two fuse together to form a deuterium nucleus of mass 2.01410 u. Calculate the velocity of the deuteron.

4.26: *(a)*. A 1 kg mass moving with a velocity of 9 m s^{-1} along $x-$axis undergoes a perfectly inelastic collision with a 2 kg mass moving along the $y-$axis at 6 m s^{-1}. Calculate the velocity of the composite mass after collision.

(b) The 3 kg composite mass in part *(a)* collides perfectly inelastically with an identical third mass, and the composite mass after collision comes to rest. Calculate the velocity of the third mass.

4.27: A 100 kg canon is mounted on wheels and rests on a firm flat ground. A 10 kg shell is fired horizontally with a muzzle velocity of 200 m s^{-1}. *(i)* Calculate the recoil velocity of the canon. *(ii)* If the canon is rigidly fixed to the deck of a ship of mass 2×10^7 kg, calculate the recoil velocity of the ship when the canon is fired. *(iii)* Discuss qualitatively the recoil in both cases if the shell was fired at 37° above horizontal.

4.28: A radium nucleus of mass 222 units decays by emitting an α−particle (mass 4 units) with a velocity of $1.5 \times 10^7 \ m \ s^{-1}$. Calculate: *(i)* the recoil velocity of the daughter nucleus, *(ii)* Ratio of the kinetic energy of the α−particle to that of the daughter nucleus, ignoring the mass defect of the decay process.

4.29: A $3M$ mass at rest spontaneously explodes into three equal parts. One piece flies off along the x−axis with velocity v, the second piece flies off along the y−axis with the same speed. Calculate the velocity of the third piece after explosion.

4.30: A railroad container of mass M initially moving with velocity u_o is being loaded with grain which is dropped into the container at a constant rate of $m \ kg \ s^{-1}$. Ignoring the vertical component of the velocity with which the grain fall into the container, find an expression for the velocity of the container as a function of time. What shall be the speed when the container has been loaded with the same mass of grain as the mass of the empty container?

Chapter 5

Work, Energy and Power

The objective of this chapter is to study the technical aspects of the three concepts; Work, Energy and Power. These terms are used in everyday language, but have precise meanings when used technically. It is common experience that when one does work one expends bodily energy which must be replenished by the consumption of food. One also hears people say that they do not have the energy to do certain work, or that the more powerful person is capable of doing more work within the same time frame. This common understanding of work, energy and power and the relationship between them is somewhat comparable to the technical definitions of these terms. In technical terms, as we shall see in this chapter, doing of work results in the consumption of energy, and if the power is higher, then work can be done at a faster rate. Technically, one defines work in terms of the external forces acting on a system and the displacement of the system, which results either in an increase or a decrease in the energy of the system. In this Chapter, after having defined work and energy in technical terms, we establish relationship between the work and the gain or loss of energy.

At the end of this chapter, the following shall have been studied

- Definitions of Work, Energy and Power

- Work-Energy Theorem

- Conservation of Energy

- Efficiency

These concepts have important applications in mechanical systems, performance of which depends on their work-energy balance. The work energy principles can also be used to analyze certain aspects of kinematical problems, provided one is able to identify the forces involved, or conversely from the work energy principles one can evaluate unknown forces which may not be so easily possible to determine.

5.1 Work

Work is a scalar quantity defined in terms of the force acting on a body and its displacement. The concept of work is introduced at a very early stage in physics curriculum at school level for a motion in one dimension, where the force acting on a body and its displacement are along parallel directions. If a constant force F_x acts on a body along the x−axis, and there is a displacement s of the body parallel to the direction of the force, then the work (W) is given by the product of force and the displacement as:

$$W = F_x \, s \tag{5.1}$$

Recognizing that force and displacement are both vectors, their vector nature in equation (5.1) is infused by introducing a sign convention such that displacement along the direction of the force is taken to be positive, and opposite to the direction of the force is taken to be negative. This results in positive and negative work respecively, which are discussed later.

Definition of work given by equation (5.1), though complete in itself, is restrictive to the extent that one can not calculate work from this expression when the force and displacement are not along the same direction. An example of such a scenario is when an object is pulled along a horizontal surface by a force applied at an angle θ above horizontal. Recalling the properties of scalar (dot) product of two vectors from Chapter 1, one notes that it satisfies all the aspects of the definition of work given by equation (5.1), namely the product is a scalar, and it involves the parallel components of the two vectors. Thus if a vector force **F** acts on a body which is displaced by a distance $\boldsymbol{\Delta r}$, the work done is given by the dot product:

$$W = \mathbf{F} \cdot \boldsymbol{\Delta r} = F\,\Delta r \cos\theta = F\,(\Delta r \cos\theta) = (F\cos\theta)\,\Delta r \tag{5.2}$$

where θ is the angle between the force and the displacement vectors as shown in Figure (5.1).

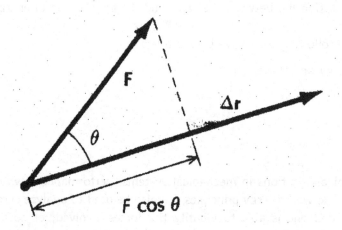

Figure 5.1: A force **F** applied to a body in a direction θ to the direction of the displacement $\boldsymbol{\Delta r}$.

From equation (5.2) and Figure (5.1) we note that work is given by the product of the parallel components of the force and displacement, the same as defined by equation (5.1). The general definition of work, given by equation (5.2) clearly displays the scalar nature of work. The SI unit of work is *Joule*, (J), which is the work done by a $1\,N$ force during a displacement of $1\,m$.

$$1J = 1N\,m = 1\,kg\,m^2\,s^{-2} \tag{5.3}$$

Once again recalling the properties of vectors, and from Figure (5.1) note that one can always define a plane containing both the vectors $\mathbf{F}$ and $\mathbf{\Delta r}$, so that the calculation of work reduces to a $2-d$ problem. However, in a $3-d$ space if the coordinates axes are chosen such that they do not lie in the plane containing the force-displacement vectors, then equation (5.2) for work can be expressed in terms of the Cartesian components of force and displacement as:

$$W = F_x\,\Delta x + F_y\,\Delta y + F_z\,\Delta x \tag{5.4}$$

Again note that each of the terms on the right side in equation (5.4) is the product of the parallel components of force and displacement. If any of the component of the force or the displacement is zero, the corresponding work is zero.

From equation (5.2), we consider the following special cases.

Case (i): $\theta = 0$, *i.e.* the force and the displacement are along the same direction, for example when an object is pulled with an external force. Equation (5.2) reduces to $W = F\,\Delta r$ which is the same as the definition of work given by equation (5.1). The work done in this case is positive. As a result of the applied external force the object accelerates along the direction of displacement, its velocity increases, and the object gains kinetic energy (defined later). This can be generalized to state that positive work done by an external force on a body results in the increase in the energy of the body.

Case (ii): $\theta = 180^o$, *i.e.*, the force and the displacement are directed opposite to each other along the same straight as is the case with the force of friction which acts opposite to the displacement. Since $cos\,180^0 = -1$, equation (5.2) reduces to $W = -F\,\Delta r$. This expression for work is also similar to equation (5.1), but with a negative sign. From experience we know that when a moving body (with kinetic energy) experiences force of friction, it experiences deceleration, and eventually having lost its kinetic energy tends to come to rest. Thus the negative work done by a body against external forces results in the loss in energy of the body.

The observations from the above two cases can be summarized as follows: *Positive work is the work done by external force(s) on the body which results in the increase in energy of the body, and negative work is the work done by the body against the external force(s) which results in the decrease in the energy of the body*

Case (iii): $\theta = 90^o$, *i.e.* the force and the displacement are perpendicular to each other, and the work done $W = 0$. Thus if the force and the displacement are perpendicular to each other, work done is zero.

Now let us consider an interesting simple example to illustrate the calculation of work. Consider a person carrying a heavy baggage on his head. The force to support the baggage against the force of gravity is applied vertically upwards, whereas the displacement from waking is along the horizontal, and hence the work done as per case *(iii)* is zero. But the question is why does the person feels exhausted when carrying the baggage? What about the increased force of friction? Students please explain!!! This example also illustrates how important it is to identify the forces correctly for any calculations.

Recipe 5.1: Calculation of work.

- Choose appropriate set of axes. Generally, but not necessary, it is convenient to choose one of the axes to be parallel to the direction of displacement.

- Identify all the forces acting on the system, and resolve them into components along the chosen set of axes.

- Calculate work for the required components of forces of interest by multiplying them with the parallel component of displacement. Work corresponding to the perpendicular components of displacement is zero.

- If the force and displacement are along the same direction, work is positive, and the object gains energy from the work done by the force.

- If the force and the displacement are anti-parallel, work done is negative, and the object loses energy in doing work against the external force.

Example 5.1: A $100\,N$ force applied at 37° above horizontal pulls a $10kg$ mass along a rough horizontal surface to a distance of $5\,m$, as illustrated in Figure (5.2). The coefficient of friction is 0.15. Calculate the work done by the external force and the force of friction. What is the net work done on the mass?

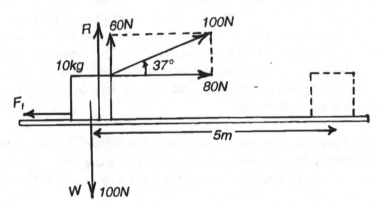

Figure 5.2: A mass being pulled along a horizontal surface by a force **F** applied at an angle of 37° above horizontal.

Solution: Following Figure (5.2) and the principles discussed in Chapter 3, the forces of interest are (students to confirm):
Forward pulling force: $F_x = 100\cos 37^o = 80\,N$
Normal reaction force: $R = W - 100\cos 37^o = 100 - 60 = 40\,N$
Force of friction: $F_f = \mu R = 0.15 \times 40 = 6\,N$

Work done by the pulling force: $W_1 = 80 \times 5 = 400 J$

Work done by the force of friction: $W_2 = 6 \times (-5) = -30 J$, (the force and displacement are anti-parallel.

Net work done: $W = W_1 + W_2 = 400 - 30 = 370 J$. As the work done is positive there is an equal increase in the kinetic energy of the mass.

Exercise 5.1: If in Exercise 5.1 the mass is pushed along the same surface by the same force, calculate the work done by the applied force and the force of friction. What is the net work done on the mass?

5.1.1   Work done on a particle moving with constant acceleration

The work W done by an external force F on a particle of mass m moving with constant acceleration a can be found by using Newton's second law $F = ma$ and the kinematic equations. Let s be the displacement. Then:

$$
\begin{aligned}
W &= Fs \\
&= mas \quad \text{where} \quad a = (v-u)/t \quad \text{and} \quad s = (u+v)t/2 \\
&= m\left(\frac{v-u}{t}\right)\left(\frac{v+u}{2}\right)t \\
&= \frac{1}{2}m(v^2 - u^2) \\
&= \frac{1}{2}mv^2 - \frac{1}{2}mu^2
\end{aligned}
\tag{5.5}
$$

where the first term $\frac{1}{2}mv^2$ is known as the *final kinetic energy* (KE_{final}) and the second term $\frac{1}{2}mu^2$ is known as the *initial kinetic energy* ($KE_{initial}$).

Equation (5.5) is an example of the *Work-Energy Theorem*, which states that the work done by the resultant force acting on a moving body is equal to the change in its kinetic energy.

$$
\begin{aligned}
W &= \Delta E \tag{5.6} \\
&= KE_{final} - KE_{initial} \tag{5.7}
\end{aligned}
$$

5.1.2   Work done on a particle moving in the gravitational field

Consider a particle of mass m moving in a gravitational field, from A to D, through B and C, which are at heights $0, h_1, h_2, h_{max}$ respectively, as illustrated in Figure (5.3). The force required to raise the mass in the gravitational field, keeping it in equilibrium is equal to its weight mg applied vertically upwards. Hence the work done W in moving the mass from B to C is given by

$$
\begin{aligned}
W &= \text{Force} \times \text{Vertical Displacement} \\
&= mg(h_2 - h_1) \quad \text{where} \quad (h_2 - h_1) \quad \text{is the distance moved from B to C} \\
&= mgh_2 - mgh_1
\end{aligned}
\tag{5.8}
$$

where the first term mgh_2 is the *final gravitational potential energy* (PE_{final}) and the second term mgh_1 is the *initial gravitational potential energy* ($PE_{initial}$).

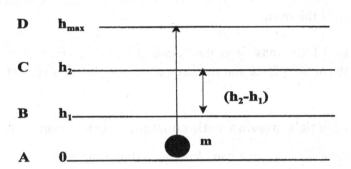

Figure 5.3: A particle of mass m moving in a gravitational field, from A to D, through B and C.

Equation (5.8) is another example of the *Work-Energy Theorem*, which in this case states that the work done by the resultant force in the gravitational field is equal to the change in gravitational potential energy.

$$W = \Delta E \qquad (5.9)$$
$$= PE_{final} - PE_{initial} \qquad (5.10)$$

In the above illustration, the external force mg used to raise the mass from h_1 to h_2 needs further explanation. The weight $W = mg$ of the mass acts vertically downwards, whereas the external force mg is applied vertically upwards, and this results in net zero force on the mass. Then the obvious question is how could then the mass be raised? However, we did state that while the mass is being raised, it must be maintained in equilibrium, meaning that there should be no change in the state of rest (or uniform motion) of the mass. The net zero force on the mass is consistent with this requirement. Let us now consider that the external applied force is larger than mg, *i.e.*, $F = mg + \Delta F$. This will result in a net upwards force ΔF on the mass which shall produce an upwards acceleration of the mass, and the kinetic energy of the mass shall increase. If one calculates the work done by the force F (see problem at the end of the chapter), one finds that the increase in the kinetic energy is just equal to the work done by the extra force ΔF. Hence, the force required to raise a mass in gravitational field keeping it in equilibrium is just equal to its weight applied vertically upwards.

5.1.3   Work done by a variable force

Suppose the force is not a constant, but a function of position, given by

$$F_x = F(x) \qquad (5.11)$$

Let us divide the displacement $x = a$ to $x = b$ into n elemental displacements Δx_i, $(i = 1$ to $n)$ such that the force over each elemental displacement can be treated to be constant. Then the total work done is obtained by summing up the work done during each elemental displacement, *i.e.*,

$$W = \sum_{i=1}^{n} F_x(x_i)\Delta x_i \qquad (5.12)$$

which is a generalisation of equation (5.1), and can also be given in terms of an integral

$$W = \int_a^b F_x(x)dx \qquad (5.13)$$

Equation (5.13) (as well as (5.1)) for the work done can be interpreted in terms of the force-displacement graph as : *Work done by an external force is equal to the area under the force-displacement graph.*

An illustration of work done by a variable force is the work done W on a spring of force constant k when the spring is extended in length from a to b (Figure 5.4), noting that the spring exerts a restoring force $F_s(x) = -kx$, given by Hooke's law.

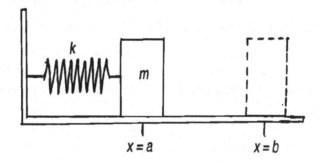

Figure 5.4: Mass-spring system.

The external force F_x required to extend the spring must be applied opposite to the restoring force, *i.e.*, $F_x = -F_s = kx$, and

$$\begin{aligned}
W &= \int_a^b F_x(x)dx \\
&= \int_a^b (kx)dx \\
&= k\left[\frac{1}{2}x^2\right]_a^b \\
&= \frac{1}{2}k(b^2 - a^2) \qquad (5.14) \\
&= EPE_{final} - EPE_{initial} \qquad (5.15)
\end{aligned}$$

where the first term $\frac{1}{2}kb^2$ is the *final elastic potential energy* (EPE_{final}) and the second term $\frac{1}{2}ka^2$ is the *initial elastic potential energy* ($EPE_{initial}$).

Equation (5.15) is another example of the *Work-Energy Theorem*, which in this case states that the work done on the spring by the external force is equal to the change in its elastic potential energy.

$$
\begin{aligned}
W &= \Delta E && (5.16) \\
  &= EPE_{final} - EPE_{initial} && (5.17)
\end{aligned}
$$

5.1.4 Work-Energy Theorem

We have already met several examples of the work-energy theorem, in earlier sections: (5.1.1), (5.1.2) and (5.1.3). The underlying principle in all these examples is that the work done is equal to a change in kinetic energy or potential energy. This can be generalised so that the work-energy theorem states that *the work done is equal to the change in energy*. Example (5.2) illustrates a much more general application of the work-energy theorem.

Recipe 5.2: Applications of Work-energy principles:

- Calculate the net work done by the external forces acting on the system

- Calculate the total mechanical energy (kinetic + potential) of the system before and after the event under consideration, and determine the change in mechanical energy.

- The magnitude of the change in mechanical energy is equal to the net work done.

Example 5.2 A particle starts from rest at a height h above the horizontal, on a rough inclined plane making an angle θ with the horizontal. The particle acquires a velocity v by the time it reaches the bottom of the inclined plane, having covered a distance s. Show, by using the work-energy theorem, that the velocity v is given by

$$
v = \left\{ 2g \left[h - \mu \sqrt{(s^2 - h^2)} \right] \right\}^{1/2}
$$

where μ is the coefficient of friction.

$$
\begin{aligned}
\text{Total energy at height } h \text{ on the inclined plane, } E_A &= KE_A + PE_A \\
&= \frac{1}{2}m \times (0)^2 + mgh \\[2em]
\text{Total energy at the bottom of the inclined plane, } E_B &= KE_B + PE_B \\
&= \frac{1}{2}mv^2 + mg \times (h = 0) \\
E_B &= \frac{1}{2}mv^2 \\[2em]
\text{Energy loss, } \Delta E &= mgh - \frac{1}{2}mv^2 \\[2em]
\text{Work done against friction, } W &= F_f s \\
&= \mu R s \\
&= \mu s m g \cos\theta
\end{aligned}
$$

By using the work-energy theorem,

$$
\begin{aligned}
W &= \Delta E \\
\mu s m g \cos\theta &= mgh - \frac{1}{2}mv^2 \\
v^2 &= \left\{ 2g\left[h - \mu\sqrt{(s^2 - h^2)} \right] \right\} \\
v &= \left\{ 2g\left[h - \mu\sqrt{(s^2 - h^2)} \right] \right\}^{1/2}
\end{aligned}
$$

The above result, arrived at by the work-energy theorem, can also be derived using Newton's laws. This is left as an exercise for the student.

Exercise 5.2: A mass m starting with velocity v at the foot of a rough plane inclined at angle θ slides up the plane. If the coefficient of friction of the plane is μ, to what height does the mass rise up the plane?

5.2   Energy

Energy is defined as an entity that imparts a body or a system the capability to do work, *i.e.*, in order to do work energy must be supplied. For example, when a body moves against the force of friction its kinetic energy is dissipated to overcome the frictional resistance, electrical energy is used for the work done by industrial and electrical machines, human energy is used when one pulls or pushes an object, and so on. In the following section we list various types of energy and examples of their applications.

5.2.1   Forms of energy

Various forms of energy known to mankind are classified into groups according to some characteristics that are common to energies within a particular group. Many classification schemes are used. Here we use one such scheme to introduce the different forms of energy. The list presented is not exhaustive, and students may add to the list as many more as they know.

- **Mechanical energy:** This is the energy possessed by mechanical systems. In this category, we have two energy types:

    - Kinetic energy (KE): This could be the translational KE or the rotational KE.
    - Potential energy (PE) which itself has many forms such as the gravitational PE, elastic PE, the electrostatic PE, and the magnetostatic PE.

    As our focus on energy here is with application to mechanics, the mechanical energies are discussed in further detail in sections that follow.

- **Renewable energies:** These are the energies that are unlimited in supply and are constantly being generated in nature. Examples of these are

- Solar energy obtained from the sun as heat and light energy used to generate electricity or used directly as thermal energy.
- Wind energy can be used to do work such as pumping of water using a wind mill.
- Tidal energy from the oceans is used to generate electricity.
- Geothermal energy from the interior of the earth brought to surface by natural or man made hot water geysers is used to generate electricity.

- **Electrical energy:** This is the man made form of energy produced in power plants from other forms of energy as input. This is one of the most important form of energy for industrial work.

- **Chemical energy:** This is the energy stored in fuels such as oil, gas and coal, and is released as heat energy from their combustion. The thermal energy in turn is used for varieties of work for example to provide motive power in the transport sector, or converted to electrical energy in power plants. The energy stored in car batteries and dry cells is also the chemical energy.

- **Thermal energy:** Received from sun or from the combustion of fuel. Steam produced from combustion of fuel is used to drive steam engines and steam turbines in power stations.

- **Nuclear energy:** Stored in the nuclei of atoms, this energy is released as thermal energy through nuclear reaction. With a controlled reaction it is used in nuclear power plants, and an uncontrolled release results in nuclear bombs.

- **Light energy:** It is also known that electromagnetic energy includes not only the known visible light, but also the invisible spectrum of ultraviolet, infrared, x–ray, radio waves, microwaves, γ-ray. Their applications are varied and numerous in communication, broadcasting, medical diagnostics and treatment etc.

- **Sound energy:** This energy transmitted through a medium as longitudinal waves include not only the audible sound but also the inaudible spectrum of ultra and infra sonic waves. The importance of audible sound in communication and entertainment is well known, whereas the inaudible sound has medical and industrial applications.

5.2.2   Kinetic energy

The kinetic energy (KE) of a body is the energy by virtue of its motion. There are several forms of kinetic energy, for example

- Translational kinetic energy (TKE): An object of mass m moving with a velocity v has TKE given by

$$TKE = \frac{1}{2}mv^2 \qquad (5.18)$$

as was also defined in equation (5.5), and sometimes referred to as KE .

It is sometimes convenient to express the translational kinetic energy in terms of momentum, in the form

$$TKE = \frac{p^2}{2m} \qquad (5.19)$$

where $p = mv$ is the momentum.

- Rotational kinetic energy (RKE): An object of moment of inertia I rotating with a angular velocity ω has RKE given by

$$RKE = \frac{1}{2}I\omega^2 \qquad (5.20)$$

which is dealt with further in Chapters 6 and 9.

5.2.3 Potential energy

The potential energy (PE) of a body is the energy by virtue of its position and state. There are several forms of potential energy, for example

- Gravitational potential energy: A mass m at a height h has a gravitational potential energy (GPE) given by

$$GPE = mgh \qquad (5.21)$$

where g is the acceleration due to gravity. Note that at the zero level $GPE = 0$.

- Elastic potential energy: When a spring of force constant k is extended or compressed by a distance x, it acquires an elastic potential energy (EPE) given by

$$EPE = \frac{1}{2}kx^2 \qquad (5.22)$$

such that at zero extension or compression $EPE = 0$.

- Electrostatic potential energy: A pair of charges q_1 and q_2 at a separation r apart has electrostatic potential energy, U, given by

$$U = \frac{1}{4\pi\epsilon_0}\frac{q_1 q_2}{r} \qquad (5.23)$$

where

$$\frac{1}{4\pi\epsilon_0} = 9 \times 10^9 Nm^2C^{-2} \qquad (5.24)$$

is the *electric force constant*.

- Atomic and Nuclear potential energies are possessed by atoms and nuclei by virtue of their binding energies. This energy is released as thermal energy during chemical and nuclear reactions.

5.3 Conservation of Energy

The laws of conservation of energy are expressed in three different manners; the application depends on the system and its environment under consideration. These are:

- Universal law of conservation of energy.

- Mass - energy conservation.

- Conservation of mechanical energy.

5.3.1   Universal law of conservation of energy

According to this law of conservation of energy, *"energy can neither be destroyed nor created. It can only be converted from one form to another such that total energy input during a process is equal to the total energy output at the end of the process."* In order to apply this law, one must carefully account for all possible forms of energy consumed or generated through conversion process, no matter how small it is. Some examples are:

- In a coal thermal power plant, coal is used to produce electrical energy. During the process some thermal energy escapes the plant through radiation, with exhaust gases, exhaust steam from the turbines, and residues from the furnace, energy is lost to friction in the moving parts of the machines as heat and sound energy etc. If all energies recovered and lost are carefully accounted, the total energy of combustion of coal shall equal the electrical energy produced plus the energies lost.

- When an object moves against friction, its kinetic energy is apparently lost, but an equal amount of heat energy is produced. In case they are found not to balance perfectly, some energy might have been used to produce sound if the motion is not smooth, and some energy might have been used to chip and damage the contact surfaces.

- When electricity is used by a consumer for lighting, all of it does not convert to light. Some is lost to heating of the wires and the light source. The fluorescent light sources sometimes also have some humming. Thus one must account for all possible energies to balance an energy equation.

5.3.2   Mass-energy conservation

According to Einstein's mass-energy relationship, mass can be converted to energy and vice versa. The relationship between the mass m and the energy E is given as;

$$E = m\,c^2 \tag{5.25}$$

where $c = 3 \times 10^8\,m\,s^{-1}$ is the speed of light in free space. Such kind of mass-energy conversion takes place only through nuclear reaction in which changes are introduced in the nuclei of atoms, and only an extremely small fraction of mass is involved in the conversion. The examples of mass-energy conversion are the nuclear fuel used in nuclear power plants, and the nuclear bombs. The ratio $\frac{E}{m} = c^2 = 9 \times 10^{16}$ shows that an extremely large amount of energy is produced from an exceeding small mass, or conversely a large amount of energy shall be converted to an exceeding small mass. This is what makes the nuclear power plants and the nuclear bombs so extremely energy intensive. The energy-mass conservation equation, where mass-energy conversion is involved is written as follows:

$$\left(\sum_i E_i + \sum_i m_i\,c^2 \right)_{initial} = \left(\sum_j E_j + \sum_j m_j\,c^2 \right)_{final} \tag{5.26}$$

where on the left side of the equation E_i's and m_i's are the total energy and mass inputs and on the left side they are the total outputs.

5.3.3 Conservation of mechanical energy

This is the simplest form of the energy conservation law that applies only to the mechanical energy of mechanical systems. The total mechanical energy (ME) of a system is given by the sum of all its kinetic energies (KE_i) and potential energies (PE_i). Thus:

$$ME = \sum_i KE_i + \sum_i PE_i \qquad (5.27)$$

In order to understand the conditions of the applicability of the law, and to define it, we must first revisit forces and their nature.

Conservative and non-conservative forces

Recalling the work-energy theorem in context of the gravitational potential energy $m\,g\,h$ of a mass m raised to a height h in the gravitational field, we note that the gravitational potential energy, and hence the work done to raise the mass depends on the initial and final position of the mass, and it does not depend on the path followed between the two positions. One can raise the mass either directly vertically between the two positions, or take any possible path starting from the initial position to reach the final position. The work done, and the change in potential energy shall be the same. A proof of this is left as an exercise for students. A force field in which work done does not depend on the path, and depends only on the initial and final positions, is known as the *conservative force field*. Other examples of conservative force fields are, the spring or the elastic force, the electrostatics force, the magnetostatic force etc.

Now consider the force of friction. Suppose an object with a fixed kinetic energy starts from some initial position in a horizontal plane. Moving along a straight line, it reaches a final position where after having used-up all its kinetic energy in doing work against the force of friction it comes to rest. Next consider other possible paths between the initial and final positions. All such paths shall be longer than the straight-line path, and the object shall not be able to reach the final point, because the work done against friction along all other paths is larger than the kinetic energy of the body. Thus the work done against the force of friction depends not on the initial and final position of the object, but on the path followed. Such a force is known as the *non-conservative force*. Examples of such forces are the solid friction, and the fluid friction.

When an object is projected vertically upwards, its kinetic energy decreases and its potential energy increases such that the decrease in KE is equal to the increase in the PE. The reverse processes takes place when the object falls freely under gravity. Thus when an object moves freely under gravity its total mechanical energy remains constant. In contrast to this when an object moves against the force of friction, it does so at the expense of its kinetic energy which is converted to the thermal energy. Although the total energy of the system is conserved, but not the mechanical energy. We can now state the law of conservation of mechanical energy as:

When an object moves freely in a conservative force filed, its mechanical energy is conserved.

$$ME_i = \left(\sum_j KE_j + \sum_j PE_j\right)_{initial} = \left(\sum_k KE_k + \sum_k PE_k\right)_{final} = ME_f \qquad (5.28)$$

5.4   Examples and Applications of Energy Conservation

The following examples illustrate the law of conservation of energy and the work-energy principle.

5.4.1   Vertical motion under gravity

Consider a particle of mass m moving in a gravitational field, from A to D, through B and C, as illustrated in Figure (5.5), such that A, B, C, D are at heights $0, h_1, h_2, h_{max}$ respectively, and the respective velocities are $u, v_1, v_2, 0$. Using kinematic equations, the velocities and heights are related as:

$$v_1^2 = u^2 - 2gh_1$$
$$v_2^2 = u^2 - 2gh_2$$
$$0 = u^2 - 2gh_{max}$$

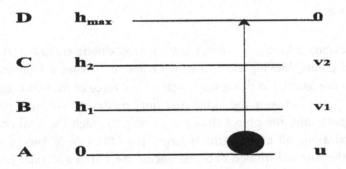

Figure 5.5: Vertical motion under gravity.

The total mechanical energies at the various positions are given by

$$\text{Total energy at, } A, E_A = KE + PE$$
$$= \frac{1}{2}mu^2 + mg \times 0$$
$$E_A = \frac{1}{2}mu^2 \qquad (5.29)$$

$$
\begin{aligned}
\text{Total energy at}, B, E_B &= KE + PE \\
&= \frac{1}{2}mv_1^2 + mgh_1 \\
&= \frac{1}{2}m(u^2 - 2gh_1) + mgh_1 \qquad (5.30) \\
E_B &= \frac{1}{2}mu^2 \qquad (5.31)
\end{aligned}
$$

$$
\begin{aligned}
\text{Total energy at}, C, E_C &= KE + PE \\
&= \frac{1}{2}mv_2^2 + mgh_2 \\
&= \frac{1}{2}m(u^2 - 2gh_2) + mgh_2 \qquad (5.32) \\
E_C &= \frac{1}{2}mu^2 \qquad (5.33)
\end{aligned}
$$

$$
\begin{aligned}
\text{Total energy at}, D, E_D &= KE + PE \\
&= \frac{1}{2}m \times (0)^2 + mgh_{max} \\
E_D &= mgh_{max} \qquad (5.34)
\end{aligned}
$$

From the energy conservation principle, the total energies at A, B, C, D are conserved, that is

$$
\begin{aligned}
E_A = E_B = E_C &= \frac{1}{2}mu^2 = \text{Constant} \\
&= E_D = mgh_{max}
\end{aligned}
$$

which gives

$$
u = \sqrt{2gh_{max}} \qquad (5.35)
$$

Although total energy is conserved, it can be *converted* or *transformed* from one form to another, for example, while the total energy at A is totally kinetic, that at D is totally potential. The kinetic energy at A has been converted to potential energy at D.

5.4.2 Motion in the gravitational field along a smooth track

Consider a particle of mass m moving in a gravitational field, from A to C, through B, along a smooth track as illustrated in Figure (5.6), such that A, B, C are at heights $h_A, h_B = 0$ and h_C respectively, and the respective velocities are $v_A = 0, v_B$ and v_C,

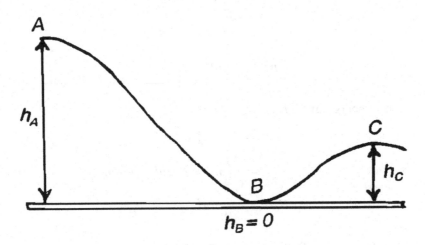

Figure 5.6: Motion under gravity along a smooth track.

The total mechanical energies at the various positions are given by

$$\text{Total energy at}, A, E_A = KE_A + PE_A$$
$$= \frac{1}{2}mv_A^2 + mgh_A$$
$$= \frac{1}{2}m \times (0)^2 + mgh_A$$
$$E_A = mgh_A \qquad\qquad (5.36)$$

$$\text{Total energy at}, B, E_B = KE_B + PE_B$$
$$= \frac{1}{2}mv_B^2 + mgh_B$$
$$= \frac{1}{2}mv_B^2 + mg \times (0)$$
$$E_B = \frac{1}{2}mv_B^2 \qquad\qquad (5.37)$$

$$\text{Total energy at}, C, E_C = KE_C + PE_C$$
$$E_C = \frac{1}{2}mv_C^2 + mgh_C$$

From the energy conservation principle, the total energies A, B, C are conserved, that is

$$E_A = E_B \;\Rightarrow\; mgh_A = \frac{1}{2}mv_B^2 = \text{Constant}$$
$$= E_C = \frac{1}{2}mv_C^2 + mgh_C$$

which gives

$$v_C = \sqrt{2g(h_A - h_C)} \tag{5.38}$$

This is another illustration of *energy conversion*. It can be noted again that although the total energy is conserved, it can be *converted* from one form to another, for example, while the total energy at A is totally potential, that at B is totally kinetic, and that at C is both kinetic and potential.

5.4.3 Roller coaster

Consider a roller coaster whereby a particle of mass m moving from A to D, through B, and a gap at C which is at the top of the loop, as illustrated in Figure (5.7), such that A, B, C, D are at heights $h_A = h, h_B = 0, h_C = 2R$ and h_D respectively, and the respective velocities are $v_A = 0, v_B, v_C$ and v_D. The question is, at what height $h_A = h$ must the mass start if it is to loop the path without leaving the track at the gap?

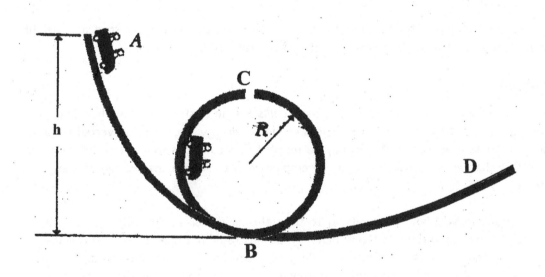

Figure 5.7: Roller coaster.

The total mechanical energies at the various positions are given by

$$
\begin{aligned}
\text{Total energy at,} A, E_A &= KE_A + PE_A \\
&= \frac{1}{2}mv_A^2 + mgh_A \\
&= \frac{1}{2}m \times (0)^2 + mgh \\
E_A &= mgh
\end{aligned} \tag{5.39}
$$

$$
\begin{aligned}
\text{Total energy at the gap,} C, E_C &= KE_C + PE_C \\
&= \frac{1}{2}mv_C^2 + mgh_C \\
&= \frac{1}{2}mv_C^2 + mg2R \quad\quad\quad\quad\quad (5.40) \\
E_C &= \frac{1}{2}mgR + 2mgR \text{ where at the gap } mg = \frac{mv_C^2}{R} (5.41)
\end{aligned}
$$

The total energies E_A and at the gap, E_C are conserved, that is

$$
\begin{aligned}
E_A = E_C \Rightarrow mgh &= \frac{1}{2}mgR + 2mgR \\
&= \frac{5}{2}mgR
\end{aligned}
$$

which gives

$$
h = \frac{5}{2}R \quad\quad\quad\quad\quad (5.42)
$$

In order to complete the loop without leaving the track at the gap, the mass must start at a height h given by equation 5.42. This is the principle used in the design of roller coasters.

5.4.4 Collisions

Collisions are classified as *elastic collisions* and *inelastic collisions*. In elastic collisions, there is no loss of kinetic energy. In inelastic collisions, there is loss of kinetic energy since it is converted to other forms of energy such as heat, sound etc. It may be recalled that collisions were discussed earlier in Chapter 4. In this chapter we revisit this topic to emphasize the role of conservation of energy in elastic collisions.

An object of mass m_1 moving with a velocity u_1 to the right collides elastically with another object of mass m_2 also moving to the right with a velocity u_2, as illustrated in Figure (5.8). After collision, the two masses m_1 and m_2 have velocities v_1 and v_2 respectively. The velocities v_1 and v_2 after collision can be expressed in terms of the initial parameters u_1, u_2, m_1 and m_2 using the conservation principles.

By using the law of conservation of linear momentum, the total momentum *before* collision is equal to the total momentum *after* collision.

$$
\begin{aligned}
m_1u_1 + m_2u_2 &= m_1v_1 + m_2v_2 \\
m_1(u_1 - v_1) &= m_2(v_2 - u_2) \quad\quad\quad\quad\quad (5.43)
\end{aligned}
$$

By using the law of conservation of energy (kinetic), the total energy *before* collision is equal to the total energy *after* collision.

$$
\begin{aligned}
\frac{1}{2}m_1u_1^2 + \frac{1}{2}m_2u_2^2 &= \frac{1}{2}m_1v_1^2 + \frac{1}{2}m_2v_2^2 \\
m_1(u_1^2 - v_1^2) &= m_2(v_2^2 - u_2^2) \\
m_1(u_1 - v_1)(u_1 + v_1) &= m_2(v_2 - u_2)(v_2 + u_2) \quad\quad\quad\quad\quad (5.44)
\end{aligned}
$$

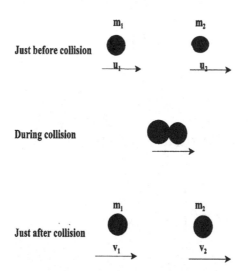

Figure 5.8: Elastic collision of two masses.

Dividing equations (5.44) by (5.43), and after algebraic simplification, expressions for v_1 and v_2 are obtained as

$$v_1 = \left(\frac{m_1 - m_2}{m_1 + m_2}\right) u_1 + \left(\frac{2m_2}{m_1 + m_2}\right) u_2 \tag{5.45}$$

$$v_2 = \left(\frac{2m_1}{m_1 + m_2}\right) u_1 + \left(\frac{m_2 - m_1}{m_1 + m_2}\right) u_2 \tag{5.46}$$

5.4.5 Simple pendulum

Consider a simple pendulum consisting of a particle of mass m suspended by a massless and un-streachable string of length l, and swinging side to side in a vertical plane, as illustrated in Figure (5.9). When the mass is at positions A, B, C, its respective heights from the lowest position B are $h_A, h_B = 0$ and h_C, and the respective velocities are $v_A = 0, v_B$ and $v_C = 0$. When the mass is at position C, the string makes an angle θ with the vertical.

$$
\begin{aligned}
\text{Total energy at, } C, E_C &= KE_C + PE_C \\
&= \frac{1}{2}mv_C^2 + mgl(1 - \cos\theta) \\
&= \frac{1}{2}m \times (0)^2 + mgl(1 - \cos\theta) \\
E_C &= mgl(1 - \cos\theta) \tag{5.47}
\end{aligned}
$$

$$\text{Total energy at, } B, E_B = KE_B + PE_B$$

$$= \frac{1}{2}mv_B^2 + mgh_B$$

$$= \frac{1}{2}mv_B^2 + mg \times (0) \tag{5.48}$$

$$E_B = \frac{1}{2}mv_B^2 \tag{5.49}$$

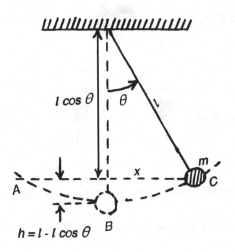

Figure 5.9: The simple pendulum.

The total energies at B, C are conserved, that is

$$E_B = E_C \Rightarrow \frac{1}{2}mv_B^2 = mgl(1 - \cos\theta)$$

which gives

$$v_B = \sqrt{2gl(1 - \cos\theta)} \tag{5.50}$$

This is yet another illustration of *energy conversion*. It can be noted that although the total energy is conserved, it can be *converted* from one form to another, for example, while the total energy at A is totally potential (why?), that at B is totally kinetic, and that at C is totally potential.

5.4.6   Ballistic pendulum

The ballistic pendulum consists of a block of mass m_2 supported by a light rod of length l, as illustrated in Figure (5.10). A bullet of mass m_1 moving horizontally with a velocity v_1 is shot into a stationary block and is embedded in it. After impact, the bullet-block system acquires a velocity v_s, and is displaced by a distance $h = l(1 - \cos\theta)$, where θ is the angle the rod makes with the vertical.

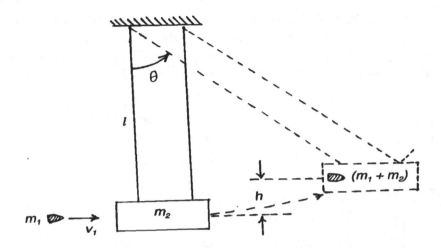

Figure 5.10: The ballistic pendulum.

By using the law of conservation of linear momentum, the total momentum *before* impact is equal to the total momentum *after* impact.

$$
\begin{aligned}
m_1 v_1 + m_2 \times (0) &= (m_1 + m_2)v_s \\
v_s &= \frac{m_1}{m_1 + m_2} v_1
\end{aligned}
\tag{5.51}
$$

By using the law of conservation of mechanical energy, the total energy immediately *after* impact is equal to the total energy of the block when it rises to height h.

$$
\begin{aligned}
\frac{1}{2}(m_1 + m_2)v_s^2 &= (m_1 + m_2)gh \\
v_s &= \sqrt{2gh}
\end{aligned}
\tag{5.52}
$$

Equating equations (5.51) and (5.52), and rearranging we obtain the velocity of the bullet as

$$
v_1 = \frac{(m_1 + m_2)}{m_1}\sqrt{2gh} \quad \text{or}
\tag{5.53}
$$

$$
v_1 = \frac{(m_1 + m_2)}{m_1}\sqrt{2gl(1 - \cos\theta)}
\tag{5.54}
$$

It is important to note that in the ballistic pendulum there are two events. First, there is inelastic collision for which the energy conservation does not apply. After impact, the block with the bullet rises freely under gravity, and in this case conservation of mechanical energy is applicable.

5.4.7   A mass dropped on a vertically held spring

Consider a mass-spring system illustrated in Figure (5.11). A mass m is dropped from rest, from a vertical height h on a vertically held spring of force constant k. When the spring relaxes to its neutral state, the mass is projected vertically upwards to a height h'. The four successive stages of the system are shown in Figure (5.11) A, B, C and D respectively. The system is conservative, and the total mechanical energy in various stages of the system is conserved.

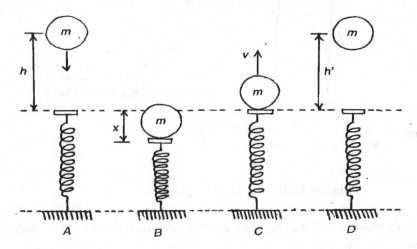

Figure 5.11: A mass dropped on a vertically held spring.

The total mechanical energies of the system at stages A, B, C and D are given as:

$$
\begin{aligned}
E_A &= mgh \\
E_B &= \frac{1}{2}kx^2 - mgx \\
E_C &= \frac{1}{2}mv^2 \\
E_D &= mgh'
\end{aligned}
\tag{5.55}
$$

and from the conservation of energy:

$$
\begin{aligned}
E_A &= E_B = E_C = E_D \Rightarrow \\
&= mgh = \frac{1}{2}kx^2 - mgx = \frac{1}{2}mv^2 = mgh'
\end{aligned}
\tag{5.56}
$$

These are trivial exercises left for students to show that (i) the mass rises to the same height h, i.e., $h' = h$, (ii) velocity with which the mass leaves the spring is the same with which it lands upon it when dropped, and is given by $v = \sqrt{2gh}$. Here we calculate the length x by which the spring is compressed when the mass is at its lowest point. From equation (5.56), we have:

$$
mgh = \frac{1}{2}kx^2 - mgx \Rightarrow kx^2 - 2mgx - 2mgh = 0
\tag{5.57}
$$

Solving the quadratic equation (5.57) gives:

$$x = \frac{mg \pm \sqrt{m^2 g^2 + 2kmgh}}{k} \qquad (5.58)$$

Considering the figure, and the expression for total energy at B, (E_B), the magnitude of x must be positive. Hence, the length by which the spring is compressed is given by:

$$x = \frac{1}{k}\left(mg + \sqrt{m^2 g^2 + 2kmgh}\right)$$

5.4.8 Bullet-block system

Consider a bullet-block system. A bullet of mass m_1 moving with a velocity v_1 is fired vertically upwards into a stationary block of mass m_2. After impact, the bullet-block system acquires a velocity v_s, and rises to a height h. It may be recalled that this problem was discussed in chapter 4. We give a complete solution here.

By using the law of conservation of linear momentum, the total momentum *before* impact is equal to the total momentum *after* impact.

$$\begin{aligned} m_1 v_1 + m_2 \times (0) &= (m_1 + m_2)v_s \\ v_1 &= \frac{m_1 + m_2}{m_1}v_s \end{aligned} \qquad (5.59)$$

By using the law of conservation of mechanical energy, the total energy immediately after impact is equal to the total energy *after* the block rises to a height h.

$$\begin{aligned} \frac{1}{2}(m_1 + m_2)v_s^2 &= (m_1 + m_2)gh \\ v_s &= \sqrt{2gh} \end{aligned} \qquad (5.60)$$

From equations (5.59) and (5.60), we obtain

$$v_1 = \frac{(m_1 + m_2)}{m_1}\sqrt{2gh} \qquad (5.61)$$

This is also a two event process like the case of the ballistic pendulum discussed earlier.

5.5 Power

The *power* delivered by a force to a body is the rate at which the force does work on that body. If there is an amount of work ΔW done in an interval of time Δt, then the *average power*, $\bar{P}$ is given by

$$\bar{P} = \frac{\Delta W}{\Delta t} \qquad (5.62)$$

and the *instantaneous power*, P, is given by

$$P = \frac{dW}{dt} \tag{5.63}$$

Using the definition of work given by equation (5.2), the instantaneous power can also be given in the form

$$
\begin{aligned}
P &= \frac{dW}{dt} \\
&= \frac{\mathbf{F} \cdot \mathbf{dr}}{dt} \\
&= \mathbf{F} \cdot \frac{\mathbf{dr}}{dt} \\
&= \mathbf{F} \cdot \mathbf{v} \tag{5.64}
\end{aligned}
$$

where $\mathbf{v}$ is velocity.

It can be noted that power is a *scalar quantity*. The unit for power is the $Watt$, which is the rate of work of 1 Joule per second.

$$1W = 1Js^{-1} \tag{5.65}$$

From equation (5.65) one can express the unit of energy as $1J = W\,s$.

The SI units of power W or $J\,s^{-1}$, and the unit of energy $W\,s$ are very small units. They are not convenient to use for commercial and industrial applications. For such applications larger units such as kW $(1\,kW = 1000\,W)$, MW $(1\,MW = 10^6\,W)$, and GW $(1\,GW = 10^9\,W)$ are used for power. For example the capacity of a power supply line to a house hold and large house hold appliances such as cooking range, washing machines are rated in kW. The generation capacity of a medium sized commercial power plant is rated in MW. Likewise, the commercial unit of energy consumption, for the purpose of billing is $kW\,h$ (kilowatt hour). As an exercise convert $1\,kW\,h$ of energy to the SI units of energy.

As work and energy are intimately related, work done results either in the dissipation (as against the force of friction) or in the creation (as the gain in kinetic or potential energy) of an equal amount of energy. The power is therefore in general defined *as the rate of energy consumption or the rate of energy production*. For example a light bulb of $100\,W$ power consumes $100\,J$ of energy per second, and a electric generator of $1\,kW$ power produces electricity at a rate of $1000\,J$ per second.

5.6   Efficiency

Although as per the energy conservation law, the total energy in any process is always conserved, but when energy is used to perform a specific task, all the energy input to the system is not directed to the required task. A part of the energy is lost to other processes. For example, when electricity is used for light, part of it is lost due to heating, in a car engine when petrol is ignited, only a part of it provides the motive power to the car, and the remaining is lost to heating, and in a coal fired power plant some of the heat of combustion of coal is lost to environment with only a fraction of it being converted to electricity. The efficiency (η) of a energy consuming or a energy generating device is defined as the *fraction of total input energy converted to useful work or energy*.

$$\eta = \frac{\text{Useful output work or energy}}{\text{Total input work or energy}} \tag{5.66}$$

Sometimes the efficiency is also expressed as a percentage given by:

$$\text{Percentage efficiency} = \eta\,\% = \eta \times 100 = \frac{\text{Useful output work or energy}}{\text{Total input work or energy}} \times 100 \tag{5.67}$$

The efficiency is a fraction less than one or less than 100%. As per the laws of nature, no system can ever be designed to have a perfect 100% efficiency, as it violates the laws of thermodynamics.

5.7 Problems

5.1. A $5\,kg$ mass placed on a horizontal surface is pulled to a distance of $3\,m$ by a $50\,N$ force applied at an angle of 45° above horizontal. Identify all the forces acting on the mass and calculate the work done by each force if *(i)* the surface if smooth, and *(ii)* its coefficient of kinetic friction is 0.3. What is the gain in the kinetic energy of the mass in each case?

5.2. A $5\,kg$ mass placed on a horizontal surface is pushed to a distance of $3\,m$ by a $50\,N$ force applied at an angle of 45° above horizontal. Identify all the forces acting on the mass and calculate the work done by each force if *(i)* the surface if smooth, and *(ii)* its coefficient of kinetic friction is 0.3. What is the gain in the kinetic energy of the mass in each case?

5.3. A mass M is pulled up a smooth plane inclined at angle θ above horizontal with a force F applied at an angle α above the plane. Apply work-energy principles to calculate the acceleration of the mass up the plane.

5.4. A mass M is pushed up a smooth plane inclined at angle θ above horizontal with a force F applied at an angle α above the plane. Apply work-energy principles to calculate the acceleration of the mass up the plane.

5.5. A mass M is pulled up a rough plane inclined at angle θ above horizontal with a force F applied at an angle α above the plane. The coefficient of kinetic friction is μ. Apply work-energy principles to calculate the acceleration of the mass up the plane.

5.6. A mass M is pushed up a rough plane inclined at angle θ above horizontal with a force F applied at an angle α above the plane. The coefficient of kinetic friction is μ. Apply work-energy principles to calculate the acceleration of the mass up the plane.

5.7.A mass m at rest is lifted vertically up to a height h with a force $F = (mg + \Delta F)$ Calculate:
(a) Work done by the force F.
(b) Upward acceleration of the block, and the gain in its kinetic energy.
(c) What is the net force responsible for the acceleration of the block, and the gain in its kinetic energy?
(d) What is the gain in the potential energy of the mass, and what force is responsible for it?
(e) What conclusion can you draw about the force required to raise a mass in gravitational field while maintaining it in equilibrium?

5.8. A $3\,kg$ mass is projected vertically upwards with a velocity of $10\,m\,s^{-2}$ from the top of a $50\,m$ high building, and after rising above the building, the mass falls to the ground. Use work energy principles to calculate:
(a) The maximum height to which the mass rises above ground.
(b) Velocity with which the mass hits the ground.
(c) After hitting the ground, the mass penetrated the soft ground below to a depth of $10\,cm$. Calculate the average force of friction of the ground.

5.9. A $5\,kg$ mass placed on a rough plane inclined at 30° above horizontal slides down freely from a height of $2\,m$. The coefficient of friction of the plane is 0.2. Using work energy principles, calculate the velocity of the mass on reaching the foot of the plane.

5.10. A mass m starting from rest slides down a hill of height h and length L.
(a) If the average friction force is f, prove that the velocity of the mass on reaching the foot of the hill is given by:

$$v = \left(\frac{2}{m} \left(m\,g\,h - f\,L \right) \right)^{\frac{1}{2}}$$

(b)If the coefficient of kinetic friction is μ. show that the expression for velocity reduces to:

$$v = \left(2g \left(h - \mu \sqrt{L^2 - h^2} \right) \right)^{\frac{1}{2}}$$

5.11. A skier of $80\,kg$ mass skies down a hill of slope 20° at a constant speed. How much energy per km of distance is used up against the force of friction?

5.12. Brakes of a $1800\,kg$ car traveling at $60\,km\,h^{-1}$ fail. The driver switches off the engine, and drives on the graveled sandy shoulder of the road. The car comes to rest after traveling $150\,m$. Calculate the average friction force.

5.13. A car driving along a level road climbs a $20\,m$ high hill of $500\,m$ length with its engine switched off. What should be the minimum speed of the car on the level road, if *(i)* there is no friction, and *(ii)* average coefficient of friction is 0.25.

5.14. A $0.5\,kg$ point mass is suspended freely with a light, inextensible string. The mass is pulled to the side keeping the string tight till the string make an angle of 37° with the vertical. Calculate:

(a) The work done to pull the mass to the side, and its mechanical energy in that position.
(b) Velocity of the ball when it passes the lowest point when released.
(c) How far up the mass rises when it continues to swing to the other side?

5.15 A simple pendulum of length l is set into oscillations from a small angle θ to the vertical. Show that the maximum speed of the pendulum bob is given by

$$\sqrt{2gl(1-\cos\theta)}$$

5.16. A $1\,kg$ mass is resting on a vertically held spring of force constant $k = 1.2 \times 10^4\,N\,m^{-1}$ and the spring is compessed by $0.5\,cm$. When the mass is released, it is projected up by the spring force. Calculate:
(a) The velocity with which the mass leaves the spring.
(b) Height to which the mass rises in the air from the neutral position of the spring.
(c) If the mass again falls back on the spring, by what length it shall be compressed?

5.17. A $100\,g$ mass freely suspended by an ideal spring stretches it by $2\,cm$. Calculate the change in the elastic potential energy of the spring if the same spring is compressed with a force of $2.5\,N$

5.18. In a toy gun a compressed spring is used to shoot balls of $50\,g$ mass. If the force constant of the spring is $50\,N\,cm^{-1}$, and it is compressed by $5\,cm$, what is the muzzle velocity of the ball.

5.19. Two masses M_1 and M_2 are suspended freely with a light unstreachable string passing over a light, frictionless pulley. Use work-energy principles and the kinematic equations to determine the acceleration of the masses when released from rest. Use the expression for acceleration to find the tension in the string.

5.20. A mass M_1 placed on a rough plane inclined at angle θ above horizontal, is attached to another mass M_2 with a light, inextensible string. The string passes over a smooth light pulley, and the mass M_2 hangs freely. The coefficient of kinetic friction for the plane is μ. Use work-energy principles to find an expression for the acceleration of the mass if the mass M_2 *(i)* ascends, and *(ii)* descends.

5.21. A $10\,g$ bullet traveling at $200\,m\,s^{-1}$ penetrates $10\,cm$ in a wooden block. Calculate average force of friction from the work energy principles. Verify your answer by calculating the force using kinematic equations and the Newton's laws of motion.

5.22. A 10 g bullet is fired into a 3 kg block from below where it is embedded, and the block-bullet system rises by a height of 2m. Calculate the initial velocity of the bullet.

5.23. A bullet of $15\,g$ traveling at $300\,m^{-1}$ penetrates a wooden block of $5\,kg$ mass, which then slides on a rough horizontal surface to a distance of $20\,cm$ before coming to rest. Calculate the average force of friction, and the coefficient of friction for the plane. If the same plane was inclined at 37° above horizontal, how far up the plane shall the block slide?

5.24. In your home you have the following electrical appliances: *(i)* a $40\,W$ refrigerator which is on 50% of the time, *(ii)* a $1.5\,kW$ cooking range used for 2 hours per day, *(iii)* a $2\,kW$ hot water system used for 3 hours per day. *(iv)* a $100w$ music system used for 15 hours per day, *(v)* a $60W$ television used for 3 hours per day, *(vi)* two $100W$ light bulbs each used for 3 hours per day, *(vii)* four $60W$ bulbs each used for 4 hours per day, *(viii)* miscellaneous gadgets $200W$ used for 2 hours per day. Calculate the cost of your electricity consumption for the month of July if the cost per unit of electricity is $55\,t$ per unit including VAT if (a) all appliance operated at 100% efficiency, and (b) if the average efficiency of all appliances is 90%.

5.25. A $80\,kg$ person ascends the stairs of a $20\,m$ building in $10\,s$. Calculate the power of the person.

5.26. A pump lifts water from a $50\,m$ deep well at a rate of $180\,l$ per minute. If the pump operates at 70% efficiency, calculate the daily electricity consumption of the pump in commercial units, if it is used for 8 hours per day. If the water pumped is used to fill-up an overhead water storage tank at a $20\,m$ height how much additional electricity shall be consumed by another pump with 65% efficiency.

Chapter 6

Circular Motion including Rotational Kinematics

The objective of this chapter is to study circular motion, including rotational kinematics. Circular motion occurs in various situations in nature as well as in scientific, technological and consumer devices. The existence of rotational motion in nature includes the orbital motion of planets and the electronic motion in the atoms. In technology, circular and rotational motions are observed in the motion of the wheels, pulleys, gears, motors, spinning tops etc. Engineers and technologists who design these products, and scientists who propose related models must all be well conversant with all aspects of circular and rotational motions. This makes the study of rotational motion in mechanics as important and relevant as any other area of mechanics. The simplest representation of a rotational motion is the circular motion of a point mass about a fixed axis, which can explain the main general features of rotational motions, and be applied to many a number of cases. In this Chapter we focus on the circular motion of point masses to lay foundation for this important area of mechanics. At the end of the chapter students are expected to have understood the following concepts which will be applied later in the study of rigid body dynamics in Chapter 9.

- Angular displacement, angular velocity and angular acceleration

- Period of rotation

- Rotational kinematic equations of motion

- Angular momentum

- Moment of inertia

- Torque

- Rotational kinetic energy

- Examples illustrating circular motion: Roller coasters, Motion in a vertical circle, Conical pendulum, Banking of a road, Planet-satellite system (eg Earth-Moon system), Electronic orbits in an atom and an electric charge in a uniform magnetic field.

6.1   Rotational Kinematics

Rotational kinematics is the study of rotational motion in which the only parameters considered are angular displacement, angular speed and angular acceleration.

Consider the motion of a particle in a circle of radius r as illustrated in Figure (6.1).

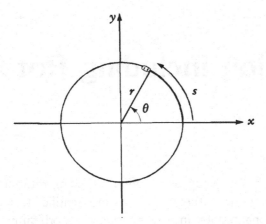

Figure 6.1: Geometry of circular motion showing an angular displacement θ, radius r of the circle and the arc of length s.

The centre of the circle is known as the *centre of rotation*, and an axis perpendicular to the plane of the circle through the center about which the particle rotates is called the *axis of rotation*. The circular path of the particle is the *trajectory* of motion, and instantaneous linear velocity v is tangential to the trajectory. The radius of the circle r, joining the particle to the center of rotation is the *radius vector* In analogy with the translational motion, we define the rotational kinematic quantities namely angular displacement, angular velocity, and angular acceleration, and relate them to corresponding translational quantities.

6.1.1   Angular displacement

The angular displacement is the angle moved by the radius vector in a specified time interval. It is measured in *radians* where:

$$\pi \,\text{radian} = 3.1415 \,\text{radian} = 180^o \Rightarrow 1 \,\text{radian} = 57.3^o \tag{6.1}$$

The angular displacement, denoted by θ in Figure (6.1), is related to the linear displacement s during the same interval of time as:

$$\text{Angular displacement } \theta = \frac{s}{r} \,\text{ radian} \tag{6.2}$$

6.1.2   Angular velocity

The angular velocity ω is the rate of change of angular displacement with time. If $\Delta\theta$ is the angular displacement in time interval Δt, then the angular velocity is given as:

$$\text{Angular velocity, } \omega = \frac{\Delta\theta}{\Delta t} \Rightarrow \frac{d\theta}{dt} \text{ radian per second} \qquad (6.3)$$

where the two terms in equation (6.3) can be interpreted as the average and instantaneous angular velocities as were defined in the case of translational motion.

The angular velocity is a vector quantity of magnitude ω and it is in the direction of the axis of rotation as illustrated in Figure (6.2). From the figure it is related to the translational velocity vector V, and the radius vector r as:

$$\mathbf{V} = \omega \wedge r \qquad (6.4)$$

A further discussion of this topic is given in Chapter 9 dealing with rigid body dynamics.

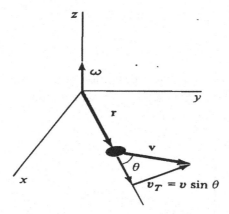

Figure 6.2: Angular velocity, ω, of a particle executing circular motion in the xy plane is along the axis of rotation along the z axis.

In case of the circular motion r and ω are perpendicular to each other, and equation (6.4) reduces to:

$$v = r\,\omega \Rightarrow \omega = \frac{v}{r} \qquad (6.5)$$

6.1.3 Period of rotation

The period of one complete rotation T is the time taken for one complete circle of rotation. For a uniform circular motion it can be obtained either from the angular velocity ω recognizing that the angle included by a circle is 2π radians, or from the circumference $2\pi r$ of the circle divided by the linear velocity v, $i.e.$,

$$T = \frac{2\pi}{\omega} = \frac{2\pi r}{v} \qquad (6.6)$$

The equation (6.6) also reflects the relationship between v and ω given by equation (6.5).

6.1.4   Angular acceleration

The angular acceleration α is the rate of change of angular velocity with time. If $\Delta\omega$ is the change in angular velocity in time interval Δt, then the angular acceleration is given as:

$$\text{Angular acceleration, } \alpha = \frac{\Delta\omega}{\Delta t} \Rightarrow \frac{d\omega}{dt} = \frac{d^2\theta}{dt^2} \; rad \; s^{-2} \qquad (6.7)$$

where the two terms in equation (6.7) can be interpreted as the average and instantaneous angular accelerations.

The angular acceleration is a vector quantity of magnitude α and it is in the direction of the axis of rotation as illustrated in Figure (6.3). From the figure it is related to the (translational) acceleration vector a, and the radius vector r as:

$$\mathbf{a} = \alpha \wedge r \qquad (6.8)$$

A further discussion of relation between angular acceleration and translational vectors is deferred to a later section where we introduce the tangential and normal components of acceleration, and is also given in Chapter 9 dealing with rigid body dynamics.

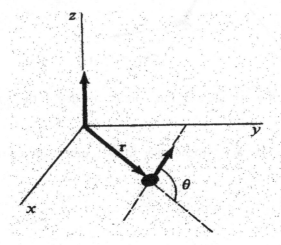

Figure 6.3: Angular acceleration, α, of a particle executing circular motion in the xy plane is along the axis of rotation along the z axis.

In case of the circular motion, r and the angular acceleration α are perpendicular to each other, and equation (6.8) reduces to:

$$a = r\alpha \Rightarrow \alpha = \frac{a}{r} \qquad (6.9)$$

6.1.5 Rotational kinematic equations

Upon integrating equation (6.7), we obtain

$$\begin{aligned} \omega &= \int \alpha dt \\ &= \alpha t + C_1 \end{aligned} \tag{6.10}$$

where C_1 is a constant of integration, which can be found by noting the initial conditions that when $t = 0, \omega(t = 0) = \omega_0 \Rightarrow C_1 = \omega_0$, and hence

$$\omega = \omega_0 + \alpha t \tag{6.11}$$

We can also integrate equation (6.3) to obtain

$$\begin{aligned} \theta &= \int \omega dt \\ &= \int (\omega_0 + \alpha t) dt \\ &= \omega_0 t + \frac{1}{2} \alpha t^2 + C_2 \end{aligned} \tag{6.12}$$

where C_2 is a constant of integration, which can be found by noting the initial conditions that when $t = 0, \theta(t = 0) = 0 \Rightarrow C_2 = 0$, and hence

$$\theta = \omega_0 t + \frac{1}{2} \alpha t^2 \tag{6.13}$$

If both sides of equation (6.11) are squared, we obtain

$$\begin{aligned} \omega^2 &= (\omega_0 + \alpha t)^2 \\ &= \omega_0^2 + \alpha^2 t^2 + 2\omega_0 \alpha t \\ &= \omega_0^2 + 2\alpha \left(\omega_0 t + \frac{1}{2} \alpha t^2 \right) \\ \omega^2 &= \omega_0^2 + 2\alpha \theta \end{aligned} \tag{6.14}$$

where equation (6.13) has been used in reaching the final equation given in (6.14).

From equation (6.13), algebraic simplification leads to

$$\begin{aligned} \theta &= \omega_0 t + \frac{1}{2} \alpha t^2 \\ &= \frac{t}{2} (2\omega_0 + \alpha t) \\ &= \frac{t}{2} (\omega_0 + \omega_0 + \alpha t) \\ \theta &= \frac{1}{2} (\omega_0 + \omega) t \end{aligned} \tag{6.15}$$

Equations (6.11), (6.13), (6.14) and (6.15) constitute what are known as *rotational kinematic equations* for uniformly accelerated motion, which are collected below for convenience.

$$\omega = \omega_0 + \alpha t$$
$$\theta = \omega_0 t + \frac{1}{2}\alpha t^2$$
$$\omega^2 = \omega_0^2 + 2\alpha\theta$$
$$\theta = \frac{1}{2}(\omega_0 + \omega)t$$

The rotational kinematic equations can be applied to several cases of interest discussed below.

Recipe 6. 1: Rotational kinematics

- Read the given problem carefully and list the known values, expresing them in SI units
- Identify the unknown values
- When rotational motion is from rest, $\omega_0 = 0$
- When a particle stops, $\omega = 0$
- Express the unknown values in terms of known values using one or more of the rotational kinetic equations and solve

Example 6. 1: A body moving with a uniform angular acceleration of 0.2 rev/s^2 has an angular velocity of 1 rev/s at time $t = 0$. Calculate the following quantities after 10s.
(a) Angular velocity.
(b) Angular displacement.
(c) Average angular velocity over the 10s period.

Solution
(a)

$$\omega_0 = 1\text{rev/s} = 2\pi\text{rad/s}$$
$$\alpha = 0.2\text{rev/s}^2 = 0.4\pi\text{rad/s}^2$$
$$\omega = ?$$

Using

$$\omega = \omega_0 + \alpha t$$
$$\omega = (2\pi + 0.4\pi \times 10)\ \text{rad/s}$$
$$\omega = 6\pi\text{rad/s}$$

(b)

$$\theta = \omega_0 t + \frac{1}{2}\alpha t^2$$
$$= 2\pi \times 10 + \frac{1}{2} \times 0.4\pi \times 10^2$$
$$= 40\pi\ \text{rad}$$

(c)

$$
\begin{aligned}
\omega_{average} &= \frac{\omega_0 + \omega}{2} \\
&= \frac{2\pi + 6\pi}{2} \\
&= 4\pi \text{rad/s}
\end{aligned}
$$

Exercise 6. 1: A body moving with uniform angular acceleration of 0.4 rev/s^2 undergoes 120 rev in 20 s. Calculate the following quantities
(a) Angular velocity at $t = 0$.
(b) Angular velocity at $t = 20$ s.
(c) Average angular velocity over the 20s period.

6.1.6 Tangential and Centripetal aceleration

Consider a particle moving in a circle along an arc PP', of length $\Delta s = ds$, as illustrated in Figure 6.4. The particle has a tangential velocity $\mathbf{v} = \mathbf{v}\hat{\theta}$, where v is the magnitude in the tangential direction of unit vector $\hat{\theta}$. Since the velocity changes direction, the particle must have an acceleration, $\mathbf{a} = \mathbf{dv}/\mathbf{dt}$.

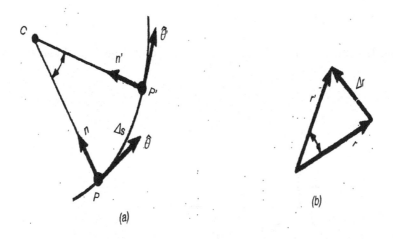

Figure 6.4: Unit tangential vector $\hat{\theta}$ and unit normal vectors $\mathbf{n}$ for a particle moving along a circular trajectory.

$$
\begin{aligned}
\text{The arc } PP' &= \Delta s = ds = r\delta\theta \\
\text{The velocity } v &= \frac{ds}{\delta t} \\
\text{Hence, } r\delta\theta &= v\delta t \\
\text{Angular velocity, } \omega &= \frac{\delta\theta}{\delta t} = \frac{v}{r}
\end{aligned}
$$

or

$$v = r\omega = r\dot{\theta} \tag{6.16}$$

The acceleration is given by

$$
\begin{aligned}
\mathbf{a} &= \frac{d\mathbf{v}}{dt} \\
&= \frac{d(v\hat{\theta})}{dt} \\
&= \dot{v}\hat{\theta} + v\frac{d\hat{\theta}}{dt} \quad \text{where} \quad \frac{d\hat{\theta}}{dt} = \frac{d\hat{\theta}}{d\theta}\frac{d\theta}{dt} = \mathbf{n}\frac{d\theta}{dt} = \mathbf{n}\frac{v}{r} \\
\mathbf{a} &= \dot{v}\hat{\theta} + \frac{v^2}{r}\mathbf{n} \\
\mathbf{a} &= a_\theta\hat{\theta} + a_n\mathbf{n} \tag{6.17}
\end{aligned}
$$

where $a_\theta = \dot{v}$ is the *tangential acceleration* and $a_n = v^2/r$ is the *centripetal acceleration*. The centripetal acceleration is along the radial direction directed towards the centre. For a uniform motion (constant speed), the tangential acceleration $a_\theta = 0$ and there is only the centripetal acceleration a_n. The force responsible for producing a_n is known as the centripetal force given by

$$F_n = ma_n = m\frac{v^2}{r} = mr\omega^2 \tag{6.18}$$

6.2   Angular Momentum

The angular momentum $\mathbf{L}$ is a vector quantity along the direction of the axis of rotation as illustrated in Figure (6.5).

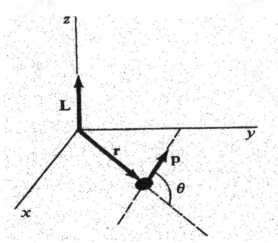

Figure 6.5: The angular momentum $\mathbf{L}$ is along the z axis, normal to r and p which lie in the xy plane.

The angular momentum $\mathbf{L}$ is defined as the moment of linear momentum:

$$\mathbf{L} = \mathbf{r} \wedge \mathbf{p} \tag{6.19}$$

or

$$\mathbf{L} = \mathbf{rp}\sin\theta \qquad (6.20)$$

Note that when $\theta = 90^0$,

$$
\begin{aligned}
L &= rp \\
&= rmv \\
&= rm(r\omega) \\
&= mr^2\omega \\
L &= I\omega \qquad (6.21)
\end{aligned}
$$

where $I = mr^2$ is known as the *moment of inertia*. A further discussion of this topic is given in Chapter 9 dealing with rigid body dynamics.

6.3 Torque

The torque τ is a vector quantity in the direction of the axis of rotation as illustrated in Figure (6.6).

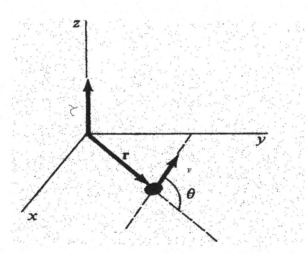

Figure 6.6: The torque τ is along the z axis, normal to r and F which lie in the xy plane.

The torque τ is defined as the moment of a force.

$$\tau = \mathbf{r} \wedge \mathbf{F} \qquad (6.22)$$

or

$$\tau = \mathbf{rF}\sin\theta \qquad (6.23)$$

Note that when $\theta = 90^0$,

$$\tau = rF$$

$$
\begin{aligned}
&= rma \\
&= rm(r\alpha) \\
&= mr^2\alpha \\
\tau &= I\alpha
\end{aligned}
\tag{6.24}
$$

where as we have seen earlier $I = mr^2$ is the *moment of inertia*. Equation (6.24) is the rotational motion equivalent of Newton's second law of motion ($\mathbf{F} = m\mathbf{a}$). It can be noted that the moment of inertia $I = mr^2$ plays the same role in rotational motion as the mass does in translational motion, that is, it represents the resistance of a rotating body to a change in rotation. A further discussion of this topic is given in Chapter 9 dealing with rigid body dynamics.

6.4   Rotational Kinetic Energy

The Rotational Kinetic Energy (RKE) of a particle executing rotational motion is defined as

$$
\begin{aligned}
RKE &= \frac{1}{2}mv^2 \\
&= \frac{1}{2}m(r\omega)^2 \\
&= \frac{1}{2}(mr^2)\omega^2 \\
&= \frac{1}{2}I\omega^2
\end{aligned}
\tag{6.25}
$$

where again $I = mr^2$ is the *moment of inertia*. A further discussion of this topic is given in Chapter 9 dealing with rigid body dynamics.

6.5   Examples illustrating Circular Motion

6.5.1   Roller Coaster

The principle of roller coaster was discussed in chapter 5 using energy conservation principles. Here we revisit the problem from the circular motion perspective. A trolley of mass m (including riders) starting from rest at A slides down a presumed smooth track (friction force neglected) from a height h, loops a circle of radius R, through B and C, and continues on to the track to D as illustrated in Figure 6.7. There is an optimum height h for the trolley to complete the circle, which can be related to the radius of the circle as follows.

When the trolley is at the highest point of the circle, it must have a certain minimum velocity so that it shall not fall under gravity. For motion in the vertical circle, the minimum speed at the highest point should be such that the weight of the trolley acting towards the center of the circle provides the necessary centripetal force. Thus, denoting the minimum speed by v_{min}:

$$
\frac{m\,v_{min}^2}{R} = m\,g
\tag{6.26}
$$

Equation (6.26) can be simplified to obtain the kinetic energy of the trolley at this point as:

$$KE = \frac{1}{2}\, m\, v_{min}^2 = \frac{1}{2}\, m\, g\, R \tag{6.27}$$

and the potential energy is:

$$PE = m\, g \times 2R \tag{6.28}$$

This gives the total mechanical energy of the trolley at the highest point of the circle as:

$$E = \frac{1}{2}\, m\, g\, R + 2\, m\, g\, R = \frac{5}{2}\, m\, g\, R \tag{6.29}$$

From energy conservation principle, this must be equal to its potential energy $m\,g\,h$ at the starting point at height h, *i.e.*,

$$\frac{5}{2}\, m\, g\, R = m\, g\, h \tag{6.30}$$

Equation (6.30) on simplification gives:

$$h = \frac{5}{2}\, R \tag{6.31}$$

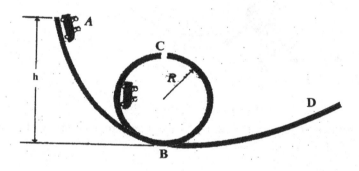

Figure 6.7: The roller coaster.

This is the minimum height from where the trolley must start, same as obtained in chapter 5 where the focus was on energy considerations. Note that it does not depend on the mass of the trolley. In actual practice as additional safety margin, the trolley starts at a higher level so that its speed at the top of the circle is higher than the minimum required. If, under certain circumstances, the speed falls below the minimum required, the weight of the trolley downward becomes larger than the required centripetal force at that speed, resulting in its fall under gravity before reaching the highest point of the circle.

6.5.2 Motion in a Vertical Circle

Consider a small sphere of mass m attached to the end of a string of length R, which is rotated in a vertical circle about a centre O as illustrated in Figure (6.8). At any instant the string makes an angle θ with the vertical and experiences a tension T. Let us answer the following questions.

(a) What is the tangential acceleration at any position in the path?
(b) What is the tension T at any point in the path?
(c) At what position is the tension maximum?
(d) At what position is the tension minimum?
(e) What is the minimum velocity of the mass for it to complete the circle?
(f) At what position would the string most likely break if the speed increases?

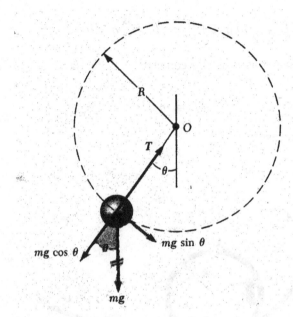

Figure 6.8: Motion in a vertical circle about a centre O for a system consisting of a small sphere of mass m attached to the end of a string of length R.

(a) The only forces acting on the sphere are the weight, mg and the tension T. The weight can be resolved into two components, the tangential component $mg\sin\theta$ and the radial component $mg\cos\theta$. The tangetial component satisfies

$$
\begin{aligned}
ma_t &= \sum F_t \\
&= mg\sin\theta \\
a_t &= g\sin\theta
\end{aligned}
\tag{6.32}
$$

where a_t is the tangential acceleration which causes the velocity v to change.

(b) The tension T at any point can be obtained by applying Newton's second law in the radial direction, noting that the radial acceleration is v^2/R.

$$
\begin{aligned}
m\frac{v^2}{R} &= \sum F_r \\
&= T - mg\cos\theta
\end{aligned}
$$

$$T = m\left(\frac{v^2}{R} + g\cos\theta\right) \qquad (6.33)$$

(c) The maximum tension T_{max} occurs when T given in equation (6.33) is maximum, and this is when $\cos\theta = 1$, which occurs when $\theta = 0$, which is at the *bottom* of the path, and hence

$$T_{max} = m\left(\frac{v_{bot}^2}{R} + g\right) \qquad (6.34)$$

From equation (6.23) we note that the tension at the lowest point of the circle comprises two terms. At this point the tension in the string must not only provide the necessary centripetal force, which is the first term on the right in the equation, but also support the weight acting vertically downwards which is the second term.

(d) The minimum tension T_{min} occurs when T given in equation (6.33) is minimum, and this is when $\cos\theta = -1$, which occurs when $\theta = 180^0$, which is at the *top* of the path, and hence

$$T_{min} = m\left(\frac{v_{top}^2}{R} - g\right) \qquad (6.35)$$

At the top point of the circle, weight of the sphere acting vertically downwards towards the center of the circle provides part of the centripetal force, and tension provides only the remainder of the required centripetal force. Hence the negative sign.

(e) The minimum possible tension for the circle to complete at the top is $T_{min} = 0$, which from equation (6.24) correspond to the velocity:

$$v_{min} = \sqrt{Rg} \qquad (6.36)$$

This is the minimum required velocity of the sphere at the top so that the circle can be complete. If the velocity is smaller than this, the weight of the sphere shall be larger than the required centripetal force and the mass shall simply slump (fall) under gravity without completing the circle. Note that this velocity does not depend on the mass of the sphere, and depends only on the radius of the circle.

(f) The string is most likely break when the tension is maximum, that is, at the bottom of the path.

6.5.3   Conical Pendulum

Consider a small sphere of mass m revolving in a horizontal circle of radius r with a constant velocity v at the end of a string of length L, as illustrated in Figure (6.9). At any instant the string makes an angle θ with the vertical and experiences a tension T.

The only forces acting on the sphere are the weight, mg and the tension T. The tension T can be resolved into two components, the radial component $T_r = T\sin\theta$ and the vertical component $T_z = T\cos\theta$. The vertical component satisfies

$$T\cos\theta = mg \qquad (6.37)$$

The radial component supplies the centripetal force and satisfies

$$
\begin{aligned}
m\frac{v^2}{r} &= \sum F = T_r \\
&= T\sin\theta
\end{aligned}
\tag{6.38}
$$

Dividing equation (6.38) by (6.37) gives

$$
\begin{aligned}
\tan\theta &= \frac{v^2}{rg} \\
v^2 &= rg\tan\theta \\
v &= \sqrt{rg\tan\theta}
\end{aligned}
\tag{6.39}
$$

from which the period T of the conical pendulum is obtained as

$$
T = \frac{2\pi r}{v} = \frac{2\pi r}{\sqrt{rg\tan\theta}} = 2\pi\sqrt{\frac{r}{g\tan\theta}}
\tag{6.40}
$$

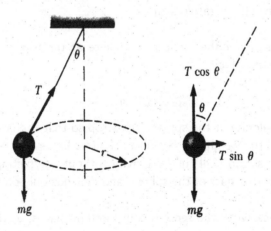

Figure 6.9: The conical pendulum and its free body diagram.

But $r = L\sin\theta$, so that the period T can also be given as

$$
T = 2\pi\sqrt{\frac{L\cos\theta}{g}}
\tag{6.41}
$$

Note that the period of the conical pendulum does not depend on the mass.

6.5.4 Banking of a Road

Speed tracks are normally in the form of a curve of radius R and banked at an angle θ as shown in Figure (6.10). Banking allows the driver in the racing car not to rely only on friction to achieve a maximum speed possible without skiding. This can be understood as follows.

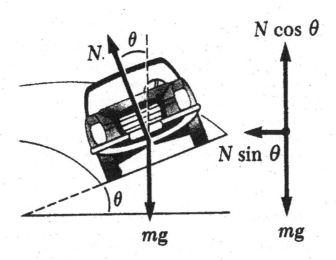

Figure 6.10: Motion of a car on a road banked at an angle θ to the horizontal. The centripetal force is provided by the horizontal component of the normal reaction.

Let us first consider a horizontal flat curved road of radius R, which a driver is negotiating at speed v. As the car tends to go off the track, a force of friction $F_f = \mu N = m g \mu$ acts towards the center of the circle where m is the mass of the car, and μ is the coefficient of friction. This force provides the necessary centripetal force to maintain the car on track. Thus:

$$\frac{m v^2}{R} = F_f = m g \mu \Rightarrow v = \sqrt{R g \mu} \tag{6.42}$$

Thus from equation (6.42), the safe speed to negotiate a curve depends on the coefficient of friction which depends on many external parameters such as the condition of the tyres, condition of the road, whether it is dry or wet or there is an oil spill etc. Thus for a flat curve a safe speed can not be predicted, and to negotiate the curve unharmed at estimated safe speed simply depends on chance. To avoid this 'game of chance' in driving along curved roads, the curved sections of highways and racing track are banked.

The car on the banked road experiences two forces, the weight mg and the normal reaction N. The normal reaction N can be resolved into two components: the vertical component $N \cos \theta$ which is balanced by the weight mg, and the horizontal component $N \sin \theta$ which provides the necessary centripetal force for the centripetal acceleration v^2/R, and hence

$$N \cos \theta = mg$$

$$N \sin \theta = m \frac{v^2}{R}$$
$$\tan \theta = \frac{v^2}{Rg}$$
$$v^2 = Rg \tan \theta$$
$$v = \sqrt{Rg \tan \theta} \qquad (6.43)$$

Equation (6.43) gives the safe speed on a banked curve, which depends only on the radius of the curve and the angle of banking which are fixed construction parameter of the road. Thus a car negotiating such a curve is able to do so safely at this speed irrespective of external parameters such as friction. The friction which we have ignored here is of course present which provides an additional safety margin, and a skillful driver may be able to take the curve at a higher speed. But it is not advisable, and one must particularly follow the speed signs along the curved roads. Most serious accidents of car going off the road happen when this basic principle of physics is violated by a driver by going at a high speed.

6.5.5   Planet-Satellite System

Consider the motion of a satellite around a planet, for example the moon of mass m around the earth of mass M_E, as illustrated in Figure 6.11. Suppose that the orbit is approximately a circle of radius r. The satellite moving in a circular orbit experiences a centripetal acceleration $a = v^2/r$. The required centripetal force for the circular motion is supplied by the gravitational interaction between the planet (e.g. earth) and the satellite (e.g. moon).

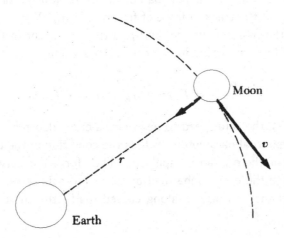

Figure 6.11: Earth-Moon system approximated as a circular orbit of the moon around the earth. The centripetal force is provided by the gravitational force.

Thus we have:

$$
\begin{aligned}
m\frac{v^2}{r} &= \sum F \\
&= \frac{GM_E m}{r^2} \\
v &= \sqrt{\frac{GM_E}{r}} = \sqrt{\frac{GM_E}{R_E + h}}
\end{aligned}
\tag{6.44}
$$

where $r = R_E + h$ with R_E being the radius of the earth and h the height of the satellite (moon) above the earth.

6.5.6 Electron orbits in an atom

Although the dynamics of an electron in an atom obey quantum mechanical laws, it is possible to get a reasonable understanding if we approximate the dynamics using classical equations of motion governing the circular motion (from Newton's second law) and Bohr's theory of the atom as below.

In *Bohr's theory of the atom*, the following basic assumptions are made:

(i) In an atom, an electron of mass m_e moves with a velocity v around a nucleus in circular orbits of radii r such that the angular momentum L of the electron is *quantised* in multiples of $\hbar$. While in these orbits, the electron does not radiate.

$$
L = m_e v r = n\hbar \text{ where } n = 1, 2, 3, \cdots
\tag{6.45}
$$

and hence

$$
r = \frac{n\hbar}{m_e v}
\tag{6.46}
$$

(ii) When an electron makes a transition from a higher energy level to a lower energy level, it is accompanied by *emission* of radiation, and when an electron *absorbs* a photon it moves from a lower energy level to a higher energy level. The change in energy between the initial energy level E_i and the final energy level E_f is related to the frequency ν of the associated photon as:

$$
E_i - E_f = h\nu
\tag{6.47}
$$

From classical considerations, the electron moving in a circular orbit experiences a centripetal acceleration $a = v^2/r$. The required centripetal force for the circular motion is supplied by the Coulomb interaction between the electron and the nucleus. The electron-nucleus system is illustrated in Figure 6.12.

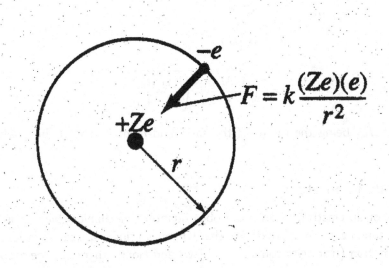

Figure 6.12: An atom approximated as a circular orbit of an electron around the nucleus. The centripetal force is provided by the electrostatic force, F.

Thus we have:

$$m_e a = \sum F$$
$$m_e \frac{v^2}{r} = \frac{kZe^2}{r^2}$$
$$v^2 = \frac{Ze^2}{4\pi\epsilon_0 m_e r} \tag{6.48}$$

where Z is the atomic number of the atom, $k = 1/(4\pi\epsilon_o)$ is the Coulomb's force constant, and $\epsilon_o = 8.854 \times 10^{-12} C^2 N^{-1} m^{-2}$ is the permittivity of free space. From Bohr's theory, the radius r_n for an electron in an orbit n can be obtained as below. From equation (6.46), we have

$$v = \frac{n\hbar}{m_e r} \tag{6.49}$$

Using equations (6.48) and (6.49),
$$\frac{Ze^2}{4\pi\epsilon_0 m_e r} = \frac{n^2\hbar^2}{m_e^2 r^2}$$

and hence the radius of the nth orbit ($r = r_n$) is given as

$$r_n = \frac{4\pi\epsilon_0 n^2 \hbar^2}{m_e Z e^2} \tag{6.50}$$

where for $n = 1$ and for the hydrogen atom $(Z = 1)$, we have

$$r_1(\text{hydrogen}) = \frac{4\pi\epsilon_0\hbar^2}{m_e e^2} = a_0 \tag{6.51}$$

known as the *Bohr's radius*, and $r_n = (a_0/Z)n^2$. From equations (9.48) and (9.50), the velocity v_n for an electron in an orbit n can be obtained as below.

$$
\begin{aligned}
v_n^2 &= \left(\frac{Ze^2}{4\pi\epsilon_0 n\hbar}\right)^2 \\
v_n &= \frac{Ze^2}{4\pi\epsilon_0 n\hbar}
\end{aligned}
\tag{6.52}
$$

where for $n = 1$ and for the hydrogen atom $(Z = 1)$, we have

$$v_1(\text{hydrogen}) = \frac{e^2}{4\pi\epsilon_0\hbar} = v_0$$

as the speed of the electron in the first Bohr's orbit of the hydrogen atom, and $v_n = v_0 Z/n$.

Thus the speed of the electron in the nth atomic orbit is inversely proportional to the quantum number n, and decreases in the ratio, $1 : \frac{1}{2} : \frac{1}{3} : \frac{1}{4} : \frac{1}{5}...$ for $n = 1, 2, 3, 4, 5, ...$

The total energy E_n for an electron in the nth orbit is the sum of its kinetic energy (KE) and potential energy (PE), and can be obtained as below.

$$
\begin{aligned}
E_n &= KE + PE \\
&= \frac{1}{2}m_e v_n^2 - \frac{1}{4\pi\epsilon_0}\frac{Ze^2}{r_n} \\
&= -\frac{m_e Z^2 e^4}{2\hbar^2(4\pi\epsilon_0)^2}\frac{1}{n^2}
\end{aligned}
\tag{6.53}
$$

When the above results are applied to the hydrogen atom $(Z = 1)$, the energy levels are obtained as

$$
\begin{aligned}
E_n &= -\frac{m_e e^4}{2\hbar^2(4\pi\epsilon_0)^2}\frac{1}{n^2} = -E_0\frac{1}{n^2} \\
&= -\frac{(9.11\times10^{-31})(1.6\times10^{-19})^4(9\times10^9)^2}{2(1.05\times10^{-34})^2}\frac{1}{n^2} \\
&= -(2.19\times10^{-18})\frac{1}{n^2} \quad \text{J} \\
&= -\frac{13.6}{n^2} \quad \text{eV}
\end{aligned}
\tag{6.54}
$$

where $E_0 = 13.6$ eV $= 2.19 \times 10^{-18}$ J is the *ionization energy* of the hydrogen atom..

6.5.7   An electric charge in a uniform magnetic field

Consider an electric charge q of mass m moving with a velocity $\mathbf{v}$ entering into a region where the magnetic field $\mathbf{B}$ is in the z-direction.

The charge will experience a magnetic force $q\mathbf{v} \wedge \mathbf{B}$. Applying Newtons's second law,

$$m\mathbf{a} = \sum \mathbf{F}$$
$$m\frac{d^2\mathbf{r}}{dt^2} = q\mathbf{v} \wedge \mathbf{B}$$

which is a vector equation, and can be separated into the three components to give

$$m\ddot{x} = qB\dot{y} \quad \text{along } x - axis \tag{6.55}$$
$$m\ddot{y} = -qB\dot{x} \quad \text{along } y - axis \tag{6.56}$$
$$m\ddot{z} = 0 \quad \text{along } z - axis \tag{6.57}$$

where equation (6.57) implies that $\dot{z} = $ constant. If this constant is taken to be zero ($\dot{z} = 0$), the motion is in the xy plane.

Upon differentiating equation (6.55),

$$m\frac{d^3x}{dt^3} = qB\frac{d^2y}{dt^2}$$
$$= qB\left(-\frac{qB}{m}\frac{dx}{dt}\right)$$

where equation (6.56) has been used. Integrating the above equation,

$$\frac{d^2x}{dt^2} = -\left(\frac{qB}{m}\right)^2 x$$
$$\frac{d^2x}{dt^2} + \left(\frac{qB}{m}\right)^2 x = 0 \tag{6.58}$$

which is a second order differential equation with a solution

$$x = A\cos\omega t \tag{6.59}$$

where

$$\omega = \frac{qB}{m} \tag{6.60}$$

Likewise, combining equations (6.55) and (6.56), we obtain

$$y = A\sin\omega t \tag{6.61}$$

and hence

$$x^2 + y^2 = A^2 = R^2$$

where we have used $A = R$, and R is the radius of the circular path of the charge in the magnetic field. The radius R can also be expressed in terms of m, v, q and B by noting that the centripetal aceleration is v^2/R and hence by Newton's second law

$$
\begin{aligned}
m\frac{v^2}{R} &= \sum F \\
&= qvB \quad \text{where we have used} \quad F = qv \wedge B = qvB \\
\Rightarrow R &= \frac{mv}{qB}
\end{aligned}
\tag{6.62}
$$

6.6 Problems

6.1 A body with an angular velocity of 1 rev/s at $t = 0$ is given an angular acceleration of 0.2 rev/s^2 for 10 s. At the end of 10 s, calculate
(a) the final angular velocity.
(b) the average angular velocity
(c) the angular displacement.

6.2. A cyclist is riding a cycle at a speed of $30\,km\,h^{-1}$. If the radius of the wheel is $0.4\,m$ at what rate are the wheels rotating, and how many rotations shall each wheel complete in covering a distance of $5\,km$.

6.3. An electric fan rotating uniformly with $1200\,rpm$ comes to rest in $30\,s$ after the power is switched off. How many rotations shall the fan complete before coming to rest?

6.4. A potter's wheel starting from rest is given an acceleration of $5\,rad\,s^{-2}$, Calculate:
(a) The time for it to reach an angular speed of $10\,rad\,s^{-1}$ and angular displacement in radians and revolutions during this period.
(b) Angular velocity after $20\,s$ and the revolutions completed during $20\,s$.
(c) Average angular velocities at *(i)* the end of period in (a), *(ii)* at the end of period in (b) *(iii)* in going from (a) to (b).

6.5 A particle of mass $0.5\,kg$ is moving clockwise in a circular orbit of radius $2.0\,m$. At some point on the trajectory its velocity is $5\,m\,s^{-1}$. If the tangential acceleration of the particle at this point is $4\,m\,s^{-2}$, calculate the total acceleration and the total force acting on the particle. Repeat the calculations if the tangential acceleration is $-4\,m\,s^{-2}$

6.6. An object moves in a circular path with a radius of 6.2 m and experiences a centripetal acceleration equal to three times that of gravity ie $3g$.
(a) What is the speed of the object?
(b) What is the period of the motion?

6.7. (a) What is the angular speed and the period of a four speed record player turntable when operating at the following speeds (i)$16\frac{2}{3}$ (ii)$33\frac{1}{3}$ (iii)45 (iv)78 revolutions per minute (rpm)?

(b) Calculate the centripetal acceleration on a particle at the end of the outer edge of a rotating 12 in LP record at $33\frac{1}{3}$ rpm?

(c) What shuld be the minimum coefficient of friction for the particle to stay in place?

6.8. A $8\,kg$ mass is rotating at $600\,rpm$ (rpm = revolutions per minute) in a circular path of radius $1.5\,m$.

(a) What is the torque required to stop the mass in $30\,s$?

(b) What is the loss in its kinetic energy?

(c) What is the average power supplied to stop the mass.

(d) What is the rate of change of its angular momentum. Compare it to the torque calculated in part (a) and comment on the two values. Can you state a law relating the two?

6.9. A rotating point mass loses $100\,J$ of energy when slowed down from $60\,rpm$ to $30\,rpm$ in $3\,s$. Calculate:

(a) Moment of inertia of the mass.

(b) The change in the angular momentum and hence the torque applied.

(c) Angular deceleration from using the value of the torque, and from the kinematic equation. Compare the two values.

6.10. A mass with moment of inertia of $4\,kg\,m^2$ about an axis of rotation has angular velocity of $5\,rad\,s^{-1}$. An accelerating torque of $10\,N\,m$ is applied to it for $10\,s$ Calculate:

(a) Change in the angular momentum and the kinetic energy.

(b) Angular displacement in $10\,s$.

(c) How long will it take for the mass to stop if the torque was retarding?

6.11. A body of mass 5 kg is placed in a bucket and is rotated in a vertical circle of radius 0.8 m. What must be the minimum speeed of the bucket at the top of its path so that the mass placed in it does not spill down?

6.12. A light string can support a maximum of $2.5\,kg$ mass without breaking. Calculate:

(a) Maximum possible speed of a $1\,kg$ mass moving in a horizontal circle of radius $1\,m$ tied to the string.

(b) Maximum possible speed of a $4\,kg$ mass moving in a horizontal circle of radius $1\,m$ tied to the string.

(c) Maximum possible speed of a $1\,kg$ mass moving in a vertical circle of radius $1\,m$ tied to the string. Will this mass be able to complete the vertical circle.

(d) Maximum possible mass tied to the string that can complete motion in a vertical circle of radius $1\,m$.

6.13. A person is standing against the wall inside a large vertical cylinder of radius R with a detachable floor. The coefficient of static friction between the person and the cylinder wall is μ. When the cylinder is rotating at certain speed, floor from below the person is removed, and the person stays in position without falling. Prove the maximum allowed period of rotation of the cylinder for this stunt to be successful is:

$$T = \sqrt{\frac{4\pi^2 R \mu}{g}}$$

6.14. A $100g$ mass placed on a smooth horizontal table is tied to a light string which passes through a hole in the table. The other end of the string is tied to a $500g$ mass which hangs freely. The mass on the table moves in a circle of $75\,cm$ radius. Calculate the speed of the mass to maintain the equilibrium.

6.15. A $60\,kg$ skater holding on to a $1.5\,m$ long rope tied to a pole is moving at $5\,m\,s^{-1}$ around the pole? Calculate the pull exerted by the rope on the skater's arms.

6.16. A point mass is moving in a horizontal circle on the inner surface of a smooth hemispherical bowl of radius R. The trajectory of motion lies vertically halfway between the bottom and the top edge of the bowl. Prove that the speed of the mass is given by: $\sqrt{1.5\,g\,R}$

6.17. A stunt pilot loops a vertical circle of radius $3\,km$ in the air. What should be the minimum speed of the plane to successfully perform the stunt? Calculate the speed of the plane when it passes the lowest point of the circle.

6.18. A body of mass m is suspended with a string of length 2m and the mass revolves about the point of suspension such that it describes a circle in the horizontal plane, constituting a conical pendulum where the string makes an angle of 30^0 with the vertcal. Calculate the period of the system.

6.19. A merry go round consists of a wheel of a radius $5\,m$ around which seats for children are suspended with strings of length 2.5. When in rotation at full speed the strings makes an angle of 30° from the vertical. If the mass of the seat and the child is $35\,kg$ what is the speed of rotation of the merry go round, and what is the tension in the string.

6.20. A car of mass 1000 kg is moving with a speed of 20 ms^{-1} and takes a curve of radius 200 m.
(a) What must be the coefficient of friction between the tyres and the road if the car is to round the curve without skidding?
(ii) What must be the angle of banking if the car is to round the curve safely without skidding at the given speed?

6.21. A car makes a curve of radius 100 m at a section of the road where the coefficient of friction between the tyres and the road is 0.8. What is the maximum speed can the car have if it is to round the curve without skidding?

6.22 A stunt man in a circus drives a motor cycle on the nearly vertical wall of a circular well of radius $100\,m$ at a speed of $300\,km\,h^{-1}$. Calculate the angle of inclination of the wall from the horizontal. If a racing car were to drive along a flat curve of the same radius with the same speed what should be the coefficient of friction for the car to safely negotiate the curve.

6.23. A car starts from rest and accelerates at $2\,m\,s^{-2}$ along a circular path of radius $250\,m$.
(a) At what speed the tangential and centripetal accelerations of the car are equal.
(b) What should be the coefficient of friction for the car to negotiate the flat curve at this speed.
(c) If instead the curve was to be banked, what should be the angle of banking for this speed.

6.24. A crate of fragile goods is placed on the floor of a truck, and the truck negotiates a flat curve of $40m$ radius in the road at $60\ kmh^{-1}$ without the crate slipping.
(a) What is the coefficient of friction for the crate?
(b) If the coefficient of friction is 0.6 what is the safe speed for the truck so that the crate inside shall not slip?

6.25. A $1000\,kg$ car drives over a speed breaker in the road which can be approximated to be a circle of radius $50\,m$. What is the maximum speed with which the car can pass the hump, without bouncing off the road.

6.26. (a) Estimate the value of the height above the earth for which a satellite would appear to be stationary above the earth.
(b) What is the minimum number of stationary satellites required so that every point along the equator is in view of at least one of the satellites?
The following constants will be needed in solving this problem.

$$\text{Radius of the earth, } R_E = 6368\text{km}$$
$$\text{Gravitational constant, } G = 6.67 \times 10^{-11}\text{Nm}^2\text{kg}^{-2}$$
$$\text{Mass of the earth, } M_E = 6 \times 10^{24}\text{kg}$$

6.27. An electron, a proton and an alpha particle each moving with a speed of $3 \times 10^6 m\,s^{-1}$ enter a uniform magnetic field of $5\,T$ (T=SI unit of magnetic field) perpendicular to the direction of the field. Calculate the radius of the trajectory of the of each particle in the magnetic field.

Chapter 7

Gravitation and Planetary Motion

Motion of the planets, the moon, the sun (apparent motion as conceived by early man) and the stars had always intrigued our ancient ancestors. Questions they had plenty, but none the answers. Such questions must have been: How the positions of the planets and stars changed in the sky? How the changes in weather took place at regular intervals? How phases of the moon were related to these motions? etc. According to the recorded history, as early as 400 B.C. Greeks were looking at the 'heavenly bodies' in the sky and wondered about their nature, how they were created, what they were made of, how they moved about in the sky, and so on. In the second century, Greek astronomers put forward the *Geocentric* or *Ptolemaic* model of planetary motion according to which the earth was at the centre of the solar system, and all the other planets including the sun and the moon moved around the earth in complex orbits. This view prevailed up to about the sixteenth century, although an increasing number of observations could not be explained by the model. In the 16th century, Nicolaus Copernicus proposed the *Heliocentric* or the *Copernican* model according to which sun is at rest at the centre of the solar system, and other planets revolve around it in closed orbits. As it turned out later, this was the correct model of planetary motion, as far as the sun being stationary at the centre of the solar system is concerned. In the late 16th century Tycho Brahe spent about twenty years in the systematic observation and recording of data of planetary motion. His data collected without the use of a telescope (telescope had not been invented yet) contain errors only to within $\frac{1}{15}$ of a degree. Such was the dedication, commitment and skill of the early scientists. Using his data and the heliocentric model, Johannes Kepler around early 17th century formulated three laws of planetary motion known as the *Kepler's laws* which describe the planetary motion accurately. It is interesting to note that when Kepler proposed these laws, the force of gravitation and the laws that govern it had not been known. It was only in 1666 that Sir Isaac Newton, using the Kepler's third law and analyzing the orbital motion of the moon, formulated the universal law of gravitation. This solved the mystery of the planetary motion once for all. In this chapter we first discuss the gravitational force which is responsible for holding the universe in place, although chronologically it was the last to be discovered, after the laws of orbital motion had been established. This is followed by the Kepler's laws of planetary motion in light of the universal law of gravitation. In the later part of the chapter terrestrial applications of gravitation are presented. The chapter is concluded with a brief remark on the effect of gravity on the path of light. The remarkable similarity between the planetary motion involving unimaginably large planetary bodies separated by astronomically large distances on one hand, and classical electronic orbits in the atom involving micro particle, the electron and the atomic nuclei, separated by exceeding small distances on the other hand point to the fact how simple the nature indeed is. Students who can foster a skill of seeing such simplicity of physics undoubtedly are likely to be more successful physicists than those who rely on *one problem-one formula* approach to learning physics.

7.1 Newton's universal law of gravitation

From the analysis of the orbital motion of the moon, Newton concluded that the force which is responsible for an apple from a tree to fall to the earth is also responsible for the orbital motion of the moon around the earth. He further concluded that the same force is responsible for the orbital motion of planets around the sun. This force, known as the gravitational force, is an attractive force, and is a universal force, *i.e.*, it acts between any pair of two masses irrespective of their sizes (masses) and separation.

Consider two masses m_1 and m_2 separated by a distance r as shown in Figure (7.1). The gravitational force between them is governed by the Newton's Universal law of Gravitation stated as follows.

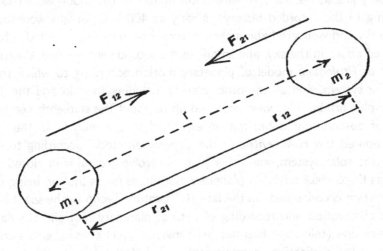

Figure 7.1: Gravitational force between a pair of masses.

Newton's Universal Law of Gravitation: *Two masses m_1 and m_2 separated by a distance r attract each other by a force which is directly proportional to the product of the masses and is inversely proportional to the square of the separation between them. The force is a central force, i.e., it acts along the line joining the masses.* Thus the magnitude of the gravitational force between the two masses is given by:

$$F = G \frac{m_1 m_2}{r^2} \tag{7.1}$$

where $G = 6.67 \times 10^{-11} \, N \, m^2 \, kg^{-2}$ is the universal gravitational constant. The direction of the force on each mass is along r, acting towards each other as shown in the figure. If we denote the force on mass m_1 due to mass m_2 as $\mathbf{F_{12}}$, and the force on mass m_2 due to mass m_1 as $\mathbf{F_{21}}$, then $\mathbf{F_{12}} = -\mathbf{F_{21}}$, and

$$|F_{12}| = G\,\frac{m_1\,m_2}{r_{12}^2} = G\,\frac{m_1\,m_2}{r_{21}^2} = |F_{21}| \Rightarrow F = G\,\frac{m_1\,m_2}{r^2} \tag{7.2}$$

In vector notation, one can write equation (7.1) as:

$$\mathbf{F} = -G\,\frac{m_1\,m_2}{r^3}\,\mathbf{r} = -G\,\frac{m_1\,m_2}{r^2}\,\hat{\mathbf{r}} \tag{7.3}$$

where $\hat{r}$ is a unit radial vector. Using appropriate notations, write down the vector expressions for the forces $\mathbf{F_{12}}$ and $\mathbf{F_{21}}$, and explicitly prove that $\mathbf{F_{12}} = -\mathbf{F_{21}}$.

Because of the very small value of G, the gravitational force is a very weak force, and it assumes significant value only when at least one of the masses involved is a large planetary body. It is this gravitational force which is responsible for the acceleration due to gravity of the planet, and for the orbital motion of the planetary bodies. The relationship between the acceleration due to gravity (g) and the universal gravitational force, and the variation of g with altitude and latitude in the case of earth were discussed in Chapter 3. In this chapter we shall be concerned with the application of the gravitational force to the orbital motion of the planetary bodies.

Example 7.1: Using the planetary data given in Table (7.1), calculate the gravitational force between the earth and a $100\,kg$ mass on the surface of the earth, and at a height h equal to the earth's radius. What are the values of the acceleration due to gravity at these locations?

Solution: Let F_0 and F_h be the magnitudes of the gravitational force, and g_0 and g_h be the accelerations due to gravity at the surface of the earth and at height h respectively. Using the data from Table (7.1):

$$F_0 = 6.67 \times 10^{-11} \times \frac{5.98 \times 10^{24} \times 100}{\left(6.37 \times 10^6\right)^2} = 983.0\,N \quad \text{and}$$

$$g_0 = \frac{F_0}{m} = \frac{983.0}{100} = 9.83\,m\,s^{-2}$$

Likewise:

$$F_h = 6.67 \times 10^{-11} \times \frac{5.98 \times 10^{24} \times 100}{\left(2 \times 6.37 \times 10^6\right)^2} = 245.7\,N \quad \text{and}$$

$$g_0 = \frac{F_0}{m} = \frac{245.7}{100} = 2.46\,m\,s^{-2}$$

The value of g_0 obtained is slightly larger than the average value of g at the equator, because we have used the average radius of the earth and not the equatorial radius which is larger than the average radius. However, these calculations clearly display the variation of g with altitude.

Exercise 7.1: Using the data from Table (7.1), calculate the force on a 100 kg mass on the surface of the moon and the value of acceleration due to gravity on moon's surface. Compare your results to those for the earth calculated in Example 7.1. What conclusions can you draw about the weight of a person on moon compared to his weight on earth?

Recipe 7.1: Gravitational force on a mass in an assembly of masses

- Determine the gravitational forces, their magnitudes and directions, on the desired mass due to each one of the remaining masses in the assembly.

- Find the vector sum of all the forces to determine the magnitude and direction of the resultant force on the mass.

- This procedure may be repeated for each mass in the assembly to determine the resultant force on it.

7.1.1   Experimental determination of G

The present accepted value of G was measured at the U.S. National Bureau of Standards in 1942 using a torsion balance shown in Figure (7.2). The experiment is known as the Cavandish experiment after Lord Cavendish who first measured G by this technique in 1798.

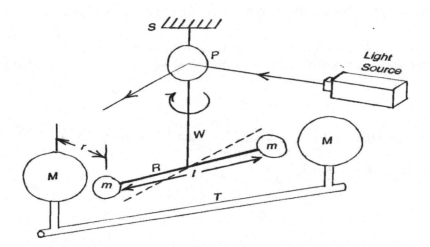

Figure 7.2: Schematic view of a torsional balance used to measure the value of G.

Two small masses m (of the order of 20 g each) are attached to the two ends of a thin, rigid rod R of length l, and are suspended in a horizontal plane from a rigid support S with a very thin, delicate wire W. The wire W has a very small torsion constant so that a small torque on the masses shall produce a large torsion in the wire. A light mirror P is attached to the wire W to measure the torsion of the wire by reading the deflection of a spot of light reflected from the mirror.

Two large masses M (of the order of 10 kg) attached to a rigid frame T are fixed to a rotating table in such a way that they lie in the same horizontal plane as the small masses m. The separation between the large masses is about the same as between the small masses, and they are symmetrically located with respect to the small masses. By rotating the table the large masses can be brought close to the small masses to within a few cm of them.

To perform the experiment, the table with the large masses is rotated such that the line joining them is perpendicular to the rod R joining the small masses, and position of the light spot on the scale reflected from mirror P is noted. In this position there is no net torque on the small masses from the gravitational forces between each pair of small and large masses. (Draw an appropriate diagram of the masses and show why?). Next the turn table with the large masses is rotated till they are within few cm (about 10 cm) of the small masses lying on opposite sides. The gravitational forces between each pair of neighboring large and small masses exert a torque which in turn produces a torsion in the wire, resulting in the deflection of the reflected light-spot on the scale. The torsion of the wire is determined by measuring the deflection of the spot, (remember from optics that a θ torsion of the wire shall cause the reflected spot of light to be rotated by 2θ), and the value of G is determined in the following manner.

Let θ be the torsion of the wire, κ be the torsion constant of the wire, r is the separation between the neighboring pair of large and small masses.

The gravitational force on each small mass due to the neighboring large mass is:

$$F = G\frac{Mm}{r^2} \tag{7.4}$$

Torque produced by the gravitational forces on the two small masses:

$$\tau_G = G\frac{Mm}{r^2}l \tag{7.5}$$

The elastic torque due to the torsion of the wire is:

$$\tau_e = \kappa\theta \tag{7.6}$$

In equilibrium the two torque are equal and opposite. Therefore, from equations (7.5) and (7.6):

$$G\frac{Mm}{r^2}l = \kappa\theta \tag{7.7}$$

This gives:

$$G = \frac{\kappa\theta r^2}{Mml} \tag{7.8}$$

The torsion constant κ of the wire is determined from the time period (T) of oscillation of the suspended pair of small masses using them as a torsion pendulum:

$$T = 2\pi\sqrt{\frac{I}{\kappa}} \tag{7.9}$$

where I is the moment of inertia of the suspended masses about the axis of rotation (the wire).

7.1.2   Measuring the mass of the earth

Once the value of G is determined, and if the radius of earth (R_e) is known, one can calculate the mass of the earth from the acceleration due to gravity (g) using the following equation (prove!):

$$M_e = \frac{g\,R_e^2}{G} \tag{7.10}$$

Cavendish was the first person to have determined the mass of the earth in this manner, and therefore, he is known to be the one who weighed the earth.

Example 7.2: In a Cavendish experiment, the small masses are 20 g each and the large masses are 10 kg each. The separation between the small masses is 50 cm, and the torsion constant of the wire is $4.5 \times 10^{-8} kg\,m^2\,s^{-2}$. Calculate the torsion in the wire due to the gravitational torque when the large masses are positioned at a distance of 10 cm from the small masses.

Solution: Using equation (7.7):

$$\theta = G\frac{M\,m}{r^2} \times \frac{l}{\kappa} = 6.67 \times 10^{-11} \times \frac{10 \times 0.02 \times 0.5}{0.1^2 \times 4.5 \times 10^{-8}} = 0.0149\,\text{radian} = 0.86^o$$

Exercise 7.2: In a Cavendish experiment, the small masses are 50 g each, the large masses are 20 kg each, and the separation between the small masses is 60 cm. When the large masses are positioned at a distance of 15 cm from the small masses, the torsion in the wire from the gravitational torque is 0.6^o. Calculate the torsion constant of the wire.

7.2   Gravitational field strength

Gravitational interaction between two masses can be expressed in terms of a gravitational field of the masses in the same way as the electrostatic interaction between charges is expressed in terms of the electrostatic field of the charges. Only difference is that for the charges there are two types of electric fields, repulsive and attractive corresponding to the positive and the negative charges respectively (electric field is defined as the force per unit positive test-charge), whereas in the case of the gravitational field there is only attractive field.

According to the *field concept* all masses are surrounded by their gravitational field. As soon as another mass enters the existing gravitational field of a mass, it interacts with the field of the mass and experiences a gravitational force of attraction towards the mass present. This interaction works for both the masses, *i.e.*, for two interacting masses M and m, mass m interacts with the gravitational field of mass M, and mass M interacts with the gravitational field of mass m. The manner in which the gravitational field is defined ensures that the forces on the two masses are equal and opposite. The gravitational field strength of a mass is defined as follows.

Definition 7.1: *The gravitational field strength of a mass M at a point in space is defined as the gravitational force per unit mass placed at the point under consideration.*

The gravitational field is a vector quantity and it acts along the direction of the gravitational force. Its SI unit is $N\,kg^{-1}$ or $m\,s^{-2}$, (prove that the two units are equivalent). We note that the unit of gravitational field strength is the same as that of the acceleration, the justification for which can be found by relating the gravitational force to gravitational acceleration using Newton's second law of motion, (see Chapter 3).

From the gravitational force between masses M and m separated by a distance r, the magnitude of the gravitational field strength of mass M at a distance r is:

$$\frac{|F|}{m} = G\frac{M}{r^2} \tag{7.11}$$

and it acts along r towards mass M. Likewise, the gravitational field strength of mass m at a distance r is:

$$\frac{|F|}{M} = G\frac{m}{r^2} \tag{7.12}$$

and it acts towards mass m along r. When the mass m is brought into the gravitational field of mass M or vice versa, they experience equal and opposite forces of attraction as given by equation (7.4).

To further elaborate on the gravitational field strength, let g be the acceleration due to gravity due to mass M. The gravitational force of mass M on mass m can be expressed in two ways either in terms of the gravitational field strength, or in terms of the acceleration due to gravity applying the Newton's second law of motion, both of which must be the same. Thus:

$$|F| = G\frac{M}{r^2}m = mg$$

or

$$g = G\frac{M}{r^2},$$

and in vector notation

$$\mathbf{g} = -G\frac{M}{r^2}\hat{r} \tag{7.13}$$

Comparing equations (7.11) and (7.13) we note that the gravitational field strength of a mass is the same in magnitude and direction as the acceleration due to gravity due to the mass under consideration. Graphically gravitational field strength is expressed by gravitational field lines converging towards the mass as shown in Figure (7.3). The direction of the field lines gives the direction of the acceleration due to gravity. As one gets further from the mass, the field lines diverge, and the field strength decreases with distance in proportion to the density of the filed lines. Thus the figure graphically displays the decrease in the value of the acceleration due to gravity with distance from the object, (altitude in the case of the earth). Using equation (7.13) students are assigned to prove that the acceleration due to earth's gravity varies with altitude according to the following equation:

$$g(h) = g_0 \frac{R_e^2}{(R_e + h)^2} \tag{7.14}$$

where g_0 is the mean acceleration due to gravity on the earth's surface, R_e is the mean radius of the earth, and h is the altitude from the earth's surface. Students should state the assumptions made in deriving the above expression. At what altitude in terms of R_e shall the acceleration be one half the acceleration on the surface?

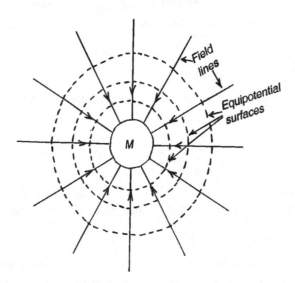

Figure 7.3: Graphical representation of gravitational field strength and the gravitational potential.

7.2.1   Gravitational potential and potential energy

Definition 7.2: In analogy with the electrostatic potential, *the gravitational potential (V_G) at a point in space due to mass M is defined as the work done to move a unit mass from infinity to the point under consideration.*

Gravitational potential is a scalar quantity, and its SI unit is $J\,kg^{-1}$. Consider two points r_1 and r_2 in the gravitational field of mass M. Then the gravitational potential difference between the two points is the work done to move a unit mass from one point to the other, given by:

$$V_{G,2} - V_{G,1} = \int_{r_1}^{r_2} G\frac{M}{r^2}\,dr = -G\,M\left(\frac{1}{r_2} - \frac{1}{r_1}\right) \tag{7.15}$$

If the initial point is taken to be at infinity ($r_1 \to \infty$) and $r_2 = r$ then the gravitational potential at point r is given by:

$$V_G(r) = -G\frac{M}{r} \tag{7.16}$$

From equation (7.16) the gravitational potential at $r = \infty$ is zero, and with reference to the zero potential at infinity , the potential at other points in space is negative. This is so because the unit mass when free to move accelerates towards mass M under the attractive gravitational force. To keep the unit mass in equilibrium, energy must be extracted from the moving mass, *i.e.*, the work done is negative.

From equations (7.13) and (7.16), the gravitational field strength ($\mathbf{g}$) and the gravitational potential (V_G) are related by the following equation in the same way as the electrostatic field and electrostatic potential are related:

$$\mathbf{g} = -\frac{dV_G}{dr}$$

The gravitational potential is the same at all equidistant points from mass M, (equation 7.16). The locus of such points is known as the equipotential surface. For a point mass or a spherical body, the equipotential surfaces are concentric spherical surfaces, as shown in Figure (7.3). The equipotential surfaces are normal to the gravitational field lines, and by definition the work done to move a mass between two points on an equipotential surface is zero.

The gravitational potential energy (U_G) of mass m in the gravitational field of mass M is simply the product of the mass m and the gravitational potential of M at the point:

$$U_G = m\,V_G = -G\frac{M\,m}{r} \tag{7.17}$$

If we regard mass M as the earth, and mass m as another mass in the gravitational field of the earth, then the expression for the gravitational potential energy given by equation (7.17) is markedly different from the commonly used expression $'mgh'$ for the gravitational potential energy of mass m raised to height h above the earth's surface. In order to relate the two expressions we must first clearly understand the terms of reference of the two expressions for the potential energy which are enumerated below:

- In equation (7.17) zero of the potential energy is taken at infinity which is physically the true zero of the potential energy as given by equation (7.16). However, one is free to choose the zero of energy at any arbitrary point as per the convenience of application, because when one considers the change in energy in going from one point in the field to another point, the difference in energy is unaffected by the choice of the zero.

- For routine applications, the zero for earth's gravitational potential energy in conventionally taken to be on the earth's surface.

- For small height h above earth's surface (compared to the radius R_e of the earth) the value of the acceleration due to gravity is treated to be constant, equal to its value on the surface of the earth, g_0.

With these considerations one can derive the conventional expression for the potential energy of mass m raised to height h above earth's surface as follows:

$$U_{G,h} = \int_0^h G\frac{M_e}{R_e^2}\,m\,dh = \int_0^h g_0\,m\,dh = m\,g_0\,h \tag{7.18}$$

where $g_0 = G\frac{M_e}{R_e^2}$ is the acceleration due to gravity on earth's surface.

Equation (7.18) for the gravitational potential energy can also be obtained using equation (7.17) under approximation and assumptions stated above. The gravitational potential energy of a mass m on the surface of earth, from equation (7.17) is:

$$U_{G,R_e} = -G\frac{M_e\,m}{R_e} \tag{7.19}$$

and the gravitational potential energy of the same mass at a height h is:

$$U_{G,R_e+h} = -G\frac{M_e\,m}{(R_e + h)} \tag{7.20}$$

The change in potential energy of the mass when raised to height $h\,(<< R_e)$ above earth's surface, from equations (7.19) and (7.20), is:

$$
\begin{aligned}
\Delta U_{G,h} &= U_{G,R_e+h} - U_{G,R_e} = -G\frac{M_e\,m}{(R_e + h)} + G\frac{M_e\,m}{R_e} \\
&= G\frac{M_e}{R_e}\,m\left(1 - \frac{1}{(1 + \frac{h}{R_e})}\right) \\
&= G\frac{M_e}{R_e}\,m\left(1 - \left(1 + \frac{h}{R_e}\right)^{-1}\right) \approx G\frac{M_e}{R_e}\,m\left(1 - \left(1 - \frac{h}{R_e}\right)\right) \\
&= G\frac{M_e}{R_e^2}\,m\,h = m\,g_0\,h
\end{aligned} \tag{7.21}
$$

This is the gravitational potential energy of mass m with respect to earth's surface at a height $h\,(<< R_e)$, same as given by equation (7.18).

Recipe 7.2: Gravitational potential energy of an assembly of masses

Consider an assembly of masses $m_1, m_2, m_3, \ldots m_n, \ldots$. The work done to assemble the masses is obtained from the gravitational potential energy of individual masses as follows.

- Calculate the gravitational potential energy for each pair of masses in the assembly.

- The work done to assemble the masses is the sum of the potential energy of each pair of masses, considering each pair only once, given as:

$$U = -\frac{1}{2}G\sum_i m_i \sum_{j\neq i}\frac{m_j}{r_{ij}} \tag{7.22}$$

- Attention is drawn to the factor $\frac{1}{2}$ included in equation (7.22). Explain why?

7.2.2 Escape velocity

Escape velocity from the earth is the minimum velocity required for an object to escape the earth's gravitational field. Consider a mass m projected vertically upwards from earth's surface with a velocity v, and it rises to a maximum height r_{max} (from the centre of the earth). At the surface of the earth, total energy E_i of the mass consists of the kinetic energy and the potential energy, given as:

$$E_i = \frac{1}{2} m v^2 - \frac{G M_e m}{R_e} \qquad (7.23)$$

At the maximum height, the kinetic energy of the mass is zero, and the total energy comprises only the potential energy, *i.e.*,

$$E_f = -\frac{G M_e m}{r_{max}} \qquad (7.24)$$

Since, the gravitational field is a conservative force field, the energy is conserved, *i.e.*, $E_i = E_f$, and from equations (7.23) and (7.24) we obtain:

$$v^2 = 2 G M_e \left(\frac{1}{R_e} - \frac{1}{r_{max}} \right) \qquad (7.25)$$

For the mass to escape the earth's gravitational field, $r_{max} = \infty$, and from equation (7.25) the escape velocity v_{esc} from the earth is obtained as:

$$v_{esc} = \sqrt{\frac{2 G M_e}{R_e}} \qquad (7.26)$$

From equation (7.26) we note that the escape velocity does not depend on mass m and depends only on the mass and radius of the earth. Thus the escape velocity for a small stone projected upwards is the same as for a large space craft. Using the values of G, M_e and R_e the escape velocity from earth is found to be $11.2 \, km \, s^{-1}$. In deriving equation (7.26) we have ignored the effect of the earth's atmosphere which results in a large dissipation of energy from the frictional forces. Therefore, the true escape velocity is significantly larger than the value given by equation (7.26).

The concept of escape velocity can be used to explain the presence or the lack of atmosphere of planets. The velocities of the molecules of the atmospheric gases have the following features:

- The gas molecules are in a constant state of random motion, and their velocities have a distribution from near zero to very large values.

- Average velocity (v) of the molecules depends on their mass (m) and the temperature $(T \, K)$ given by:

$$v = \sqrt{\frac{3 \, k \, T}{m}}$$

  where $k = 1.38 \times 10^{-23} \, J \, k^{-1}$ is the Boltzmann constant.

- The average velocity of lighter molecules is larger. The average velocity is also larger at higher temperature.

These characteristics of the velocity of gas molecules combined with the escape velocity influence the environment of the planets in the following manner:

- Due to the velocity distribution of molecules, molecules with velocity larger than the escape velocity escape the planet, and the velocity of the remaining molecules is redistributed. Thus the process of escape of the atmospheric gases from the planet continues.

- Planets with smaller escape velocity lose their atmosphere faster than those with larger escape velocity. Likewise, the hot planets lose their atmosphere faster than the cold ones.

- Lighter gas molecules are the first to escape. That is why the earth's atmosphere has almost no hydrogen.

The Mercury and the earth's moon have small escape velocities, and they both have lost their entire atmosphere. Escape velocity of Mars is one-sixth that of the earth, and it has some atmosphere left. The escape velocity for Venus is about the same as for the earth, and other planets have larger escape velocities. Therefore, they all have atmosphere, the chemical composition of which depends on other factors beyond the scope of this book.

7.3   Gravitational force between extended objects

So far we have treated the gravitational force between two masses as if the two masses were either point masses, or they were perfectly spherical (as shall be seen shortly). In case of the extended objects, one must consider elemental gravitational force between small elemental masses of the two bodies and then integrate the elemental force over the two masses as discussed below.

7.3.1   Force between an extended mass and a point mass

Let M be an extended mass, and m be a point mass. Consider a small elemental mass dM of the large mass, and let the distance between dM and m be r_1 (Figure 7.4). The gravitational force between dM and m is:

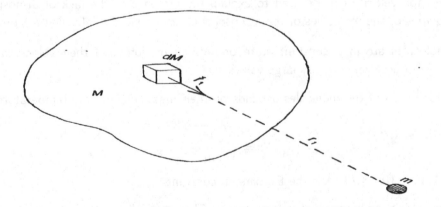

Figure 7.4: Gravitational force between an extended mass and a point mass.

$$dF = -G\frac{m\,dM}{r_1^2}\,\hat{\mathbf{r}_1} \tag{7.27}$$

The total gravitational force between M and m is obtained by integrating equation(7.27) over mass M. Thus:

$$\mathbf{F} = -Gm\int_M \frac{dM}{r_1^2}\,\hat{\mathbf{r}_1} \tag{7.28}$$

By similar integration process, students are assigned to show that the gravitational potential due to the extended mass M at point r is:

$$V_G = -G\int_M \frac{dM}{r_1} \tag{7.29}$$

7.3.2 Force between two extended masses

Let M_1 and M_2 be two extended masses. Consider small elemental masses dM_1 and dM_2 of the two masses respectively, and let the distance between dM_1 and dM_2 be r_{12} (Figure 7.5). The gravitational force between dM_1 and dM_2 is:

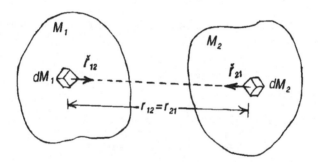

Figure 7.5: Gravitational force between two extended masses.

$$dF = -G\frac{dM_1\,dM_2}{r_{12}^2}\,\hat{\mathbf{r}_{12}} \tag{7.30}$$

The total gravitational force between M_1 and M_2 is obtained by integrating equation (7.30) over both the masses. Thus:

$$\mathbf{F} = -G\int_{M_1} dM_1 \int_{M_2} \frac{dM_2}{r_{12}^2}\,\hat{\mathbf{r}_{12}} \tag{7.31}$$

The gravitational potential energy of the two masses is:

$$U_G = -G\int_{M_1} dM_1 \int_{M_2} \frac{dM_2}{r_{12}} \tag{7.32}$$

7.3.3   Force between a point mass and a homogeneous spherical mass

The gravitational force due to homogeneous spherical masses is a special case, the integration for which leads to the following interesting results:

- For a point mass m placed outside the sphere, a homogeneous spherical mass acts as if the entire mass M of the sphere was placed as a point mass at the centre of the sphere.

- For a point mass m placed inside the sphere, away from the centre of the sphere, the entire sphere can be viewed as made up of two parts: A spherical shell lying just outside the mass m, and a sphere of smaller radius and a smaller mass $M'\,(< M)$ for which the mass m lies on the surface. The spherical shell does not interact with mass m, as if the effective mass of the shell was zero. The remaining sphere of the smaller mass and radius acts as if its mass M' was located as point mass at the centre.

Those who have done a course in electrostatics shall recall that uniform spherical charge distributions act in a similar way. Using the above results without going into the integration over spherical masses, we consider the following cases of gravitational force due to homogenous spherical masses.

Thin, uniform spherical shell of mass M and radius R

Case (i) Point mass m outside the shell at a distance $r > R$ from the centre (Figure 7.6): The gravitational force between the shell and mass m is:

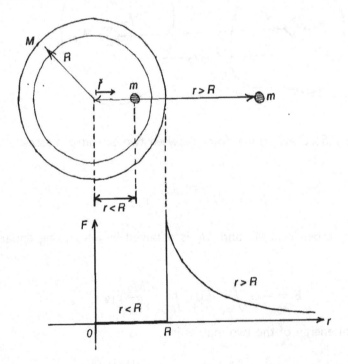

Figure 7.6: Gravitational force due to a spherical shell.

$$\mathbf{F} = -G\frac{Mm}{r^2}\,\hat{\mathbf{r}},\ (r > R) \qquad (7.33)$$

Case (ii) Point mass m inside the shell at a distance r < R from the centre (Figure 7.6): The gravitational force between the shell and the mass m is zero, *i.e.*,

$$F = 0,\ (r < R) \qquad (7.34)$$

The gravitational potential inside the shell is constant. Prove it!

Uniform solid sphere of mass M and radius R

Case (i) Point mass m outside the sphere at a distance r > R from the centre (Figure 7.7): The gravitational force between the sphere and the point mass is:

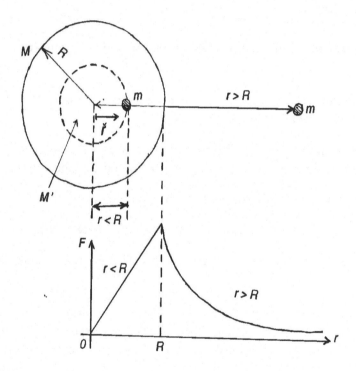

Figure 7.7: Gravitational force due to a uniform solid sphere.

$$\mathbf{F} = -G\frac{Mm}{r^2}\,\hat{\mathbf{r}},\ (r > R) \qquad (7.35)$$

Case (ii) Point mass m inside the solid sphere at a distance r < R from the centre (Figure 7.7): The gravitational force between the sphere and mass m is:

$$F = -G\frac{M'm}{r^2}\,\hat{\mathbf{r}},\ (r < R) \qquad (7.36)$$

where M' is the mass of the inner, small sphere on the surface of which the point mass m lies. If the sphere is of uniform density, then:

$$M' = M \frac{r^3}{R^3} \qquad (7.37)$$

Combining equations (7.36) and (7.37) we get:

$$F = -G \frac{M\,m}{R^3}\, r\,\hat{\mathbf{r}}, \ (r < R) \qquad (7.38)$$

Students are assigned to prove equation (7.37) and (7.38). If the density of the sphere is not constant but varies with the distance from the centre, then the mass is spherically symmetric, but not homogenous. In that case M' must be obtained by integration. As an example, assume that the density of the sphere varies linearly with distance from the centre, $i.e.$, $\rho(r) = \rho_0\, r$, where ρ_0 is a constant, then $M' = \pi\,\rho_0\, r^4$. Students with a good background in integration shall find it interesting to prove this expression for M'.

Recipe 7.3: Gravitational force between a point mass m and a symmetric extended mass M.

- Choose a suitable set of axes, consistent with the symmetry of the extended mass, and the symmetry of the configuration of the system.

- Choose a suitable elemental mass, dictated by the symmetry of the problem. Write down the elemental gravitational force between the elemental mass and the point mass.

- If necessary, express the elemental force in terms of its components.

- Integrate the elemental force/ its components over the entire extended mass.

- If necessary, combine the components of the force to determine the magnitude and direction of the force on the point mass.

Example 7.3: A point mass m is placed at a distance h from the near end of a uniform rod of mass M and length L. The mass lies on the axis along the length of the rod. Determine the gravitational force between the rod and the point mass, and the gravitational potential energy of the mass.

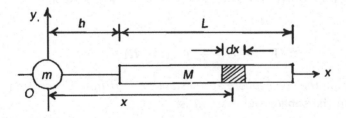

Figure 7.8: Gravitational force due to a uniform rod.

Solution: Referring to Figure (7.8) consider an elemental mass dM of the rod of width dx at a distance x from the point mass.

Mass per unit length of the rod: $\frac{M}{L}$
Mass dM of the element: $dM = \frac{M}{L} dx$
Elemental gravitational force between dM and m: $dF = G m \frac{M}{L} \frac{dx}{x^2}$, and it acts along the $x-$ axis.
The total force F is obtained by integrating dF over x between the limits h to $(L+h)$:

$$F = Gm\frac{M}{L}\int_h^{L+h}\frac{dx}{x^2} = \frac{GmM}{L}\left(-\frac{1}{x}\right)_h^{L+h}$$
$$= \frac{GmM}{h(L+h)}$$

The force F is along the $x-$ axis, as shown in the figure. In the limit $h >> L$ the expression for the force reduces to:
$$F = \frac{GmM}{h^2}$$

Likewise the gravitational potential energy is:

$$U_G = -Gm\frac{M}{L}\int_h^{L+h}\frac{dx}{x} = -\frac{GmM}{L}(ln\,x)_h^{L+h}$$
$$= \frac{GmM}{L}ln\left(\frac{h}{L+h}\right)$$

Prove that in the limit $h >> L$ the expression for the potential energy reduces to:
$$U_G = -\frac{GmM}{h}$$

Exercise 7.3: A point mass m is placed at a distance h from the centre of a uniform rod of mass M and length L. The mass lies at a distance h from the centre of the rod on the axis perpendicular to the length of the rod. Determine the gravitational force between the rod and the point mass, and the gravitational potential energy of mass m. Determine the force and the potential energy for the case when $h >> L$

7.4 Kepler's laws of Planetary motion

The orbital motion of the planets results from the gravitational force of attraction between the orbiting planet and the planet around which it orbits. Recalling circular motion, we know that for an object of mass m to move with velocity v in a circular orbit of radius r, there must be a centripetal force, $F_c = \frac{mv}{r^2}$, acting on the object towards the centre of the orbit. An application of this is encountered in the Bohr's theory of the hydrogen atom whereby the electron moves in a circular orbit around the nucleus; the centripetal force for the orbital motion being provided by the Coulomb's force of attraction on the electron due to the positive nucleus. In the case of the orbital motion of planets, the situation is nearly similar: Gravitational force of attraction on the orbiting planet due to the planet around which it orbits provides the necessary centripetal force to keep the planet moving in the orbital trajectory. Only difference is that the planetary orbits are not circular, rather they are elliptical. After years of study and analysis of data of the orbital motion of planets, Johannes Kepler deduced three empirical laws about the orbital motion of planets. These laws as applied to the solar system are stated as follows:

Kepler's First Law - The Law of Orbit: *The planets move in elliptical orbits with sun at one of the focus of the ellipse.*

Kepler's Second Law - The Law of Areal Velocity: *The line joining the planet to the sun (or the position vector of the planet) sweeps out equal areas in equal time intervals.*

Kepler's Third Law - The Law of Orbital Period: *The square of the orbital period of the planet is proportional to the cube of the semi major axis of the elliptical orbit.* In some texts this law is stated differently as: *The square of the orbital period of the planet is proportional to the cube of the mean distance of the planet from the sun.* In our discussion we shall show that these two statements of the law are equivalent except that the proportionality constants in the two cases differ by a constant factor.

7.4.1   Kepler's first law

Without going into explicit mathematical derivation, Kepler's first law is the direct consequence of the fact that the gravitational force between the sun and the planet is a central force (acting along the radius vector), and varies as the inverse of the square of the distance between them ($F_G \propto \frac{1}{r^2}$). In fact the orbit in such a case is a conic section, *i.e.*, it can either be an ellipse or a circle or a hyperbola or a parabola. The trajectory of the orbit depends on the energy, (discussed later). If the energy is just right, as in the case of planetary motion, the orbit is an ellipse (or a circle) with sun at one of the focus as shown in Figure (7.9). In the figure, F_1 and F_2 are the two foci of the ellipse with sun at F_1, and a and b respectively are the semi major and semi minor axes of the ellipse. An ellipse is defined as the locus of a point the sum of whose distances from the two foci is constant. Thus if $P = (x, y)$ is the position of the planet at any instant, then by definition: $PF_1 + PF_2 =$ constant. The equation of the ellipse with origin at the intersection of the two axes is:

$$\frac{x^2}{a^2} + \frac{y^2}{b^2} = 1 \qquad (7.39)$$

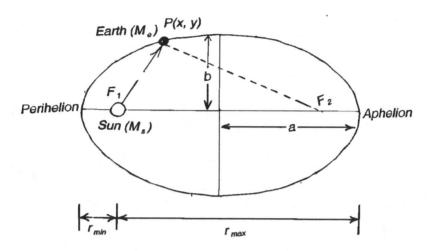

Figure 7.9: An elliptical orbit of a planet around the sun.

A circle is a special case of an ellipse with $a = b$ with equation $x^2 + y^2 = a^2$. The farthest and the closest positions of the planet from the sun are the *Aphelion* and *perihelion* respectively. In case of the earth satellite, these positions of the satellite are called *apogee* and *perigee* respectively.

7.4.2 Kepler's second law

The Kepler's second law is a consequence of the conservation of angular momentum. Consider Figure (7.10 (a)). The line of action of the gravitational force of the sun on the earth passes through the sun. If we take an axis of rotation through the centre of the sun and perpendicular to the plane of the orbit, then the torque of the force on the earth about the axis of rotation is zero. Hence the angular momentum of the earth about the chosen axis of rotation is conserved, *i.e.*,

$$\tau = \mathbf{r} \times \mathbf{F_G} = 0, \quad \text{and hence}$$
$$\mathbf{L} = \mathbf{r} \times \mathbf{p} = M_e (\mathbf{r} \times \mathbf{v}) = \text{constant} \tag{7.40}$$

In Figure (7.10 (b)) let P_1 and P_2 be the two positions of the earth over a short interval of time $dt (\rightarrow 0)$, and v is the average speed of the planet over this interval of time. Then the area dA swept by the radius vector over this interval of time is the area of the triangle $S P_1 P_2$, where $P_1 P_2 = d\mathbf{r} = \mathbf{v}\, dt$ and:

$$dA = \frac{1}{2} \mathbf{r} \times (P_1 P_2) = \frac{1}{2} (\mathbf{r} \times \mathbf{v})\, dt \tag{7.41}$$

and the areal velocity of the earth is:

$$\frac{d\mathbf{A}}{dt} = \frac{1}{2} (\mathbf{r} \times \mathbf{v}) = \frac{1}{2 M_e} (M_e \mathbf{r} \times \mathbf{v}) = \frac{\mathbf{L}}{2 M_e} = \text{constant} \tag{7.42}$$

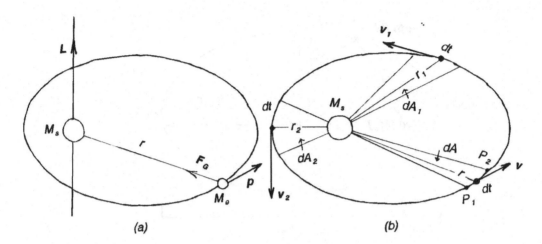

Figure 7.10: Conservation of angular momentum and the Kepler's second law.

Equation (7.42) is the statement of the Kepler's second law of the constancy of areal velocity. A constant areal velocity implies, that when a planet is closer to the sun, it must move faster, and when it is far from the sun it must move slower so that the radius vector always sweeps out the same area per unit time interval as shown in Figure (7.10(b)). Thus the earth moves fastest when it passes the perihelion, and it moves the slowest when it passes the aphelion. Logically, this is an obvious conclusion, because when the earth is closest to the centre of force, it experience the maximum centripetal force, and it therefore, must move fastest. Looking at it in simple terms of a circular orbit of radius r, we have:

$$\frac{M_e v^2}{r} = G\frac{M_e M_s}{r^2} \quad \text{or}$$

$$v^2 = G\frac{M_s}{r} \tag{7.43}$$

G and the mass of the sun M_s being constant, the smaller the value of r the larger is the velocity v.

7.4.3   Kepler's third law

Let us consider the simple case of the circular orbit of the earth of radius r with sun at the centre. If v is the orbital velocity of the earth, which in the case of a circular orbit is constant (why? - explain), then the time period T of one revolution is:

$$T = \frac{2\pi r}{v} \tag{7.44}$$

The centripetal force for the orbital motion is provided by the gravitational force, (Figure 7.11) *i.e.*,

$$F_G = G\frac{M_s M_e}{r^2} = \frac{M_e v^2}{r} = F_c \tag{7.45}$$

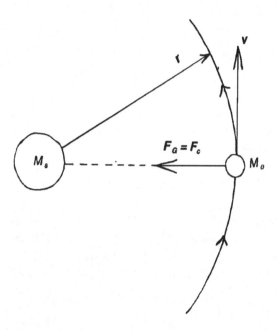

Figure 7.11: Forces in a circular planetary orbit of radius r.

Substituting the value of v in terms of T from equation (7.44) in equation (7.45), and simplifying we get:

$$T^2 = \frac{4\,\pi^2}{G\,M_s}\,r^3 \qquad\qquad (7.46)$$

Thus the square of the orbital period of earth is proportional to the cube of the average distance of the earth from the sun as stated by the Kepler's third law. The constant of proportionality does not depend on the mass of the earth, rather it depends on the mass of the sun around which the earth revolves. Hence, this constant is the same for all planets of the solar system, and its value is: $2.97 \times 10^{-19}\,s^2\,m^{-3}$. The time period of earth's orbital motion is well known, and if the average distance of earth from the sun is known from some independent measurement, equation (7.46) can be used to estimate the mass of the sun. In fact, equation (7.46) has a large potential for application, for example from the known time period of other planets in the solar system, their average distance from the sun can be calculated, and using similar equation for the orbital motion of moon around earth, if average distance of moon from the earth is known, mass of the earth can be calculated. Table (7.1) gives the planetary and orbital data for the planets in the solar system, which we shall use for problems at the end of the chapter.

Now reverting to an elliptical orbit, let a and b respectively be the semi major and semi minor axes of the ellipse and $e\,(<1)$ is the eccentricity of the ellipse. Let r be the radius of the equivalent circular orbit. Then from the properties of an ellipse:

$$r = \sqrt{ab}$$
$$b = a(1-e^2)^{\frac{1}{2}}$$
$$a = b(1-e^2)^{-\frac{1}{2}} \quad \text{so that}$$
$$r^3 = a^3(1-e^2)^{\frac{3}{4}} = b^3(1-e^2)^{-\frac{3}{4}} \tag{7.47}$$

Substituting r^3 from equation (7.47) in equation (7.46) we readily find that the square of the time period of the orbital motion is proportional to the cube of the semi major (or the semi minor) axis of the orbit as is also stated by Kepler's third law.

Table 7.1: Planetary and orbital data for the solar system

Name	Mass (kg)	Mean radius (m)	Period (s)	Mean distance from sun (m)
Mercury	3.18×10^{23}	2.43×10^{6}	7.60×10^{6}	5.79×10^{10}
Venus	4.88×10^{24}	6.06×10^{6}	1.94×10^{7}	1.08×10^{11}
Earth	5.98×10^{24}	6.37×10^{6}	3.156×10^{7}	1.496×10^{11}
Mars	6.42×10^{23}	3.37×10^{6}	5.94×10^{7}	2.28×10^{11}
Jupiter	1.90×10^{27}	6.99×10^{7}	3.74×10^{8}	7.78×10^{11}
Saturn	5.68×10^{26}	5.85×10^{7}	9.35×10^{8}	1.43×10^{12}
Uranus	8.68×10^{25}	2.33×10^{7}	2.64×10^{9}	2.87×10^{12}
Neptune	1.03×10^{26}	2.21×10^{7}	5.22×10^{9}	4.50×10^{12}
Pluto	$\approx 1.4 \times 10^{22}$	$\approx 1.5 \times 10^{6}$	7.82×10^{9}	5.91×10^{12}
Sun	1.991×10^{30}	6.96×10^{8}	—	—
Moon	7.36×10^{22}	1.74×10^{6}	2.36×10^{6} *E	3.84×10^{8} **E

*E: Period of moon around earth (s).   **E: Mean distance from earth (m).

Example 7.4: From the orbital parameters of the earth, calculate the mass of the sun.

Solution:
$$M_s = \frac{4\pi^2 r^3}{G T^2} = \frac{4 \times (3.14)^2 \times (1.496 \times 10^{11})^3}{6.67 \times 10^{-11} \times (3.156 \times 10^7)^2} = 1.99 \times 10^{30} \, kg$$

Exercise 7.4: From the orbital parameters of the moon, calculate the mass of the earth.

7.4.4   Effect of other planets on the planetary orbits

In considering the orbital motion of planets around sun, we only considered the gravitational force of the sun on the individual planet, and completely ignored the gravitational force on each planet due to all others. In view of the extremely large mass of the sun (sun is $\sim 10^3$ times heavier than the heaviest planet Jupiter, and is $\sim 3 \times 10^5$ times heavier than the earth), the gravitational forces of planets on each other are so small compared to the gravitational force of the sun, that it is a justifiable approximation to ignore interplanetary gravitational forces. The orbits so calculated are correct only to within an approximation. For precise calculations of planetary orbits, the interplanetary gravitational forces are treated as perturbations, and corrections to orbital motions due to perturbations are calculated using celestial mechanics. From the results of such calculations it is found that there are two effect of the perturbation forces on the planetary orbits: *(i)* Rotation of the axis of the ellipse. In case of the earth, the rotation of the major axis is only about 21 arc minutes per century. *(ii)* Periodic variation (oscillation) of the eccentricity of the ellipse about the average value. This variation is also extremely small, and the period of oscillation is of the order of 10^5 years. However, even these extremely small effects have lead to important and interesting discoveries. Examples are:

- At the time of Newton only six planets, Mercury, Venus, Earth, Mars, Jupiter, and Saturn were known. When the seventh planet, Uranus was discovered in 1781, its calculated orbit using perturbations from the known six planets did not precisely match the actual orbit. To account for the discrepancy, the presence of an eighth planet and its position in the sky were predicted. Soon afterwards Neptune was discovered in the predicted location in the sky.

- The calculations of the orbits of Neptune and Uranus led to the prediction, and the discovery of the ninth planet Pluto. Is there a tenth planet in the solar system? Only further precise calculations of the orbits of known planets hold an answer to this question.

7.5   Energy considerations of Planetary orbits

Consider a planet of mass m orbiting in a circular orbit of radius r about a planet of mass M. Let v be the orbital velocity of the planet. The total energy of the orbiting planet comprises it kinetic and potential energies as:

$$E = KE + PE = \frac{1}{2} m v^2 - G \frac{M m}{r} \qquad (7.48)$$

The centripetal force for the orbital motion is provided by the gravitational force, *i.e.*,

$$F_c = \frac{m v^2}{r} = G \frac{M m}{r^2} = F_G \qquad (7.49)$$

From equation (7.49), the kinetic energy of the planet is:

$$\frac{1}{2} m v^2 = \frac{1}{2} G \frac{M m}{r} \qquad (7.50)$$

Substituting equation (7.50) in equation (7.48), the total orbital energy of the planet is obtained as:

$$E = -\frac{1}{2} G \frac{M m}{r} \qquad (7.51)$$

Thus the total energy of the planet in an orbit is negative. This implies that the planet is in a bound state with the planet around which it orbits, and to separate the two planets apart to an unbound state, a minimum of this much amount of energy must be supplied externally. Here, it shall be interesting for students to recall the negative orbital energy of an electron in a bound state in an atom, and the ionization energy required to separate the electron from the atom. The necessary conditions for a planet to be in a bound orbit are:

- Its total energy must be negative, and equal to half its gravitational potential energy, (equation 7.51).

- Its kinetic energy must be equal to half the magnitude of the gravitational potential energy, (equation 7.50).

When both these conditions are satisfied, the orbit of the planet is either circular or elliptical. However, if the kinetic energy of the planet does not satisfy equation (7.50), and is either too small or too large, it results in the following trajectories of the planet.

Case (i): $E = 0$. This is the case of just unbound state of the planet. The orbit of the planet is an open orbit which is parabolic, and the orbiting planet escapes the gravitational field of the gravitating planet of mass M. From equation (7.48) this condition is satisfied when the velocity is: $v = \sqrt{2\left(\frac{GM}{r}\right)}$. Compare this to the expression for the escape velocity of an object projected from the surface of the earth, and comment on your observation.

Case (ii): $E > 0$. The kinetic energy of the planet is larger than the magnitude of its gravitational potential energy (equation 7.48), so that its total energy becomes positive. This is also the unbound state of the planet with large positive energy, and the orbit is again an open orbit, which is hyperbolic. The planet escapes the gravitating planet in a hyperbolic trajectory.

Case (iii): $E \ll 0$, less than the energy given by equation (7.51). The kinetic energy of the planet is smaller than given by equation (7.50), so that its total energy has a large negative value. The trajectory of the orbiting planet intersects with the surface of the gravitating planet. The orbit does not go fully around, and the orbiting planet falls back on to the gravitating planet. The motion of a projectile fired from the surface of the earth is a common example of this case. This case has terrestrial applications in intercontinental ballistic missiles, and in space travel.

7.6   Terrestrial applications of gravitation

7.6.1   Earth satellites

Earth satellites are the man made objects that are put in orbit around earth at a specific height h above the surface. The satellites are equipped with scientific laboratories and observation equipment which are used for, telecommunication, broadcasting and telecasting, spying, other military applications, weather and climatic data collection, environmental research, pollution monitoring, and outer space research etc. The velocity v of the satellite of mass m must satisfy equation (7.49) for it to be in a stationary, bound orbit, *i.e.*,

$$\frac{mv^2}{(R_e + h)} = G\frac{M_e m}{(R_e + h)^2} = G\frac{M_e}{R_e^2}m\frac{R_e^2}{(R_e + h)^2} = g_0 m\frac{R_e^2}{(R_e + h)^2} \qquad (7.52)$$

This gives:

$$v = R_e \sqrt{\frac{g_0}{R_e + h}} \qquad (7.53)$$

7.6.2 Space travel

Let us consider the most common route of space travel, *i.e.*, landing on the moon and return back. As the space ship gets further from the earth, and approaches moon, the earth's gravity gradually decreases and that of the moon increases. At a certain point between the two the earth's gravity completely cancels the moon's gravity. On either side of this point either the earth's gravity dominates or that of the moon dominates. The travel involves the space ship getting away from the earth's gravity, entering the moon's gravity, and landing on it. This comprises the following steps:

- The space ship is put in an orbit around the earth.

- Next its speed is increased towards the moon. The spaceship enters the moon's dominant gravitational field and is set in an orbit around the moon.

- The landing module separates from the space ship. Moving with a smaller velocity, its trajectory intercepts with the surface of moon, and it is maneuvered to land on the moon.

- All the speeds, and radii of the orbits are predetermined according to the landing site on the moon. The speed and positioning in orbit are controlled by firing of rockets at the precise moment in specific direction(s) to achieve desired results.

- The reverse process in involved in the return travel.

7.6.3 Ballistic missiles

Intercontinental ballistic missiles are fired from a certain place on the earth to strike a predetermined enemy position very far from the launching site, for example on the other side of the earth's curvature. The missile is first launched vertically upwards against the minimum air resistance. At certain height, the missile turns at an angle from the vertical in a specific direction. When the missile has reached certain velocity, its engines are turned off, and the missile travels in a orbit that intersects the surface of the earth at the point of target where it strikes, (recall the case of orbital motion when the total energy is too negative). All the flight parameters, the vertical height, angle and direction to which it is turned, and the final velocity are predetermined according to the location of launch and the location of the target, and are automatically controlled by preprogrammed computers.

7.7 Gravity and the bending of light

Light comprises massless photons of energy $h\nu$. Classically photons do not experience gravitational force, and they travel in a straight line, unaffected by gravity. In early 1900's from the theory of General Relativity, Einstein predicted that photons do indeed experience gravitational force, and the light bends from its straight-line path towards the gravitating body. Because of the nature of photons, and the extreme weak nature of gravitational force, this effect can be observed only when the light passes very close to a very large planetary body, such as the sun. The bending of light by the sun was first observed and measured during the total solar eclipse of 19 May 1919, which verified the Einstein's theory. The scientific and technical details of the theory and the experiment are beyond the scope of this book, but there is a small story attributed to the experimental discovery of the bending of light that shall interest our students. After the discovery Einstein was asked by some journalists how he would have felt if the experiment had yielded negative results. Einstein responded, *"I would have felt sorry for the dear Lord for having made a mistake in his creation, because the theory is correct"*. This goes to show how Physics, if done correctly in accordance with the laws of nature is undisputably correct.

7.8 Problems

7.1 Using the data from Table (7.1), calculate the gravitational force on a 100 kg mass placed on the surface of each of the planets excluding the earth and the moon, and including the sun (assuming that it was possible for such a mass to be placed on the sun). Determine the value of the acceleration due to gravity in each case.

7.2 Using the data from Table (7.1), calculate the gravitational force between the sun and each of the planets, and the gravitational force between the earth and the moon.

7.3 Assume that all the planets are aligned along a straight line on the same side of the sun. *(i)* Calculate the total gravitational pull of all the planets on the sun. *(ii)* Calculate the gravitational pull of each of the planets except sun on the earth, and the resultant gravitational pull of the planets on earth. In which direction does this gravitational pull acts and how does it compare to the gravitational pull of the sun on the earth?

7.4 Three masses of 10 kg, 20 kg and 50 kg are placed in the $x - y$ plane at (0, 0), (0, 5) and (10, 0) respectively, where the distances are in m. *(i)* Calculate the resultant gravitational force on each of the masses due to the other two. *(ii)* What shall be the force on each mass if the mass on the origin is moved along the z-axis to a height of 15 m, while the other two masses remain in the $x - y$ plane along the horizontal? A fourth mass of 100 g is placed at point (1, 2, 3) in the Cartesian coordinates axis, the distance being in m. *(iv)* Calculate the gravitational acceleration of the fourth mass in the gravitational field of the other three masses when released from rest at $t = 0$ for both the configurations of the other three masses as in *(i)* and *(ii)* above. *(v)* Discuss qualitatively the nature of acceleration of the mass at $t > 0$.

7.5 In a Cavendish experiment, the small masses are 40 g each, the large masses are 40 kg each, the separation between the small masses is 40 cm, and the large masses are positioned at a distance of 10 cm from the small masses. The torsion constant of the wire is $5.0 \times 10^{-8} \, kg \, m^2 \, s^{-2}$. Calculate: *(i)* The gravitational force on each small mass. *(ii)* The gravitational torque. *(iii)* The torsion in the wire. *(iv)* The linear deflection of the reflected spot of light on a scale placed at a distance of 1 m from the mirror.

7.6 Using the value of g calculated for each planet in problem (7.1), and the value of average radius for the corresponding planet, calculate the mass of each planet. Compare the calculated masses with the ones given in Table (7.1).

7.7 Using the value of g calculated for each planet in problem (7.1), and values of mass and radii of corresponding planets from Table (7.1), calculate the value of G. Compare the values of G so obtained, and comment on its universality.

7.8 Calculate the total gravitational potential energy of the solar system when aligned along a straight line as described in problem (7.3).

7.9 Calculate the work done to assemble *(i)* the three masses, and *(ii)* the four masses for all the configurations of their locations as described in problem (7.4).

7.10 A mass m is released from a height h above the earth's surface. Show that the velocity with which the mass shall hit the surface of the earth is given by:

$$v = \sqrt{2\,G\,M_e\left(\frac{1}{R_e} - \frac{1}{r}\right)}$$

With what velocity a meteorite reaching the earth shall hit at the surface? How does this velocity compare with the escape velocity for the earth, and explain why? What assumptions have you made in these calculations?

7.11 Calculate the velocity of meteorites hitting the moon's surface.

7.12 From the data given in Table (7.1), calculate the escape velocity for each of the planets, including the moon and the sun. The surface temperature of sun is 6000 K. Calculate the average velocity of helium molecules at this temperature and by comparing it with the escape velocity for the sun comment on the observation that the sun is predominately composed of the helium gas.

7.13 Prove that the escape velocity v_1 and v_2 for two planets of radii R_1 and R_2 and uniform densities ρ_1 and ρ_2 respectively are related as:

$$\frac{v_1}{v_2} = \frac{R_1}{R_2}\sqrt{\frac{\rho_1}{\rho_2}}$$

From the escape velocities of various planets calculated in problem (7.12), calculate the ratios of the density of earth to that of every other planet in the solar system including the sun and the moon.

7.14 A point mass m is placed at a distance r from the centre of a uniform thin spherical shell of mass M and radius R. Find the gravitational force between the shell and the point mass, and the gravitational potential energy of the point mass m, when *(i)* $r > R$, and *(ii)* $r < R$.

7.15 A point mass m is placed at a distance r from the centre of a uniform thick spherical shell of mass M and inner and outer radii R_1 and R_2 respectively. Find the gravitational force between the two, and the gravitational potential energy of the point mass m, when *(i)* $r > R_2$, *(ii)* $r < R_1$ and *(iii)* $R_2 > r > R_1$.

7.16 A point mass m is placed at a distance r from the centre of a uniform solid sphere of mass M and radius R. Find the gravitational force between the sphere and the point mass, and the gravitational potential energy of the point mass m, when *(i)* $r > R$, and *(ii)* $r < R$.

7.17 A point mass m is placed at a distance r from the centre of a solid sphere of mass M and radius R. The density of the sphere varies linearly with distance from the centre. Find the gravitational force between the sphere and the point mass, and the gravitational potential energy of the point mass m, when *(i)* $r > R$, and *(ii)* $r < R$.

7.18 From the orbital parameters of various planets of the solar system, Calculate: *(i)* The ratio $\frac{T^2}{r^3}$. Comment on the value for different planets. What does this represent? *(ii)* Calculate the mass of the sun, and compare the value obtained from different planetary orbits.

7.19 From the mass of the sun and average distance of each planet, calculate their orbital periods. Compare the calculated values with the values given in Table (7.1).

7.20 From the mass of the sun and orbital period of each planet, calculate their average distance from the sun. Compare the calculated values with the values given in Table (7.1).

7.21 Prove that the orbital periods of planets, and their average distances from the sun are related as:

$$\frac{T_1^2}{T_2^2} = \frac{r_1^3}{r_2^3}$$

where the symbols have their usual meanings. Using the orbital periods of various planets from Table (7.1), calculate the average distance of each planet from the sun in the units of the earth's distance, (known as the astronomical unit of distance: $1\,AU$ = Earth-Sun average distance).

7.22 Period of a earth-satellite is $5.77 \times 10^3\,s$. Calculate: *(i)* Height of the satellite orbit above earth's surface. *(ii)* Orbital velocity of the satellite. *(iii)* Its kinetic, potential and total energies in terms of the mass m of the satellite. With what velocity should the satellite be launched to put it in the orbit.

7.23 Calculate the period, velocity and acceleration of an earth satellite in orbit at 400 km above earth's surface.

7.24 The Coulomb force between an electron of mass m and charge $-e$, and a nucleus of mass M and charge $Z\,e$ is:

$$F_{Coul} = -\,k\frac{z\,e^2}{r^2}$$

where r is the radius of the electronic orbit in the atom. The electrostatic potential energy of the electron and the nucleus is:

$$U_{Coul} = -k\frac{z\,e^2}{r}$$

Apply the principles of planetary orbits to find expression for the orbital period, velocity, kinetic, potential and the total energies for the electronic orbit.

7.25. A straight tunnel is dug parallel to one of the diameters through the earth at a normal distance $d\,(< R_e)$ from the centre of the earth. Prove that a mass m dropped in the tunnel shall execute a simple harmonic motion from one end of the tunnel to the other. Find an expression for the time period of the motion, and calculate its numerical value. Ignore the frictional forces on the mass, and assume earth to be a uniform sphere.

Chapter 8

Oscillations

Oscillatory and periodic motions are very fundamental to nature and to physical systems. They permeate matter and life in the universe. Atoms and molecules in materials are in a constant state of oscillation about their mean positions. The body functions of most living beings are governed by the periodic heartbeat and the rhythm of respiration. The daily cycle of rising and setting of sun in itself is periodic with slight variation in the period. The sound and light comprise of oscillatory wave motions. There is also a wide range of applications of oscillatory phenomena in science, technology and consumer products. Time keeping by clocks is based on the periodic oscillations of an oscillatory system within the clock. At the heart of the tuning of a radio and a television is an electronic oscillatory system. Scientific instruments such as sonars and radars, and geological exploration instruments employ oscillatory fields and radiation. The list is endless, but the few given examples suffice to underline the importance of this area of mechanics. In this chapter, we present the simplest of all periodic and oscillatory motions represented by simple harmonic and related motions. Though simple in features and characteristics, one is able to explain a whole range of physical data and systems using concepts developed in the study of oscillations. At the end of the chapter the following systems and associated characteristics will have been studied

- Simple Harmonic Oscillation (SHO) or Simple Harmonic Motion (SHM)

- Damped Oscillations

- Forced Oscillations and resonance

- Coupled Oscillations and normal modes

8.1 Simple Harmonic Oscillations

A Simple Harmonic Oscillation (SHO) is a type of *periodic (or oscillatory)* motion in which the acceleration is proportional to displacement and is in the opposite direction. Alternatively, SHO is referred to as Simple Harmonic Motion (SHM). In this section, we consider some examples of SHOs (or SHMs) under ideal conditions, *i.e.*, when all forces other than the restoring force such as the frictional forces are assumed to be zero. The following system that execute SHM are discussed:

- SHO(SHM) as a projection of circular motion

- Simple pendulum

- Compound pendulum

- Torsional pendulum

- Mass-Spring System: Horizontal Oscillations

- Mass-Spring System: Vertical Oscillations

- LC-Circuit

- Liquid in a U-shaped tube

8.1.1 SHO(SHM) as a projection of circular motion

Consider a particle P executing uniform circular motion. If there is a source of light from the left in the plane of motion, the projection of the circular motion of particle P will show a shadow whose motion is periodic, as illustrated in Figure (8.1).

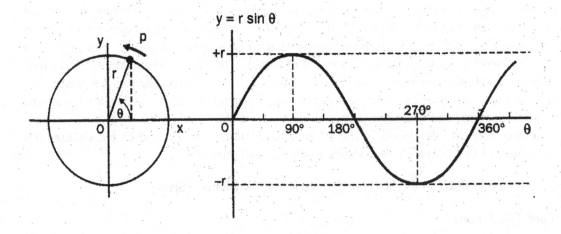

Figure 8.1: The projection of circular motion is simple harmonic motion.

If the angular displacement in time t is θ, the angular speed or the angular frequency ω is given by

$$\omega = \frac{\theta}{t} \tag{8.1}$$

and hence

$$\theta = \omega t \tag{8.2}$$

The *period*, T is the time taken to complete one complete oscillation (2π radians). Hence from equation (8.1), we obtain

$$2\pi = \omega T$$

or

$$T = \frac{2\pi}{\omega} \tag{8.3}$$

The frequency f is the number of oscillations per second given as

$$f = \frac{1}{T} = \frac{\omega}{2\pi}\ \text{s}^{-1} \tag{8.4}$$

The *displacement*, y, is periodic, that is it is a function of time, and is given by

$$\begin{aligned} y &= r\sin\theta \\ &= r\sin\omega t \\ y &= y_0\sin\theta \end{aligned} \tag{8.5}$$

where $r = y_0$ is known as an *amplitude* (maximum displacement) of the SHO.

The speed, v_c, of the particle along the circle is

$$v_c = \frac{2\pi r}{T} = r\omega \tag{8.6}$$

The speed of the shadow executing SHO is, therefore,

$$\begin{aligned} v &= v_c\cos\theta \\ &= r\omega\frac{x}{r} \\ &= \pm\omega\sqrt{r^2 - y^2} \\ v &= \pm\omega\sqrt{y_0^2 - y^2} \end{aligned} \tag{8.7}$$

where it has been shown that for a SHO, *the velocity depends on the displacement.* Note that the velocity is maximum $(v_{max} = \pm\omega y_0)$ at the equilibrium position $(y = 0)$, and with minimum value of zero $(v_{min} = 0)$ at maximum displacement $(y = y_0)$.

The centripetal acceleration, a_c, of the particle moving along the circle is

$$a_c = \frac{v_c^2}{r} \tag{8.8}$$

The acceleration of the shadow executing SHO is given by

$$\begin{aligned} a &= -a_c\sin\theta \\ &= -\frac{v_c^2}{r}\sin\theta \\ &= -\frac{v_c^2}{r^2}r\sin\theta \\ a &= -\omega^2 y \end{aligned} \tag{8.9}$$

where it has been shown that for a SHO, *the aceleration is proportional to displacement and is in the opposite direction to the displacement.*

From Newton's second law of motion, the force in a simple harmonic motion is:

$$F = m\,a = -m\,\omega^2\,y = -k\,y \qquad (8.10)$$

where we have used $m\,\omega^2 = k = $ constant. Thus the restoring force resposible for a simple harmonic motion is proportional to the displacement from the equilibrium position, and is directed opposite to the displacement (towards the equilibrium position).

Energy of SHM: The total mechanical energy of a simple harmonic oscillator comprises kinetic energy and the potential energy. The KE, using equation (8.7) for the velocity of SHM, at time t, is given as:

$$KE = \frac{1}{2}\,m\,v^2 = \frac{1}{2}\,m\,y_o^2\,\omega^2\,cos^2\,\omega\,t = \frac{1}{2}\,k\,y_o^2\,cos^2\,\omega\,t \qquad (8.11)$$

where we have used $m\,\omega^2 = k$. The potential energy at time t, using the expression for the elastic potential energy of a spring force (kx) derived in chapter 6, is given by:

$$PE = \frac{1}{2}\,k\,y^2 = \frac{1}{2}\,k\,y_o^2\,sin^2\,\omega\,t \qquad (8.12)$$

The total energy of SHM is the sum of the kinetic and the potential energies, *i.e.*,

$$E = KE + PE = \frac{1}{2}\,k\,y_o^2 = \frac{1}{2}\,m\,\omega^2\,y_o^2 \qquad (8.13)$$

From the expressions for energy the following conclusions can be drawn:

- The kinetic energy is maximum, and is equal to the total energy E when the displacement is zero, *i.e.* when the particle is passing through the equilibrium position. In this position we know that the velocity of the mass is maximum, and the restoring force and the potential energy are zero.

- The potential energy is maximum, and is equal to the total energy E when the displacement is maximum, *i.e.* when the particle is at its amplitude. In this position we know that the velocity of the mass is zero, and the restoring force and the potential energy are maximum.

- The total energy of SHM is constant, determined by its amplitude y_o. Thus SHM follows the conservation of mechanical energy, and once the system has been set in oscillation it must continue to oscillate for ever without loss in energy.

A practical oscillator actually does not conform to these observations about energy because of the presence of dissipative forces which we have neglected for an ideal oscillator. Because of the dissipative forces the energy of oscillations gradually decays, and the amplitude decreases till the oscillations have completely died out. This aspect of SHM is discussed in a later section on damped oscillations where some of the definitions and terms introduced here shall be applied.

In the above analysis of simple harmonic motion as a projection of a uniform circular motion, we implied that the particle moving along the circle starts its motion at $t = 0$ from $\theta = 0$ position along the x−axis. However, depending on the initial conditions of the motion at $t = 0$, it can start its motion at any point of the circle along its circumference, say at $\theta = \phi$. The angle ϕ is known as the initial phase of the simple harmonic motion, and it can take values either from 0 to 2π or from $-\pi$ to π (the two are equivalent).

Recipe 8.1: Oscillations and Differential equations (DE)

- Write down the equation of motion given by Newton's second law in the form $ma = \sum F$,

- Write the acceleration a as a second derivative of the displacement,

- Identify all the components to the net forces $\sum F$,

- Write down the resulting differential equation, usually Second Order DE.

- Assume a solution (periodic, damped, forced or coupled) to the differential equation depending on whether you are considering simple harmonic, damped, forced or coupled oscillations respectively and substitute it in the DE.

- Study the solution and its characteristics.

8.1.2 Simple pendulum

Consider a simple pendulum consisting of a particle of mass m suspended by a string of length l, and swinging side to side as illustrated in Figure (8.2). When the mass is at position where the string makes an angle θ with the vertical, the weight (force mg) of the pendulum bob has two components: the *radial component*, $mg \cos \theta$ balanced by the tension in the string, and the *tangential component*, $mg \sin \theta$ which provides the restoring force.

The equation of motion is given by Newton's second law, where as is known, acceleration a is proportional to the *net forces*, in the form

$$
\begin{aligned}
ma &= \sum F \\
m \frac{d^2 x}{dt^2} &= \sum F \\
&= -mg \sin \theta \\
\frac{d^2 x}{dt^2} &= -g \frac{x}{l} \\
\frac{d^2 x}{dt^2} + \frac{g}{l} x &= 0
\end{aligned}
\tag{8.14}
$$

which is a *second order differential equation*. It is common in phyics to end up with differential equations representing a physical system. What is important is not just to get the differential equation, but to obtain its solution as well. So, the question is, what is the solution, x, that satisfies the differential equation given in (8.14)?

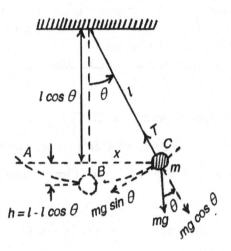

Figure 8.2: The simple pendulum.

From physical grounds, it is known that the solution x is *periodic*, and for this reason we look for periodic functions such as $\cos \omega t, \sin \omega t, e^{i\omega t}$, that is

$$x(t) = x \;=\; x_0 \sin (\omega t + \phi), \text{ or,} \qquad (8.15)$$

$$x(t) = x \;=\; x_0 \cos (\omega t + \phi), \text{ or,} \qquad (8.16)$$

$$x(t) = x \;=\; x_0 e^{i(\omega t + \phi)} \qquad (8.17)$$

where x_o and ϕ are constants, with x_0 representing the amplitude (maximum value of the displacement) and ϕ representing the phase constant (initial phase which can take values between $+\pi$ and $-\pi$). .

Without any loss of generality, for all the eaxmples of SHM presented here (sections 8.1.2 to 8.1.8) we shall only consider the cosine function with $\phi = 0$, *i.e.*, $x = x_o \cos \omega t$ as the trial solution, and derive the expression for the time period T of oscillations. However, this is not always the case. Students must try all other possible solutions given above including $x = x_o \sin \omega t$ for each of the examples and prove that they indeed also constitute the valid solutions of the respective equations of motion, and obtain the period of oscillation in each case.

Let us consider a *trial solution* to the differential equation given in (8.14).

$$x = x_0 \cos \omega t \qquad (8.18)$$

and see whether it works as a solution to the equation. Differentiating with respect to time gives

$$\frac{dx}{dt} \;=\; -\omega x_0 \sin \omega t$$

$$\frac{d^2x}{dt^2} = -\omega^2 x_0 \cos \omega t$$

$$\frac{d^2x}{dt^2} = -\omega^2 x$$

$$\frac{d^2x}{dt^2} + \omega^2 x = 0 \qquad (8.19)$$

which is the same as equation (8.14) *if and only if* we identify

$$\omega^2 = \frac{g}{l} \qquad (8.20)$$

or

$$\omega = \sqrt{\frac{g}{l}} \qquad (8.21)$$

which is known as the *natural angular frequeny of oscillation* of a simple pendulum, which gives the period T as

$$T = \frac{2\pi}{\omega} = 2\pi \sqrt{\frac{l}{g}} \qquad (8.22)$$

and thus the trial solution (8.18) is indeed a solution to the differential equation (8.14).

Equation (8.22) can be verified experimentally. A graph of T^2 against l gives a sraight line, as illustrated in Figure (8.3), of slope

$$\text{Slope} = \frac{4\pi^2}{g}$$

which can be used to find the acceleration due to gravity, g.

$$g = \frac{4\pi^2}{\text{Slope}} \qquad (8.23)$$

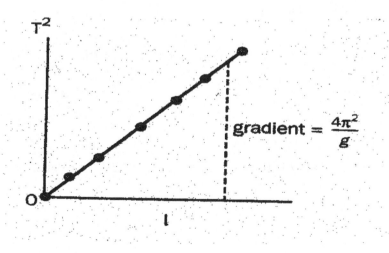

Figure 8.3: A graph of T^2 against l for a simple pendulum.

8.1.3   Compound pendulum

Consider a compound pendulum consisting of a rigid body of mass m and supported at a point A about which it oscillates in the vertical plane, as illustrated in Figure (8.4). A typical position of displacement is when the line through the axis of rotation and the centre of gravity makes an angle θ with the vertical.

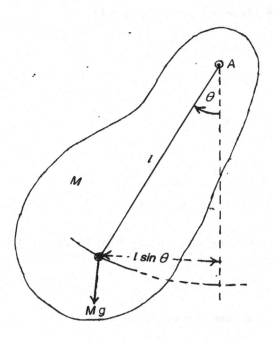

Figure 8.4: The compound pendulum.

The equation of motion is given by the torque equation where the angular acceleration $d^2\theta/dt^2$ is proportional to the *net torque*, τ, in the form

$$
\begin{aligned}
I\frac{d^2\theta}{dt^2} &= \sum \tau, \quad \text{where } I \text{ is the moment of inertia} \\
&= -Mgl\sin\theta
\end{aligned}
\tag{8.24}
$$

The moment of inertia I is given by

$$
I = M(k^2 + l^2)
\tag{8.25}
$$

where k is the radius of gyration, and l is the distance between the axis of rotation and the centre of gravity. Hence equation (8.24) becomes

$$
\begin{aligned}
M(k^2 + l^2)\frac{d^2\theta}{dt^2} &= -Mgl\sin\theta \\
\frac{d^2\theta}{dt^2} + \frac{gl}{k^2 + l^2}\theta &= 0
\end{aligned}
\tag{8.26}
$$

which is a *second order differential equation* where we have used $\sin\theta \approx \theta$ for small θ. So, the question is, what is the solution, θ, that satisfies the differential equation given in (8.26)?

As in the case of the simple pendulum considered earlier, let us consider a *trial solution*

$$\theta = \theta_0 \cos\omega t \tag{8.27}$$

and see whether it works as a solution to equation (8.26). This gives

$$\frac{d\theta}{dt} = -\omega\theta_0 \sin\omega t$$

$$\frac{d^2\theta}{dt^2} = -\omega^2\theta_0 \cos\omega t$$

$$\frac{d^2\theta}{dt^2} = -\omega^2\theta$$

$$\frac{d^2\theta}{dt^2} + \omega^2\theta = 0 \tag{8.28}$$

which is the same as equation (8.26) *if and only if* we identify

$$\omega^2 = \frac{gl}{(k^2 + l^2)} \tag{8.29}$$

or

$$\omega = \sqrt{\frac{gl}{(k^2 + l^2)}} \tag{8.30}$$

which is known as the *natural angular frequeny of oscillation* of a compound pendulum, which gives the period T as

$$T = \frac{2\pi}{\omega} = 2\pi\sqrt{\frac{(k^2 + l^2)}{gl}} \tag{8.31}$$

and thus the trial solution (8.27) is indeed a solution to the differential equation (8.26).

Equation (8.31) can be verified experimentally, and can be used to find the acceleration due to gravity, g, as discussed in chapter 9.

8.1.4   Torsional pendulum

Consider a torsional pendulum consisting of a disc suspended with a thin rigid rod. The disc experiences a torsion due to a torque applied to it as illustrated in Figure (8.5).

The equation of motion is given by the torque equation where the angular acceleration $d^2\theta/dt^2$ is proportional to the *net torque*, τ, in the form

$$I\frac{d^2\theta}{dt^2} = \sum\tau \text{ where } I \text{ is the moment of inertia}$$

$$= -\kappa\theta \text{ where } \kappa \text{ is the torsion constant of the rigid rod,}$$

$$\frac{d^2\theta}{dt^2} + \frac{\kappa}{I}\theta = 0 \tag{8.32}$$

which is a *second order differential equation* of the same form encountered earlier.

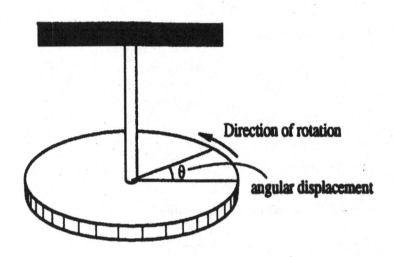

Figure 8.5: The torsional pendulum.

From physical considerations, it is known that the solution θ is *periodic*. Let us consider a *trial solution*

$$\theta = \theta_0 \cos \omega t \tag{8.33}$$

and see whether it works as a solution to equation (8.32). This gives

$$\frac{d\theta}{dt} = -\omega \theta_0 \sin \omega t$$

$$\frac{d^2\theta}{dt^2} = -\omega^2 \theta_0 \cos \omega t$$

$$\frac{d^2\theta}{dt^2} = -\omega^2 \theta$$

$$\frac{d^2\theta}{dt^2} + \omega^2 \theta = 0 \tag{8.34}$$

which is the same as equation (8.32) *if and only if* we identify

$$\omega^2 = \frac{\kappa}{I} \tag{8.35}$$

or

$$\omega = \sqrt{\frac{\kappa}{I}} \tag{8.36}$$

which is known as the *natural angular frequency of oscillation* of a torsional pendulum. This gives the period T as

$$T = \frac{2\pi}{\omega} = 2\pi\sqrt{\frac{I}{\kappa}} \tag{8.37}$$

and thus the trial solution (8.33) is indeed a solution to the differential equation (8.32).

8.1.5 Mass-Spring System: Horizontal Oscillations

Consider a mass-spring system on a horizontal smooth surface, illustrated in Figure (8.6), where the spring has a force constant k. Initially, the spring is at the equilibrium. When the spring is extended by a distance x, the spring experiences restoring force, $-kx$, according to *Hooke's law*.

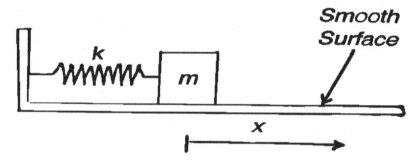

Figure 8.6: Mass-spring system.

The equation of motion is given by Newton's second law, *ie*,

$$
\begin{aligned}
ma &= \sum F \\
m\frac{d^2x}{dt^2} &= \sum F \\
&= -kx \\
\frac{d^2x}{dt^2} + \frac{k}{m}x &= 0
\end{aligned}
\tag{8.38}
$$

which is a *second order differential equation*. As before, what is the solution, x, that satisfies the differential equation given in (8.38)? From physical grounds, as before, let us consider a *trial solution*

$$x = x_0 \cos \omega t \tag{8.39}$$

and see whether it works as a solution to equation (8.38). This gives

$$
\begin{aligned}
\frac{dx}{dt} &= -\omega x_0 \sin \omega t \\
\frac{d^2x}{dt^2} &= -\omega^2 x_0 \cos \omega t \\
\frac{d^2x}{dt^2} &= -\omega^2 x \\
\frac{d^2x}{dt^2} + \omega^2 x &= 0
\end{aligned}
\tag{8.40}
$$

which is the same as equation (8.38) *if and only if* we identify

$$\omega^2 = \frac{k}{m} \tag{8.41}$$

or

$$\omega = \sqrt{\frac{k}{m}} \tag{8.42}$$

which is known as the *natural angular frequency of oscillation* of a mass-spring system. This gives the period T as

$$T = \frac{2\pi}{\omega} = 2\pi\sqrt{\frac{m}{k}} \tag{8.43}$$

and thus the trial solution (8.39) is indeed a solution to the differential equation (8.38).

Equation (8.43) can be verified experimentally. A graph of T^2 against m gives a sraight line of slope

$$\text{Slope} = \frac{4\pi^2}{k}$$

which can be used to find the force constant, k.

$$k = \frac{4\pi^2}{\text{Slope}} \tag{8.44}$$

8.1.6   Mass-Spring System: Vertical Oscillations

Consider a mass-spring system executing vertical oscillations as illustrated in Figure (8.7), where the spring has a force constant k and natural length l_0. Initially, as in the Figure 8.7 (a), the spring is unloaded. The spring is loaded wih a mass m and extends to a length $l_1 = l_0 + y'$, where y' is the extension, and the system is in the equilibrium as in Figure 8.7 (b). The spring is further extended by a length y_0, and when released, it will execute oscillations with displacement $y = y(t)$, as illustrated in Figure 8.7 (c).

In Figure 8.7 (b), the system is in equilibrium. Therefore,

$$mg = ky' \tag{8.45}$$

In Figure 8.7 (c), the spring is extended, the equation of motion is given by Newton's second law, *ie*,

$$\begin{aligned}
ma &= \sum F \\
m\frac{d^2y}{dt^2} &= \sum F \\
&= mg - k(y' + y) \\
\frac{d^2y}{dt^2} + \frac{k}{m}y &= 0
\end{aligned} \tag{8.46}$$

which is a *second order differential equation*. The student should verify that the solution is given by

$$y = y_0 \cos \omega t \qquad (8.47)$$

where, y_0 is the amplitude and ω is the *natural angular frequency of oscillation* given by

$$\omega = \sqrt{\frac{k}{m}} \qquad (8.48)$$

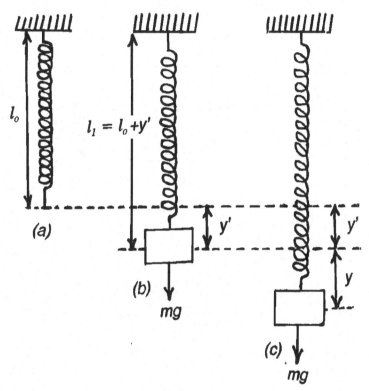

Figure 8.7: (a) The spring is unloaded with a natural length l_0. (b) The spring is loaded wih a mass m and extends to a length $l_1 = l_0 + y'$, where y' is the extension, and the system is in equilibrium. (c) Upon further extension by a length y_0, and released, the system performs vertical oscillations with displacement $y = y(t)$.

8.1.7 LC-Circuit

Consider an LC circuit consisting of an inductor of *inductance* L and a capacitor of *capacitance* C as illustrated in Figure (8.8). The potential drop across the capacitor is $V_C = q/C$, where q is the charge on the capacitor, while the potential drop across the inductor is $V_L = LdI/dt$, where $I = dq/dt$ is the electric current.

According to *Kirchoff's second law*, the potential drops in a closed circuit add up to zero, that is

$$\begin{aligned} V_L + V_C &= 0 \\ L\frac{dI}{dt} + \frac{q}{C} &= 0 \\ \frac{d^2q}{dt^2} + \frac{q}{LC} &= 0 \end{aligned} \qquad (8.49)$$

which is a *second order differential equation*. As before, what is the solution, q, that satisfies the differential equation given in (8.49)?

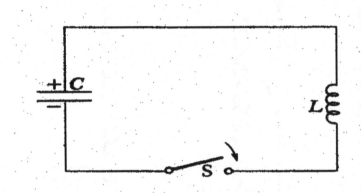

Figure 8.8: LC circuit.

From physical considerations, as before, let us consider a *trial solution*

$$q = q_0 \cos \omega t \qquad (8.50)$$

and see whether it works as a solution to equation (8.49). This gives

$$\begin{aligned} \frac{dq}{dt} &= -\omega q_0 \sin \omega t \\ \frac{d^2q}{dt^2} &= -\omega^2 q_0 \cos \omega t \\ \frac{d^2q}{dt^2} &= -\omega^2 q \\ \frac{d^2q}{dt^2} + \omega^2 q &= 0 \end{aligned} \qquad (8.51)$$

which is the same as equation (8.49) *if and only if* we identify

$$\omega^2 = \frac{1}{LC} \qquad (8.52)$$

or

$$\omega = \frac{1}{\sqrt{LC}} \qquad (8.53)$$

which is known as the *natural angular frequency of oscillation* of an LC-circuit, which gives the frequency ν as

$$\nu = \frac{1}{2\pi\sqrt{LC}} \tag{8.54}$$

and thus the trial solution (8.50) is indeed a solution to the differential equation (8.49).

8.1.8 Liquids in a U-shaped Tube

Consider a liquid of density ρ in a U-shaped tube of radius r. If l is the total length of the liquid column in the tube, the total mass of the liquid in the tube is $m = \rho\pi r^2 l$. In equilibrium, the length of liquid columns in both the limbs of the tube are equal. When the liquid in one of the limbs of the tube is displaced by x as shown in Figure (8.9), the entire column of liquid in the tube experience a force which tends to restore the liquid levels in both the limbs to equal.

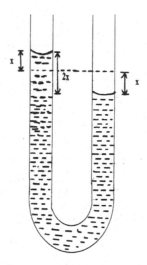

Figure 8.9: A liquid in a U-shaped tube.

The equation of motion is given by Newton's second law, where the acceleration a is proportional to the *net forces*, which is due to excess pressure and hence

$$
\begin{aligned}
\text{Net force} &= \text{Excess pressure} \times \text{Area} \\
&= \{p - (p + 2x\rho g)\}\,\pi r^2 \quad \text{where } p \text{ is the atmospheric pressure,} \\
&= -2x\rho g \pi r^2 \tag{8.55}
\end{aligned}
$$

Hence the equation of motion is

$$
\begin{aligned}
ma &= \sum F \quad \text{, (net force)} \\
\rho\pi r^2 l \frac{d^2 x}{dt^2} &= -2x\rho g \pi r^2
\end{aligned}
$$

$$l\frac{d^2x}{dt^2} + 2gx = 0$$

$$\frac{d^2x}{dt^2} + \frac{2g}{l}x = 0 \tag{8.56}$$

which is a *second order differential equation*. As before, what is the solution, x, that satisfies the differential equation given in (8.56)?

After having gone through several examples, the technique of solving differential equations should be familiar by now. Let us consider a *trial solution*

$$x = x_0 \cos \omega t \tag{8.57}$$

and see whether it works as a solution to equation (8.56). This gives

$$\frac{dx}{dt} = -\omega x_0 \sin \omega t$$

$$\frac{d^2x}{dt^2} = -\omega^2 x_0 \cos \omega t$$

$$\frac{d^2x}{dt^2} = -\omega^2 x$$

$$\frac{d^2x}{dt^2} + \omega^2 x = 0 \tag{8.58}$$

which is the same as equation (8.56) *if and only if* we identify

$$\omega^2 = \frac{2g}{l} \tag{8.59}$$

or

$$\omega = \sqrt{\frac{2g}{l}} \tag{8.60}$$

which is known as the *natural angular frequency of oscillation* of liquid oscillations in a U-shaped tube, which gives the period T as

$$T = \frac{2\pi}{\omega} = 2\pi\sqrt{\frac{l}{2g}} \tag{8.61}$$

and thus the trial solution (8.57) is indeed a solution to the differential equation (8.56).

8.2   Damped Oscillations

When dealing with simple harmonic oscillations in the previous section, dissipative forces were neglected and the resulting oscillations did not decay with time. However, in reality dissipative forces may not be negligible and the resulting oscillations have amplitudes which decay with time. The dissipative forces are due to damping mechanisms such as friction, air resistance, viscosity, electrical resistance etc, and these forces must be included in the equation of motion. The effect of damping is discussed in this section.

8.2.1 Mass-Spring System on a rough surface

Consider a mass-spring system illustrated in Figure (8.10), where the spring has a force constant k, and the mass is moving on a rough horizontal surface. Initially, the spring is at the equilibrium. When the spring is extended by a distance x, the mass experiences restoring force, $-kx$, according to *Hooke's law*, and a frictional force which is assumed to be proportional to velocity, given by $-b(dx/dt)$, where b is known as the *damping parameter* and dx/dt is velocity.

The equation of motion is given by Newton's second law, where as is known, acceleration a is proportional to the *net forces*, in the form

$$
\begin{aligned}
ma &= \sum F \\
m\frac{d^2x}{dt^2} &= \sum F \\
&= -kx - b\frac{dx}{dt} \\
\frac{d^2x}{dt^2} + \frac{b}{m}\frac{dx}{dt} + \frac{k}{m}x &= 0
\end{aligned}
\tag{8.62}
$$

which is a *second order homogeneous differential equation*. As before, our objective is to find the solution, x, that satisfies the differential equation given in (8.62).

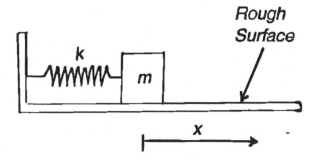

Figure 8.10: Mass-Spring System on a rough surface with damped oscillations.

From physical grounds, since the displacement must decay with time due to the presence of a frictional force, let us consider a *trial solution* which decays with time as

$$
x = x_0 e^{\lambda t}
\tag{8.63}
$$

and see whether it works as a solution to equation (8.62), where x_0 is the displacement at $t = 0$ and λ is the decay constant. This gives

$$
\begin{aligned}
\frac{dx}{dt} &= \lambda x \\
\frac{d^2x}{dt^2} &= \lambda\frac{dx}{dt} = \lambda^2 x
\end{aligned}
$$

and hence equation (8.62) becomes

$$(\lambda^2 + \frac{b}{m}\lambda + \frac{k}{m})x = 0$$

and hence

$$(m\lambda^2 + b\lambda + k) = 0 \tag{8.64}$$

which is referred to as the *auxiliary equation*, and has a solution

$$\lambda = \frac{-b \pm \sqrt{b^2 - 4km}}{2m} \tag{8.65}$$

The form of the solution (8.63) depends on the sign of $(b^2 - 4km)$ in (8.65). There are three cases of physical interest.

Case (i): Light damping (or underdamping): $(b^2 - 4km) < 0$

If $(b^2 - 4km) < 0$, the square root in equation (8.65) becomes imginary and hence λ is a complex number.

$$\begin{aligned}
\lambda &= -\frac{b}{2m} \pm \sqrt{(-1)\left(\frac{4km - b^2}{4m^2}\right)} \\
&= -\frac{b}{2m} \pm i\sqrt{\frac{k}{m} - \left(\frac{b}{2m}\right)^2} \\
&= -\frac{\Gamma}{2} \pm i\omega_1 \tag{8.66}
\end{aligned}$$

where $\Gamma = b/m$ is known as the *damping constant*, and

$$\begin{aligned}
\omega_1 &= \sqrt{\frac{k}{m} - \left(\frac{b}{2m}\right)^2} \\
&= \sqrt{\frac{k}{m} - \frac{\Gamma^2}{4}} \\
&= \sqrt{\omega_0^2 - \frac{\Gamma^2}{4}} \tag{8.67}
\end{aligned}$$

where

$$\omega_0 = \sqrt{\frac{k}{m}}$$

is the *resonant frequency* of a mass-spring system. Substituting equation (8.66) in (8.63), the displacement becomes

$$\begin{aligned}
x &= x_0 e^{\lambda t} \\
&= x_0 e^{-\frac{\Gamma}{2}t} e^{\pm i\omega_1 t}
\end{aligned}$$

whose real part is

$$
\begin{aligned}
Re\ x &= Re\ x_0 e^{-\frac{\Gamma}{2}t} e^{\pm i\omega_1 t} \\
&= x_0 e^{-\frac{\Gamma}{2}t} \cos\omega_1 t \qquad\qquad (8.68) \\
&= A(t) \cos\omega_1 t
\end{aligned}
$$

where $A(t) = x_0 e^{-\frac{\Gamma}{2}t}$. A graph of $Re\ x$ against t is shown in Figure (8.11). This is a decaying oscillatory function of angular frequency ω_1, bounded by an envelope of exponentially decaying amplitude.

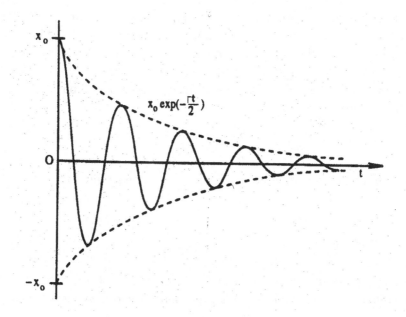

Figure 8.11: $Re\ x$ against t illustrating light damped Oscillations.

From equation (8.68), we note the following features of underdamped motion.

- The motion of an underdamped oscillator is harmonic, given by a cosine (sine) function of time in the same way as the motion of a simple harmonic oscillator.

- The angular frequency of oscillation ω_1 as given by equation (8.67) is less than the natural frequency ω_o of a corresponding simple harmonic oscillator in the absence of damping force ($\Gamma = 0$).

- The amplitude $A(t)$ of oscillation decays exponentially with time. Consequently the displacement of oscillations gradually decreases, and finally the oscillator comes to stop in equilibrium position.

- The time period of oscillation and frequency of oscillations are given by:

$$
T = \frac{2\pi}{\omega_1}, \text{ and } f = \frac{1}{T} = \frac{\omega_1}{2\pi} \qquad\qquad (8.69)
$$

Here, it is to be noted that the period of damped oscillations is larger than the period of undamped oscillations, and correspondingly the frequency of oscillations is smaller. This is to be expected because the damping force slows down the oscillations, and the oscillator must take longer to complete an oscillation.

Generalizing the expression for energy in terms of the force constant k and the amplitude derived earlier, the time dependent energy of an underdamped oscillator is given by:

$$E(T) = \frac{1}{2}\,k\,A(t)^2 = \frac{1}{2}\,k\,x_o^2 e^{-\Gamma t} = E_o\,e^{-\Gamma t} \qquad (8.70)$$

where $E_o = \frac{1}{2}\,k\,x_o^2$ is the energy of the oscillator at $t = 0$. The energy of a damped oscillator decays exponentially with time. This is consistent with the fact that a damping force being a dissipative force results in the decay in energy of the system which in turn results in the decay of the oscillations.

Case (ii): Critical damping: $(b^2 - 4km) = 0$. The square root in equation (8.65) becomes zero and hence λ is a real negative number given as:

$$\lambda = -\frac{b}{2\,m} = -\frac{\Gamma}{2} \qquad (8.71)$$

and the displacement of the critically damped oscillator decays exponentially as:

$$x = x_o\,e^{-\frac{\Gamma}{2}} \qquad (8.72)$$

without going through oscillations.

Case (iii): Heavy damping (or overdamping): $(b^2 - 4km) > 0$. If $(b^2 - 4km) > 0$, the square root in equation (8.65) is a positive quantity, and hence λ is a real number with two possible negative roots.

Let $(b^2 - 4km) = \Delta^2$. Then:

$$\lambda = -\frac{b}{2\,m} \pm \Delta \Rightarrow -\lambda_1 \text{ and } -\lambda_2 \qquad (8.73)$$

and the displacement in this case also decays exponentially without going through oscillations as:

$$x = x_o\,e^{-\lambda t} \qquad (8.74)$$

Although, in both cases of critical and heavy damping, the displacement of the oscillator decays exponentially without going through oscillations, the difference between two cases is that in a critically damped oscillator the systems reaches the equilibrium state in a minimum time, where as in a heavily damped system it takes longer to reach equilibrium. This is so because when the oscillator is displaced from equilibrium, presence of a large friction force heavily slows down its movement towards the equilibrium position. This property of critical damping is used in designing of oscillatory systems such as a coil galvanometer so that the coil returns to equilibrium position without oscillation in the minimum of time when the current producing the deflection of the coil is withdrawn. The three cases of damping are shown graphically in Figure (8.12).

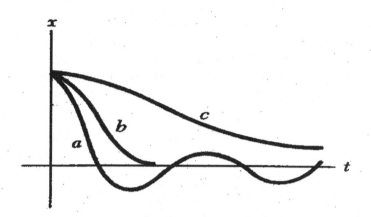

Figure 8.12: A graph of displacement against time for (a) light, (b) critical and (c) heavy damping.

8.2.2 LCR-**Circuit**

Consider an LCR circuit consisting of an inductor of *inductance* L, a capacitor of *capacitance* C and a resistor of *resistance* R, as illustrated in Figure (8.13). The potential drop across the capacitor is $V_C = q/C$, where q is charge, the potential drop across the inductor is $V_L = LdI/dt$, where $I = dq/dt$ is the electric current, and the potential drop across the resistor according to Ohm's law is $V_R = IR$.

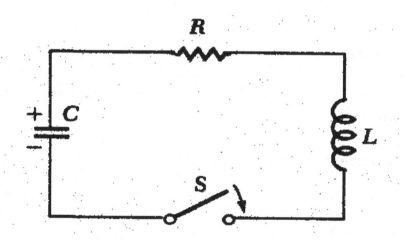

Figure 8.13: LCR circuit for damped charge oscillations.

According to *Kirchoff's second law*, the potential drops in a closed circuit add up to zero, that is

$$V_L + V_R + V_C = 0$$

$$L\frac{dI}{dt} + IR + \frac{q}{C} = 0$$

$$L\frac{d^2q}{dt^2} + \frac{dq}{dt} + \frac{q}{C} = 0 \qquad (8.75)$$

which is a *second order homogeneous differential equation* for charge q. As before, what is the solution, q, that satisfies the differential equation given in (8.75)? It can be noted that the mathematical structure of equation (8.75) is the same as that of equation (8.62), and hence their solutions are identical. We need not repeat the solution procedure which was discused in section (8.2.1), and one can note the following identities:

$$m \longleftrightarrow L$$
$$b \longleftrightarrow R$$
$$k \longleftrightarrow \frac{1}{C}$$
$$x \longleftrightarrow q$$

and hence the solution for equation (8.75) under light damping condition $\left(\text{small } R, < \sqrt{\frac{4L}{C}}\right)$ is:

$$q = q_0 e^{-\frac{\Gamma}{2}t} \cos \omega_1 t \qquad (8.76)$$

where $\Gamma = R/L$ is the *damping constant* for an LCR circuit, and

$$\omega_1 = \sqrt{\frac{1}{LC} - \left(\frac{R}{2L}\right)^2} \qquad (8.77)$$

$$= \sqrt{\omega_0^2 - \frac{\Gamma^2}{4}} \qquad (8.78)$$

where

$$\omega_0 = \frac{1}{\sqrt{LC}}$$

is the *resonant frequency* of an LC-circuit. The electrical energy of the circuit decays exponentially.

If the resistance of the circuit is large $\left(R \geq \sqrt{\frac{4L}{C}}\right)$, the decay of the energy on the capacitor simulates critical or heavy damping depending on the value of R.

8.3   Forced Oscillations

In an oscillating system, energy is proportional to the square of the amplitude. It has been shown in section (8.3) that in damped oscillations the amplitude decays with time, which means that energy is lost as the amplitude decays. However, it is possible to supply energy into the oscillating system so as to compensate for the energy loss. The supplied energy must be from an external oscillating source, for example a vibrator, oscillator or some other periodic function. There exists a definite relation between the frequency of the source of energy and the frequency of the damped oscillating system. For the maximum efficiency of the interaction, the frequency ω of the external source must be equal to the natural frequency ω_0 of oscillation of the oscillating system. When $\omega = \omega_0$, *resonance* is said to have occured.

8.3.1 Mass-spring system driven by a vibrator

Consider a mass-spring system driven by a vibrator with a periodic force $F_0 e^{-i\omega t}$, as illustrated in Figure (8.14), where the spring has a force constant k, and the mass is moving on rough surface. Initially, the spring is at the equilibrium. When the spring is extended by a distance x, the mass experiences a restoring force, $-kx$, according to *Hooke's law*, and a frictional force proportional to velocity, given by $-bdx/dt$, where b is known as the *damping parameter* and the external driving force. The vibrator provides a driving force $F_0 e^{-i\omega t}$.

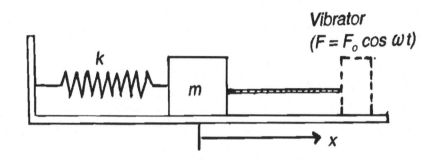

Figure 8.14: Mass-string system driven by a vibrator.

The equation of motion is given by Newton's second law, where as is known, acceleration a is proportional to the *net forces*, in the form

$$
\begin{aligned}
ma &= \sum F \\
m\frac{d^2x}{dt^2} &= \sum F \\
&= -kx - b\frac{dx}{dt} + F_0 e^{-i\omega t} \\
m\frac{d^2x}{dt^2} + b\frac{dx}{dt} + kx &= F_0 e^{-i\omega t}
\end{aligned}
\tag{8.79}
$$

which is a *second order inhomogeneous differential equation*. As before, what is the solution, x, that satisfies the differential equation given in (8.79)? To answer this question, we proceed as follows.

The general solution $x(t)$ of an inhomogeneous equation is the sum of a complementary function $x_c(t)$ (also known as the *transient response*) plus a particular integral $x_p(t)$ (also known as the *driven response*), that is

$$
x(t) = x_c(t) + x_p(t)
\tag{8.80}
$$

where $x_c(t)$ is a solution of the homogeneous equation and $x_p(t)$ is *any* solution of the inhomogeneous equation. This means

$$
m\frac{d^2x_c}{dt^2} + b\frac{dx_c}{dt} + kx_c = 0
\tag{8.81}
$$

$$
m\frac{d^2x_p}{dt^2} + b\frac{dx_p}{dt} + kx_p = F_0 e^{-i\omega t}
\tag{8.82}
$$

The complementary function has already been studied in section (8.2) when we were dealing with damped oscillations. Let us now study the particular integral satisfying equation (8.82). What is the solution, x_p, that satisfies the differential equation given in (8.82)?

From physical grounds, the driven response must have the same frequency as the external driving force, and hence x_p is of the form

$$x_p = x_0 e^{-i\omega t} \tag{8.83}$$

and see whether it works as a solution to equation (8.82). This gives

$$\frac{dx_p}{dt} = -i\omega x_p$$

$$\frac{d^2 x_p}{dt^2} = -\omega^2 x_p$$

$$(-m\omega^2 - i\omega b + k)x_0 e^{-i\omega t} = F_0^{-i\omega t}$$

$$x_0 = \frac{F_0}{m\left(\frac{k}{m} - \omega^2 - i\omega\frac{b}{m}\right)}$$

$$x_0 = \frac{F_0}{m\left(\omega_0^2 - \omega^2 - i\omega\Gamma\right)} \tag{8.84}$$

where, as before, $\omega_0 = \sqrt{k/m}$ and $\Gamma = b/m$.

The driven response has an interesting frequency dependence as we shall see below. There are two cases of physical interest: zero damping ($\Gamma = 0$) and finite damping ($\Gamma \neq 0$) as discussed below.

Case 1: Zero damping ($\Gamma = 0$)

With *zero damping*, the amplitude of the driven response given in equation (8.84) reduces to

$$x_0 = \frac{F_0}{m\left(\omega_0^2 - \omega^2\right)} \tag{8.85}$$

It can be noted that the amplitude is infinite at the resonant frequency ω_0. This is illustrated by another example in section (8.3.2).

Case 2: Finite damping ($\Gamma \neq 0$)

With *finite damping*, the amplitude of the driven response given in equation (8.84) reduces to

$$x_0 = \frac{F_0}{m\left(\omega_0^2 - \omega^2 - i\omega\Gamma\right)}$$

$$= \frac{F_0}{m\left(\omega_0^2 - \omega^2 - i\omega\Gamma\right)} \times \frac{\left(\omega_0^2 - \omega^2 + i\omega\Gamma\right)}{\left(\omega_0^2 - \omega^2 + i\omega\Gamma\right)}$$

$$= \frac{F_0\left(\omega_0^2 - \omega^2\right)}{m\left[\left(\omega_0^2 - \omega^2\right)^2 + \omega^2\Gamma^2\right]} + i\frac{F_0\omega\Gamma}{m\left[\left(\omega_0^2 - \omega^2\right)^2 + \omega^2\Gamma^2\right]}$$

$$= \frac{F_0}{m}[A + iB] \tag{8.86}$$

$$x_0 = \frac{F_0}{m}G(\omega) \tag{8.87}$$

where

$$G(\omega) = A + iB \tag{8.88}$$

$$A = \frac{(\omega_0^2 - \omega^2)}{\left[(\omega_0^2 - \omega^2)^2 + \omega^2\Gamma^2\right]} \tag{8.89}$$

$$B = \frac{\omega\Gamma}{\left[(\omega_0^2 - \omega^2)^2 + \omega^2\Gamma^2\right]} \tag{8.90}$$

where $G(\omega)$ is known as the *Response function*, A is known as the *Elastic amplitude*, and B is known as the *Absorptive amplitude*.

It is useful to define a *quality factor*, Q, of a damped forced oscillator as:

$$Q = \frac{\omega_0}{\Gamma} \tag{8.91}$$

which is a number representing the strength of damping. The smaller the damping, the larger is the value of Q, implying an oscillator of a higher quality. For an ideal simple harmonic oscillator $\Gamma = 0$, and $Q = \infty$.

Graphs of the elastic amplitude A against frequency ω, and the absorptive amplitude B against frequency ω are illustrated in Figures (8.15) (a) and (b) respectively for various values of $Q = 1, 3$ and 10. Graphs of the magnitude of amplitude $|x_0|$ against frequency ω and phase are illustrated in Figure (8.16)(a) and (b) respectively. It can be noted that the elastic amplitude is zero at ω_0 and the absorptive amplitude is very large at the resonant frequency ω_0.

If there is a phase constant, δ, the displacement is given by

$$u = u_0 e^{-i(\omega t - \delta)} \tag{8.92}$$

If the above displacement satisfies the equation of motion

$$m\frac{d^2u}{dt^2} + b\frac{du}{dt} + ku = F_0^{-i\omega t}$$

Separating the real and imaginary parts, we get

$$(\omega_0^2 - \omega^2)u_0 = \frac{F_0}{m}\cos\delta$$

$$\omega\Gamma u_0 = \frac{F_0}{m}\sin\delta$$

which give

$$\tan \delta = \frac{\omega \Gamma}{(\omega_0^2 - \omega^2)} \tag{8.93}$$

The frequency dependence of the phase constant is illustrated in Figure 8.16 (b).

Power absorption by a forced oscillator

The instantaneous power absorbed by the oscillator from the external periodic force can be calculated as follows.

$$
\begin{aligned}
\text{Instantaneous power input,} \quad P(t) \quad &= \quad \frac{\text{Work done}}{\text{Time}} \\
&= \quad \text{Driving force} \times \frac{\text{Displacement}}{\text{Time}} \\
&= \quad \text{Driving force} \times \text{Velocity} \\
&= \quad Re \left\{ F_0^{-i\omega t} \right\} Re \ \frac{dx_p}{dt} \\
&= \quad F_0 \cos \omega t \ \ Re \ \frac{dx_p}{dt} \\
&= \quad F_0 \cos \omega t \ \ \frac{d}{dt} Re \ x_p \tag{8.94}
\end{aligned}
$$

But using equation (8.83) we have

$$
\begin{aligned}
Re \ x_p \quad &= \quad Re \left(x_0^{-i\omega t} \right) \\
&= \quad Re \left\{ \frac{F_0}{m} [A + iB] [\cos \omega t - i \sin \omega t] \right\} \\
&= \quad \frac{F_0}{m} (A \cos \omega t + B \sin \omega t) \tag{8.95}
\end{aligned}
$$

Inserting equation (8.95) in (8.94), the instateneous power becomes

$$
\begin{aligned}
P(t) \quad &= \quad F_0 \cos \omega t \ \ \frac{d}{dt} \left\{ \frac{F_0}{m} (A \cos \omega t + B \sin \omega t) \right\} \\
&= \quad \frac{F_0^2}{m} \omega \left(-A \cos \omega t \sin \omega t + B \cos^2 \omega t \right) \tag{8.96}
\end{aligned}
$$

Let us now calculate the *time averaged power input*, $< P(\omega) >$, over a period T using the definition

$$< P(\omega) > = \frac{1}{T} \int_0^T P(t) dt \tag{8.97}$$

and noting that

$$< \cos \omega t \sin \omega t > \quad = \quad \frac{1}{T} \int_0^T \cos \omega t \sin \omega t \, dt$$

$$= 0 \tag{8.98}$$

$$<\cos^2 \omega t> = \frac{1}{T}\int_0^T \cos^2 \omega t\, dt$$

$$= \frac{1}{2} \tag{8.99}$$

From equations (8.97), (8.98) and (8.99), the time averaged power input becomes

$$
\begin{aligned}
<P(\omega)> &= \frac{1}{2}\frac{F_0^2}{m}\omega B \\
&= \frac{1}{2}\frac{F_0^2}{m}\omega \; \mathrm{Im}G(\omega) \\
&= \frac{1}{2}\frac{F_0^2}{m\Gamma}\frac{\omega^2\Gamma^2}{\left[(\omega_0^2-\omega^2)^2+\omega^2\Gamma^2\right]} \\
&= P_0 P_r(\omega)
\end{aligned} \tag{8.100}
$$

where P_0 and $P_r(\omega)$ can be identified as

$$P_0 = \frac{F_0^2}{2m\Gamma} \tag{8.101}$$

$$P_r(\omega) = \frac{\omega^2\Gamma^2}{\left[(\omega_0^2-\omega^2)^2+\omega^2\Gamma^2\right]} \tag{8.102}$$

and $P_r(\omega)$ is known as the *Resonance function*.

When the frequency is close to the resonant frequency, the following approximation can be made to equation (8.100).

$$
\begin{aligned}
\lim_{\omega\to\omega_0}<P(\omega)>=<P_L> &= \frac{1}{2}\frac{F_0^2}{m\Gamma}\frac{\omega_0^2\Gamma^2}{(\omega_0-\omega)^2(\omega_0+\omega)^2+\omega_0^2\Gamma^2} \\
&= \frac{1}{2}\frac{F_0^2}{m\Gamma}\frac{\omega_0^2\Gamma^2}{\left\{4\omega_0^2(\omega_0-\omega)^2+\omega_0^2\Gamma^2\right\}} \\
<P_L> &= \frac{P_0}{\left\{\frac{4(\omega_0-\omega)^2}{\Gamma^2}+1\right\}}
\end{aligned} \tag{8.103}
$$

where an approximation $\omega_0+\omega\approx 2\omega_0$ has been used, and P_L is known as the *Lorentzian lineshape*.

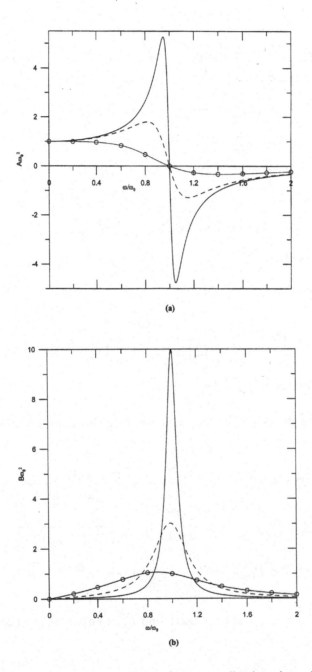

(a)

(b)

Figure 8.15: (a) Frequency dependence of the elastic amplitude, plotted as $A\omega_0^2$ vs ω/ω_0, with circles ($Q = 1$), dashes ($Q = 3$) and smooth curve ($Q = 10$). (b) Frequency dependence of the absorptive amplitude, plotted as $B\omega_0^2$ vs ω/ω_0, with circles ($Q = 1$), dashes ($Q = 3$) and smooth curve ($Q = 10$).

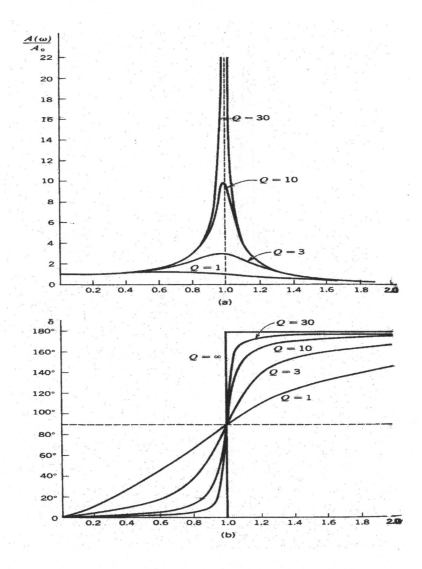

Figure 8.16: Graphs of (a) the magnitude of amplitude $|x_0|$ against reduced frequency ω/ω_0 and (b) the phase constant against reduced frequency ω/ω_0.

The frequency dependence of the time averaged power input is illustrated in figure (8.17) for varying values of the quality factor Q (and hence the damping constant, Γ). It can be noted that the sharpness of the resonance function depends on Q (and hence Γ): higher Q (low Γ) values correspond to sharper resonance.

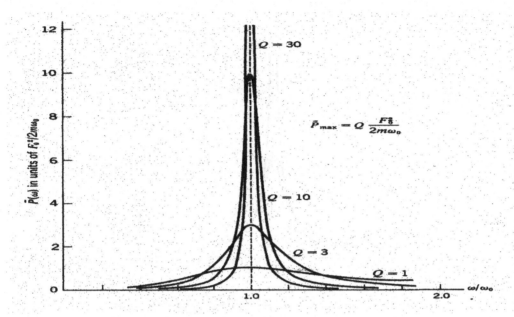

Figure 8.17: Frequency dependence of the time averaged power inpute, plotted as $< P(\omega_0) >$ vs ω/ω_0, with $Q = 1, 3, 10$ and 30.

8.3.2 Child's swing

A child's swing is essentially a driven simple pendulum consisting of a child of mass m supported by a string of length l, and driven by a periodic force $F = F_0 \cos \omega t$, as illustrated in Figure (8.18). When the string of length l makes an angle θ with the vertical, the weight (force mg) of the child has two components: the *radial component*, $mg \cos \theta$ balanced by the tension in the string, and the *tangential component*, $mg \sin \theta$ which provides the restoring force. We neglect the frictional forces which arise due to air resistance, friction etc.

The equation of motion is given by Newton's second law, where as is known, acceleration a is proportional to the *net forces*, in the form

$$m\frac{d^2x}{dt^2} = \sum F$$
$$= -mg \sin \theta + F_0 \cos \omega t$$
$$m\frac{d^2x}{dt^2} = -mg\frac{x}{l} + F_0 \cos \omega t$$
$$m\frac{d^2x}{dt^2} + m\frac{g}{l}x = F_0 \cos \omega t$$
$$m\frac{d^2x}{dt^2} + m\omega_0^2 x = F_0 \cos \omega t \quad \text{where} \quad \omega_0^2 = \frac{g}{l} \tag{8.104}$$

which is a *second order inhomogeneous differential equation*. As before, what is the solution, x, that satisfies the differential equation given in (8.104)?

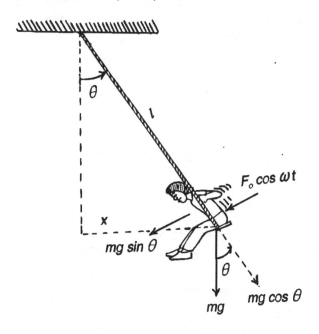

Figure 8.18: Child's swing.

Let us consider a *trial solution*

$$x = x_0 \cos \omega t \qquad (8.105)$$

and see whether it works as a solution to equation (8.104). This gives

$$\frac{dx}{dt} = -\omega x_0 \sin \omega t$$

$$\frac{d^2x}{dt^2} = -\omega^2 x_0 \cos \omega t$$

$$\frac{d^2x}{dt^2} = -\omega^2 x$$

and hence equation (8.104) becomes

$$\left(-\omega^2 + \omega_0^2\right) x = \frac{F_0}{m} \cos \omega t$$

$$x_0 = \frac{F_0}{m(\omega_0^2 - \omega^2)} \qquad (8.106)$$

which is the amplitude of the forced oscillations of the child's swing. Thus the trial solution (8.105) is indeed a solution to the differential equation (8.104). This is the same solution as equation (8.85).

8.3.3 Barton's pendulums

Bartons's pendulums consist of a system of pendulums labelled 1 to 5 and P, constructed such that their repective lengths are $l_n, n = 1, 2, 3, 4, 5$ and $l_P = l_2$, as illustrated in Figure (8.19). When pendulum P is let go, all the other pendulums labelled 1 to 5 will be driven to oscillate. The question is: which of the pendulums 1 to 5 wil have the largest amplitude?

According to the theory of the simple pendulum, each of the pendulums 1 to 5 will have a frequency

$$\omega_n = \sqrt{\frac{g}{l_n}} \quad \text{where} \quad n = 1, 2, 3, 4, 5$$

and pendulum P will have a frequency

$$\omega_P = \sqrt{\frac{g}{l_P}} = \sqrt{\frac{g}{l_2}} \quad \text{since} \quad l_P = l_2$$

Since pendulum 2 has the same length as the driving pendulum P, it will have the largest amplitude because the resonance condition $\omega_2 = \omega_P$ is satisfied.

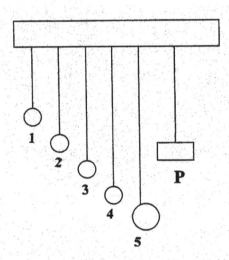

Figure 8.19: Barton's pendulums.

8.3.4 Forced LCR-circuit

Consider a forced LCR circuit consisting of an inductor of *inductance* L, a capacitor of *capacitance* C, a resistor of *resistance* R, and an oscilltor, as illustrated in Figure (8.20). The potential drop across the capacitor is $V_C = q/C$, where q is charge, the potential drop across the inductor is $V_L = LdI/dt$, where $I = dq/dt$ is the electric current, and the potential drop across the resistor is $V_R = IR$ according to Ohm's law. The circuit is druven by a power source $V = V_0 e^{-i\omega t}$. According to *Kirchoff's second law*, the potential drops in a closed circuit add up to zero, that is

$$V_L + V_R + V_C - V_0 e^{-i\omega t} = 0$$

$$L\frac{dI}{dt} + IR + \frac{q}{C} - V_0 e^{-i\omega t} = 0$$

$$L\frac{d^2q}{dt^2} + R\frac{dq}{dt} + \frac{q}{C} = V_0 e^{-i\omega t} \qquad (8.107)$$

which is a *second order inhomogeneous differential equation* in q. As before, what is the solution, q, that satisfies the differential equation given in (8.107)? It can be noted that the mathematical structure of equation (8.107) is the same as that of equation (8.79), and hence their solutions are identical.

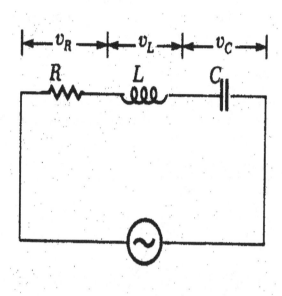

Figure 8.20: Forced LCR circuit.

There are several other examples which illustrate the principle of resonance, such as

- Radio receiver:
  A radio receiver acts on the principle of resonance between an LC-circuit and an external oscillating radio frequency. The frequency of the tuned LC-circuit is $1/\sqrt{LC}$ and the frequency of the external radio wave is ω_0. The resonance condition is achieved when $\omega = \omega_0$. The frequency of the LC circuit is varied by changing the variable capacitor in the radio circuit.

- Nuclear Magnetic Resonance (NMR):
  The expression for the lineshape in Nuclear magnetic Resonance (NMR) is analogous to that given in equation (8.103) in the form

$$\sigma(E) = \frac{\sigma(E_0)}{\left\{\frac{4(E_0 - E)^2}{\Gamma^2} + 1\right\}} \qquad (8.108)$$

- Infrared absorption by a crystal (Students should do some library search to see why crystals absorb infrared radiation and how resonance is important in this phenomenon).

- Atmospheric tides (Students should do some library search to see how the earth's motion produces atmospheric tides).

8.4   Coupled Oscillations

When there is more than one oscillator interacting with each other, we have what are known as *coupled oscillators*. In coupled oscillations, there are more than one degree of freedom. Another important concept that is introduced in the study of coupled oscillations is what is referred to as normal modes.

8.4.1   Two coupled pendulums

Consider two pendulums, each of length l and mass m, with displacements u_a and u_b. The pendulums are coupled by a rigid body of length a and a spring of force constant k, as illustrated in Figure (8.21).

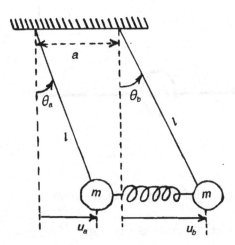

Figure 8.21: Coupled pendulums.

Initially, the system is in equilibrium when the spring has length a. At a given instant u_a and u_b are the displacements of the two masses from the equilibrium position as illustrated in Figure 8.21, and θ_a and θ_b are the respective angular displacements. After displacement from equilibrium, mass a is at u_a and mass b is at $a + u_b$ and hence the spring has a length $a + u_b - u_a$, that is its length has changed by $(u_b - u_a)$. We consider *normal mode solutions* where the displacements u_a and u_b are of the form

$$u_a \sim e^{-i\omega t} \quad \text{and} \quad u_b \sim e^{-i\omega t} \tag{8.109}$$

The equations of motion for each of the masses are given by:

For mass A,

$$m\frac{d^2u_a}{dt^2} = \sum F$$
$$= -mg\sin\theta_a + k(u_b - u_a)$$
$$-m\omega^2 u_a = -mg\frac{u_a}{l} + k(u_b - u_a)$$
$$\left(\frac{g}{l} - \omega^2 + \frac{k}{m}\right)u_a - \frac{k}{m}u_b = 0 \tag{8.110}$$

and for mass B,

$$m\frac{d^2u_b}{dt^2} = \sum F$$
$$= -mg\sin\theta_b - k(u_b - u_a)$$
$$-m\omega^2 u_b = -mg\frac{u_b}{l} - k(u_b - u_a)$$
$$\left(\frac{g}{l} - \omega^2 + \frac{k}{m}\right)u_b - \frac{k}{m}u_a = 0 \tag{8.111}$$

Equations (8.110) and (8.111) can be solved as simultaneous equations or by matrix methods. We llustrate the solutions by matrix methods where it can be noted that equations (8.110) and (8.111) can be put in matrix form as

$$\begin{bmatrix} \left(\frac{g}{l} - \omega^2 + \frac{k}{m}\right) & -\frac{k}{m} \\ -\frac{k}{m} & \left(\frac{g}{l} - \omega^2 + \frac{k}{m}\right) \end{bmatrix} \begin{bmatrix} u_a \\ u_b \end{bmatrix} = \begin{bmatrix} 0 \\ 0 \end{bmatrix} \tag{8.112}$$

A solution exists if the determinant of the 2×2 matrix vanishes, that is

$$\left(\frac{g}{l} - \omega^2 + \frac{k}{m}\right)^2 - \left(\frac{k}{m}\right)^2 = 0$$
$$\left(\frac{g}{l} - \omega^2 + \frac{k}{m} - \frac{k}{m}\right)\left(\frac{g}{l} - \omega^2 + \frac{k}{m} + \frac{k}{m}\right) = 0 \tag{8.113}$$

which gives two normal mode frequencies

$$\omega^2 = \frac{g}{l} \quad \text{and} \quad \omega^2 = \frac{g}{l} + \frac{2k}{m} \tag{8.114}$$

which are referred to as the *lower mode* and *upper mode* respectively. The two coupled masses have two normal modes.

From equation (8.110), the ratio of the displacements u_a and u_b is obtained as

$$\frac{u_a}{u_b} = \frac{k/m}{\frac{g}{l} - \omega^2 + \frac{k}{m}} \tag{8.115}$$

and this ratio *differs* for the two normal modes, giving

$$\left[\frac{u_a}{u_b}\right]_{Lower\ \ mode} = 1 \ \text{for} \ \omega^2 = \frac{g}{l} \tag{8.116}$$

$$\left[\frac{u_a}{u_b}\right]_{Upper\ \ mode} = -1 \ \text{for} \ \omega^2 = \frac{g}{l} + \frac{2k}{m} \tag{8.117}$$

The physical interpretation of these two modes is summarised below.

Mode 1 (Lower mode): $\omega^2 = \frac{g}{l}$, $\frac{u_a}{u_b} = 1$
The two masses are vibrating with displacements in the *same* direction.

Mode 2 (Upper mode): $\omega^2 = \frac{g}{l} + \frac{2k}{m}$, $\frac{u_a}{u_b} = -1$
The two masses are vibrating with displacements in *opposite* directions.

8.4.2   Two coupled masses - Longitudinal oscillations

Consider two bodies A and B, each of mass m, coupled by springs as illustrated in Figure (8. 22), where each of the springs has a force constant k. Initially, the system is in equilibrium when each spring has length a. After displacement from equilibrium, mass A is at $a + u_a$ and mass B is at $2a + u_b$ and hence the length of the left spring is $a + u_a$, the right spring has length $a - u_b$, and the middle spring has changed length by $(u_b - u_a)$. We consider *normal mode solutions* where the displacements u_a and u_b are of the form

$$u_a \sim e^{-i\omega t} \ \text{and} \ u_b \sim e^{-i\omega t} \tag{8.118}$$

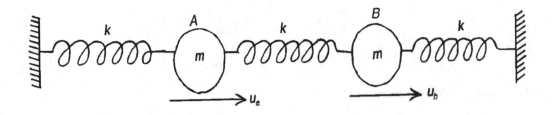

Figure 8.22: Coupled longitudinal oscillations.

The equations of motion for each of the masses are given by:

For mass A,

$$m\frac{d^2 u_a}{dt^2} = \sum F$$
$$= -ku_a + k(u_b - u_a)$$

$$-\omega^2 u_a = -2ku_a + ku_b$$

$$\left(\frac{2k}{m} - \omega^2\right) u_a - \frac{k}{m} u_b = 0 \tag{8.119}$$

and for mass B,

$$m\frac{d^2 u_b}{dt^2} = \sum F$$

$$= -k(u_b - u_a) + k(a - u_b - a)$$

$$-\omega^2 u_b = -2ku_b + ku_a$$

$$\left(\frac{2k}{m} - \omega^2\right) u_b - \frac{k}{m} u_a = 0 \tag{8.120}$$

Equations (8.119) and (8.120) can be solved as simultaneous equations or by matrix methods. We llustrate the solutions by matrix methods where it can be noted that equations (8.119) and (8.120) can be put in matrix form as

$$\begin{bmatrix} \left(\frac{2k}{m} - \omega^2\right) & -\frac{k}{m} \\ -\frac{k}{m} & \left(\frac{2k}{m} - \omega^2\right) \end{bmatrix} \begin{bmatrix} u_a \\ u_b \end{bmatrix} = \begin{bmatrix} 0 \\ 0 \end{bmatrix} \tag{8.121}$$

A solution exists if the determinant of the 2×2 matrix vanishes, that is

$$\left(\frac{2k}{m} - \omega^2\right)^2 - \left(\frac{k}{m}\right)^2 = 0$$

$$\left(\frac{2k}{m} - \omega^2 - \frac{k}{m}\right)\left(\frac{2k}{m} - \omega^2 + \frac{k}{m}\right) = 0 \tag{8.122}$$

which gives two normal mode frequencies

$$\omega^2 = \frac{k}{m} \text{ and } \omega^2 = \frac{3k}{m} \tag{8.123}$$

which are referred to as the *lower mode* and *upper mode* respectively. The two coupled masses have two normal modes.

But from equation (8.119), the ratio of the displacements u_a and u_b is obtained as

$$\frac{u_a}{u_b} = \frac{k/m}{\frac{2k}{m} - \omega^2} \tag{8.124}$$

and this ratio *differs* for the two normal modes, giving

$$\left[\frac{u_a}{u_b}\right]_{Lower \ mode} = 1 \text{ for } \omega^2 = \frac{k}{m} \tag{8.125}$$

$$\left[\frac{u_a}{u_b}\right]_{Upper \ mode} = -1 \text{ for } \omega^2 = \frac{3k}{m} \tag{8.126}$$

The physical interpretation of these two modes is summarised below.

Mode 1 (Lower mode): $\omega^2 = \frac{k}{m}$, $\frac{u_a}{u_b} = 1$
The two masses are vibrating with displacements in the *same* direction.

Mode 2 (Upper mode): $\omega^2 = \frac{3k}{m}$, $\frac{u_a}{u_b} = -1$
The two masses are vibrating with displacements in *opposite* directions.

8.4.3 Two coupled masses - Transverse oscillations

Consider two masses A and B coupled by springs as illustrated in Figure (8.23), where each of the springs has a force constant k. Initially, the system is in equilibrium when each spring has length a. After a *transverse* displacement from equilibrium, mass A is displaced by u_a and mass B is by u_b and the left spring makes an angle θ_1, the middle spring makes an angle θ_2, and the right spring makes an angle θ_3, from the horizontal, as illustrated in figure (8.23). T is the tension in each of the springs, which for small oscillations is taken to be the same.

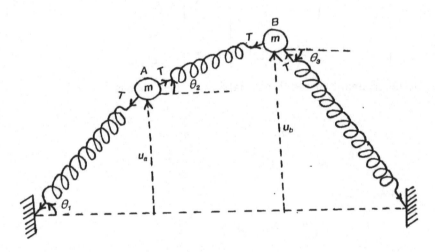

Figure 8.23: Two coupled masses showing transverse oscillations.

We consider *normal mode solutions* where the displacements u_a and u_b are of the form

$$u_a \sim e^{-i\omega t} \quad \text{and} \quad u_b \sim e^{-i\omega t} \tag{8.127}$$

The equations of motion for each of the masses are given by:

For mass A,

$$
\begin{aligned}
m\frac{d^2 u_a}{dt^2} &= \sum F \\
&= -T\sin\theta_1 + T\sin\theta_2 \\
&= -T\tan\theta_1 + T\tan\theta_2 \quad \text{where for small angles, } \sin\theta \approx \tan\theta
\end{aligned}
$$

$$-m\omega^2 u_a = -T\frac{u_a}{a} + T\frac{u_b - u_a}{a}$$

$$\left(\frac{2T}{ma} - \omega^2\right) u_a - \frac{T}{ma} u_b = 0 \tag{8.128}$$

and for mass B,

$$m\frac{d^2 u_b}{dt^2} = \sum F$$
$$= -T\sin\theta_2 + T\sin\theta_3$$
$$= -T\tan\theta_2 + T\tan\theta_3 \quad \text{where for small angles, } \sin\theta \approx \tan\theta$$
$$-m\omega^2 u_b = -T\frac{u_b - u_a}{a} - T\frac{u_b}{a}$$
$$\left(\frac{2T}{ma} - \omega^2\right) u_b - \frac{T}{ma} u_a = 0 \tag{8.129}$$

Equations (8.128) and (8.129) can be solved as simultaneous equations or by matrix methods. We llustrate the solutions by matrix methods where it can be noted that equations (8.128) and (8.129) can be put in matrix form as

$$\begin{bmatrix} \left(\frac{2T}{ma} - \omega^2\right) & -\frac{T}{ma} \\ -\frac{T}{ma} & \left(\frac{2T}{ma} - \omega^2\right) \end{bmatrix} \begin{bmatrix} u_a \\ u_b \end{bmatrix} = \begin{bmatrix} 0 \\ 0 \end{bmatrix} \tag{8.130}$$

A solution exists if the determinant of the 2×2 matrix vanishes, that is

$$\left(\frac{2T}{ma} - \omega^2\right)^2 - \left(\frac{T}{ma}\right)^2 = 0$$
$$\left(\frac{2T}{ma} - \omega^2 - \frac{T}{ma}\right)\left(\frac{2T}{ma} - \omega^2 + \frac{T}{ma}\right) = 0 \tag{8.131}$$

which gives two normal mode frequencies

$$\omega^2 = \frac{T}{ma} \quad \text{and} \quad \omega^2 = \frac{3T}{ma} \tag{8.132}$$

which are referred to as the *lower mode* and *upper mode* respectively. The two coupled mases have two normal modes.

But from equation (8.128), the ratio of the displacements u_a and u_b is obtained as

$$\frac{u_a}{u_b} = \frac{T/ma}{\frac{2T}{ma} - \omega^2} \tag{8.133}$$

and this ratio *differs* for the two normal modes, giving

$$\left[\frac{u_a}{u_b}\right]_{Lower\ mode} = 1 \ \text{for} \ \omega^2 = \frac{T}{ma} \tag{8.134}$$

$$\left[\frac{u_a}{u_b}\right]_{Upper\ mode} = -1 \ \text{for} \ \omega^2 = \frac{3T}{ma} \tag{8.135}$$

The physical interpretation of these two modes is summarised below.

Mode 1 (Lower mode): $\omega^2 = \frac{T}{ma}$, $\frac{u_a}{u_b} = 1$
The two masses are vibrating with displacements in the *same* direction.

Mode 2 (Upper mode): $\omega^2 = \frac{3T}{ma}$, $\frac{u_a}{u_b} = -1$
The two masses are vibrating with displacements in *opposite* directions.

8.4.4   Infinite linear chain of coupled masses - Longitudinal oscillations

Consider an infinite linear chain of $N(\to \infty)$ masses, each of mass m, coupled by springs each of force constant k, as illustrated in figure (8.24). The infinite linear chain of coupled masses exhibiting longitudinal oscillations is a good model for studying longitudinal oscillations of a 1-D *monatomic lattice*. We only consider nearest neighbour interactions.

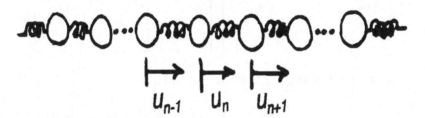

Figure 8.24: Infinite linear chain of coupled masses exhibiting longitudinal oscillations.

When the nth mass is displaced by u_n, the nearest neighbours are dislaced by u_{n-1} and u_{n+1}.

The equation of motion for the nth mass is given by Newton's second law.

$$
\begin{aligned}
m\frac{d^2 u_n}{dt^2} &= \sum F \\
&= -k(u_n - u_{n-1}) + k(u_{n+1} - u_n) \\
&= k(u_{n+1} - 2u_n + u_{n-1})
\end{aligned} \tag{8.136}
$$

which is a *second order differential equation* where we have considered only the nearest neighbour interaction. What are the solutions, u_n, u_{n-1} and u_{n+1} which satisfy the differential equation given in (8.136)?

From physical grounds, as before, let us consider *trial solution* whereby all masses are assumed to have the same amplitude of oscillations.

$$
\begin{aligned}
u_n &= A e^{i(qna - \omega t)} \\
u_{n+1} &= A e^{i[q(n+1)a - \omega t]} \\
u_{n-1} &= A e^{i[q(n-1)a - \omega t]}
\end{aligned}
$$

and see whether they work as solutions to equation (8.136), q is the wavenumber, and a is the equilibrium separation between neighbouring masses. This gives

$$
\begin{aligned}
-m\omega^2 u_n &= k\left\{ A e^{i[q(n+1)a - \omega t]} - 2A e^{i(qna - \omega t)} + A e^{i[q(n-1)a - \omega t]} \right\} \\
&= k\left\{ e^{iqa} - 2 + e^{-iqa} \right\} A e^{i(qna - \omega t)} \\
&= k\left\{ 2\cos qa - 2 \right\} u_n \\
\omega^2 &= \frac{2k}{m}[1 - \cos qa] \\
&= \frac{4k}{m}\sin^2 \frac{qa}{2} \quad \text{and hence} \\
\omega &= 2\sqrt{\frac{k}{m}}\sin \frac{qa}{2}
\end{aligned}
\tag{8.137}
$$

which is known as the *dispersion relation* for longitudinal oscillations of a 1-D monatomic lattice.

8.4.5 Infinite linear chain of coupled masses - Transverse oscillations

Consider an infinite linear chain of $N(\to \infty)$ masses, each of mass m, coupled by springs each of force constant k, as illustrated in Figure (8.25). The infinite linear chain of coupled masses exhibiting transverse oscillations is a good model for studying transverse oscillations of a 1-d *monatomic lattice*. Again, we only consider nearest neighbour interactions.

When the nth mass is displaced by u_n, the nearest neighbours are dislaced by u_{n-1} and u_{n+1}.

The equation of motion for the nth mass is given by Newton's second law.

$$
\begin{aligned}
m\frac{d^2 u_n}{dt^2} &= \sum F \\
&= -T\sin(\theta_{n-1}) - T\sin\theta_n \\
&= -T\left[\frac{u_n - u_{n-1}}{a} - \frac{u_n - u_{n+1}}{a} \right] \\
&= \frac{T}{a}[u_{n+1} - 2u_n + u_{n-1}]
\end{aligned}
\tag{8.138}
$$

which is a *second order differential equation*, where we have only considered the nearest neighbour interactions, and for small oscillations, the tension T in the springs are taken to be the same. As before, what are the solutions, u_n, u_{n-1} and u_{n+1} which satisfy the differential equation given in (8.138)?

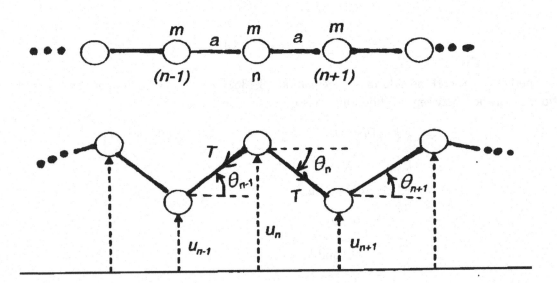

Figure 8.25: Infinite linear chain of coupled masses exhibiting transverse oscillations.

From physical grounds, as before, let us consider *trial solutions*

$$u_n = Ae^{i(qna-\omega t)}$$
$$u_{n+1} = Ae^{i[q(n+1)a-\omega t]}$$
$$u_{n-1} = Ae^{i[q(n-1)a-\omega t]}$$

and see whether these work as solutions to equation (8.138), q is the wavenumber, and a is the equilibrium separation between neighbouring masses. These give

$$-m\omega^2 u_n = \frac{T}{a}\left\{ Ae^{i[q(n+1)a-\omega t]} - 2Ae^{i(qna-\omega t)} + Ae^{i[q(n-1)a-\omega t]} \right\}$$
$$= \frac{T}{a}\left\{ e^{iqa} - 2 + e^{-iqa} \right\} Ae^{i(qna-\omega t)}$$
$$= \frac{T}{a}\left\{ 2\cos qa - 2 \right\} u_n$$
$$\omega^2 = \frac{2T}{ma}[1-\cos qa]$$
$$= \frac{4T}{ma}\sin^2\frac{qa}{2} \quad \text{and hence}$$
$$\omega = 2\sqrt{\frac{T}{ma}}\sin\frac{qa}{2} \qquad (8.139)$$

which is known as the *dispersion relation* for transverse oscillations of a 1-d monatomic lattice.

8.4.6 Transverse Oscillations of a Continuous String

Consider transverse oscillations of a continuous string. This is essentially the limit of the case studied in an earlier section. when N is very large and a very small. Consider an element of the string studied in detail in Figure (8.26).

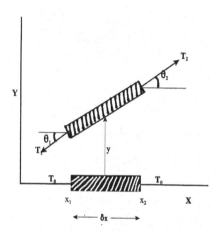

Figure 8.26: A string exhibiting transverse oscillations.

Let the element when in equilibrium have a length $\delta x = x_2 - x_1$, mass δm, linear density $\rho_l = \delta m / \delta x$ and equilibrium tension T_0.

When the string is displaced by a distance y, it is stretched and experiences tensions T_1 and T_2 in the directions θ_1 and θ_2 respectively, as shown in Figure (8.26). Approximately, $T_1 \cos \theta_1 \approx T_0$ and $T_2 \cos \theta_2 \approx T_0$.

The equation of motion for the element is given by Newton's second law.

$$
\begin{aligned}
\delta m \frac{\partial^2 y}{\partial t^2} &= \sum F \\
&= T_2 \sin \theta_2 - T_1 \sin \theta_1 \\
&= T_2 \cos \theta_2 \tan \theta_2 - T_1 \cos \theta_1 \tan \theta_1 \\
&= T_0 \tan \theta_2 - T_0 \tan \theta_1 \\
&= T_0 \left\{ \left(\frac{\partial y}{\partial x} \right)_2 - \left(\frac{\partial y}{\partial x} \right)_1 \right\} \quad \text{where} \quad \tan \equiv \frac{\partial}{\partial x} \\
&= T_0 \frac{\partial}{\partial x} \{ y_2 - y_1 \} \quad \text{where} \quad y = f(x) \\
&= T_0 \frac{\partial}{\partial x} \{ f(x_2) - f(x_1) \} \tag{8.140}
\end{aligned}
$$

Using Taylor's expansion of a function $f(x)$ about x_1,

$$f(x) = f(x_1) + \frac{\partial y}{\partial x}(x - x_1) + \cdots$$

and hence for $x = x_2$,

$$f(x_2) = f(x_1) + \frac{\partial y}{\partial x}(x_2 - x_1) + \cdots$$

which gives

$$\begin{aligned} f(x_2) - f(x_1) &\approx \frac{\partial y}{\partial x}(x_2 - x_1) \\ &= \frac{\partial y}{\partial x}\delta x \end{aligned} \tag{8.141}$$

which when it is used in equation (8.140), the equation of motion becomes

$$\begin{aligned} \delta m\frac{\partial^2 y}{\partial t^2} &= T_0\frac{\partial}{\partial x}\left(\frac{\partial y}{\partial x}\right)\delta x \\ \delta m\frac{\partial^2 y}{\partial t^2} &= T_0\delta x\frac{\partial^2 y}{\partial x^2} \\ \frac{\partial^2 y}{\partial x^2} &= \left(\frac{\delta m}{\delta x}\right)\frac{1}{T_0}\frac{\partial^2 y}{\partial t^2} \\ &= \frac{\rho_l}{T_0}\frac{\partial^2 y}{\partial t^2} \tag{8.142} \\ \frac{\partial^2 y}{\partial x^2} &= \frac{1}{v^2}\frac{\partial^2 y}{\partial t^2} \tag{8.143} \end{aligned}$$

which is known as the *wave equation*, and comparing (8.142) and (8.143), we identify

$$v^2 = \frac{T_0}{\rho_l} \implies v = \sqrt{\frac{T_0}{\rho_l}} \tag{8.144}$$

where v is the velocity of transverse waves of a continuous string.

8.4.7 Sound Waves-Longitudinal Oscillations

Consider longitudinal oscillations of a sound waves propagating in a tube of cross-sectional area A, as illustrated in Figure 8.27(a), where there are compressions (C) and extensions (E). Consider an element in Figure 8.27(a) and let us study it in detail as magnified in Figure 8.27(b).

Let the element when in equilibrium have a length $\delta x = x_2 - x_1$, mass $\delta m = \rho_0 A\delta x$, and equilibrium density ρ_0.

When the system is extended to length $\delta x' = x_2 + u(x_2) - x_1 - u(x_1) = \delta x + u(x_2) - u(x_1)$, it has density ρ, as shown in Figure 8.27(b).

From $\delta x' = \delta x + u(x_2) - u(x_1)$, and using the Taylor's expansion of $u(x)$ about x_1,

$$u(x) = u(x_1) + \frac{\partial u}{\partial x}(x - x_1) + \cdots$$

and hence for $x = x_2$,

$$u(x_2) = u(x_1) + \frac{\partial u}{\partial x}(x_2 - x_1) + \cdots$$

which gives

$$
\begin{aligned}
u(x_2) - u(x_1) &= \frac{\partial u}{\partial x}(x_2 - x_1) \\
&= \frac{\partial u}{\partial x}\delta x
\end{aligned}
$$

Using *conservation of mass*

$$
\begin{aligned}
A\rho_0\delta x &= A\rho\delta x' \\
\rho_0\delta x &= \rho\left[\delta x + u(x_2) - u(x_1)\right] \\
\rho_0 &= \rho\left(1 + \frac{\partial u}{\partial x}\right) \quad \text{where} \quad \rho = \rho_0 + \rho_e \\
\rho_0 &= \rho_0 + \rho_e + (\rho_0 + \rho_e)\frac{\partial u}{\partial x} \\
\rho_e &= -\rho_0\frac{\partial u}{\partial x} - \rho_e\frac{\partial u}{\partial x} \quad \text{where the second term is very small, and} \\
\rho_e &\approx -\rho_0\frac{\partial u}{\partial x} \quad\quad\quad\quad\quad\quad\quad\quad\quad\quad\quad (8.145)
\end{aligned}
$$

The pressure changes and density changes are related, that is

$$
\begin{aligned}
P &= P(\rho) &&(8.146) \\
&= P(\rho_0 + \rho_e) \\
&= P(\rho_0) + \frac{\partial P}{\partial \rho}(\rho - \rho_0) \\
&= P(\rho_0) + P_e &&(8.147)
\end{aligned}
$$

where

$$P_e = \frac{\partial P}{\partial \rho}\rho_e = \alpha\rho_e, \tag{8.148}$$

and

$$\alpha = \frac{\partial P}{\partial \rho} \tag{8.149}$$

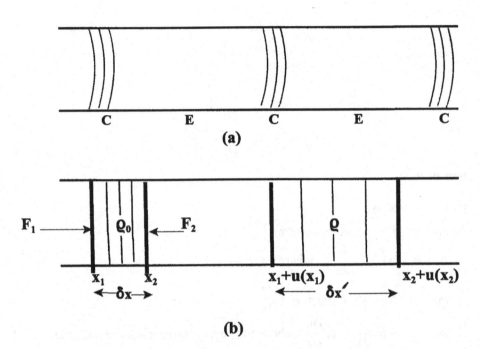

Figure 8.27: Sound waves propagating in a tube, with (a) showing compressions (C) and extensions (E), and (b) an element with length δx during equilibrium and length $\delta x'$ when extended.

The equation of motion for the element is given by Newton's second law.

$$
\begin{aligned}
\delta m \frac{\partial^2 u}{\partial t^2} &= \sum F \\
\rho_0 A \delta x \frac{\partial^2 u}{\partial t^2} &= (F_1 - F_2) \\
&= A\left[P(x_1) - P(x_2)\right]
\end{aligned}
\tag{8.150}
$$

Using Taylor's expansion of $P(x)$ about x_1,

$$
P(x) = P(x_1) + \frac{\partial P}{\partial x}(x - x_1) + \cdots
$$

and hence for $x = x_2$,

$$
P(x_2) = P(x_1) + \frac{\partial P}{\partial x}(x_2 - x_1) + \cdots
$$

which gives

$$P(x_1) - P(x_2) \approx -\frac{\partial P}{\partial x}(x_2 - x_1)$$

$$= -\frac{\partial P}{\partial x}\delta x$$

and using $P = P(\rho_0) + P_e$ from equation 8.148, assuming first term almost constant

$$= -\frac{\partial P_e}{\partial x}\delta x \tag{8.151}$$

which when used in equation (8.150), the equation of motion becomes

$$\rho_0 A \delta x \frac{\partial^2 u}{\partial t^2} = -A\frac{\partial P_e}{\partial x}\delta x$$

$$\rho_0 \frac{\partial^2 u}{\partial t^2} = -\frac{\partial P_e}{\partial x} \quad \text{where} \quad P_e = \alpha \rho_e$$

$$= -\alpha \frac{\partial \rho_e}{\partial x} \quad \text{where} \quad \alpha = \frac{\partial P}{\partial \rho}, \rho_e = -\rho_0 \frac{\partial u}{\partial x} \tag{8.152}$$

$$= \alpha \rho_0 \frac{\partial^2 u}{\partial x^2}$$

$$\frac{\partial^2 u}{\partial x^2} = \frac{1}{\alpha}\frac{\partial^2 u}{\partial t^2} \tag{8.153}$$

$$\frac{\partial^2 u}{\partial x^2} = \frac{1}{v_s^2}\frac{\partial^2 u}{\partial t^2} \tag{8.154}$$

which is known as the *wave equation* for longitudinal waves, and comparing (8.153) and (8.154), we identify

$$v_s^2 = \alpha = \frac{\partial P}{\partial \rho} \implies v_s = \sqrt{\frac{\partial P}{\partial \rho}} \tag{8.155}$$

where v_s is the velocity of sound waves.

Equation (8.155) is a general relation, and the actual velocity of sound is dependent on the material of the medium, whether it is a solid, liquid or gas.

(i) Sound waves in a solid

From

$$v_s = \sqrt{\frac{\partial P}{\partial \rho}}$$

and noting that for solids, the bulk modulus, B is defined as

$$B = -\frac{\partial P}{\partial V/V} \tag{8.156}$$

$$= -V\left(\frac{\partial P}{\partial \rho}\right)\frac{\partial \rho}{\partial V}$$

$$= -V\left(\frac{\partial P}{\partial \rho}\right)\left(-\frac{\rho}{V}\right)$$

$$= \rho\left(\frac{\partial P}{\partial \rho}\right) \tag{8.157}$$

and hence

$$v_s = \sqrt{\frac{\partial P}{\partial \rho}} \Longrightarrow v_s = \sqrt{\frac{B}{\rho}} \tag{8.158}$$

(ii) Sound waves in a liquid

As in the above derivation, the velocity of sound waves in a liquid is given by

$$v_s = \sqrt{\frac{\partial P}{\partial \rho}} \Longrightarrow v_s = \sqrt{\frac{B}{\rho}} \tag{8.159}$$

(iii) Sound waves in a gas

The velocity is given by

$$v_s = \sqrt{\frac{\partial P}{\partial \rho}} \Longrightarrow v_s = \sqrt{\frac{\gamma P}{\rho}} \tag{8.160}$$

where γ is the ratio of specific heats of gases, C_p/C_v, and the relation $PV^\gamma = constant$ for adiabatic processes in an gases has been used.

8.4.8   Electromagnetic Waves-Transverse Oscillations

According to electromagnetic theory, light is composed of the oscillating electric field $\mathbf{E}$ and the oscillating magnetic field $\mathbf{B}$. The electric field $\mathbf{E}$ and magnetic field $\mathbf{B}$ are perpendicular (*orthogonal*) to each other and are perpendicular (*transverse wave*) to the direction of propagation, as illustrated in Figure (8.28).

The electric field $\mathbf{E}$ and magnetic field $\mathbf{B}$ satisfy Maxwell's equations which are given below.

$$\nabla \cdot \mathbf{E} = \frac{\rho}{\epsilon_0} \tag{8.161}$$

$$\nabla \cdot \mathbf{B} = 0 \tag{8.162}$$

$$\nabla \wedge \mathbf{E} = -\frac{\partial \mathbf{B}}{\partial t} \tag{8.163}$$

$$\nabla \wedge \mathbf{B} = \mu_0 \mathbf{j} + \mu_0 \frac{\partial}{\partial t}[\epsilon_0 \epsilon(\omega)\mathbf{E}] \tag{8.164}$$

where

$$\rho \quad \text{is} \quad \text{the charge density,} \tag{8.165}$$

$$\epsilon_0 \quad = \quad 8.854 \times 10^{-9} Fm^{-1} \text{ is the permittivity of free space,} \tag{8.166}$$

$$\mu_0 \quad = \quad 4\pi \times 10^{-7} Hm^{-1} \text{ is the permeability of free space,} \tag{8.167}$$

$$\mathbf{j} \quad \text{is} \quad \text{the current density vector, and} \tag{8.168}$$

$$\epsilon(\omega) \quad \text{is} \quad \text{the dielectric function} \tag{8.169}$$

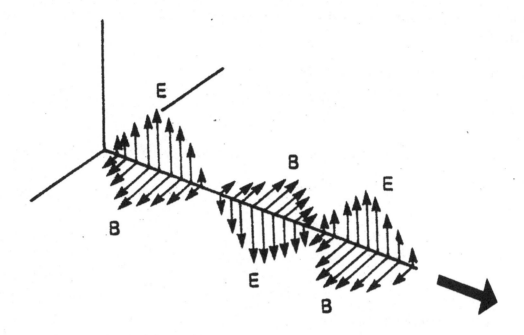

Figure 8.28: The electric field **E** and magnetic field **B** are perpendicular to each other and are perpendicular to the direction of propagation of the electromagnetic wave.

In free space, $\rho = 0$ and $j = 0$, $\epsilon(\omega) = 1$, Maxwell's equations reduce to

$$\nabla \cdot \mathbf{E} = 0 \tag{8.170}$$

$$\nabla \cdot \mathbf{B} = 0 \tag{8.171}$$

$$\nabla \wedge \mathbf{E} = -\frac{\partial \mathbf{B}}{\partial t} \tag{8.172}$$

$$\nabla \wedge \mathbf{B} = \mu_0 \epsilon_0 \frac{\partial}{\partial t} \mathbf{E} \tag{8.173}$$

From equations (8.172) and (8.173), we obtain

$$\nabla \wedge (\nabla \wedge \mathbf{E}) = -\mu_0 \epsilon_0 \frac{\partial^2}{\partial t^2} \mathbf{E} \tag{8.174}$$

Using the identity

$$\nabla \wedge (\nabla \wedge \mathbf{E}) = \nabla(\nabla . \mathbf{E}) - \nabla^2 \mathbf{E} \tag{8.175}$$

and using the definition of the velocity of light, c, given by

$$c = \frac{1}{\sqrt{\mu_0 \epsilon_0}}$$

leads to the wave equation

$$\nabla^2 \mathbf{E} = \frac{1}{c^2} \frac{\partial^2 \mathbf{E}}{\partial t^2} \tag{8.176}$$

where equation (8.170) has been used, and a similar equation for **B** can be obtained, and this is left as an exercise for the student.

The one dimensional version of equation (8.176), say along x-axis is

$$\frac{\partial^2 \mathbf{E}}{\partial x^2} = \frac{1}{c^2} \frac{\partial^2 \mathbf{E}}{\partial t^2}$$

which has a solution of a wave propagating along the x - axis given by any of the following forms:

$$E = E_0 e^{i(kx - \omega t)} \tag{8.177}$$

or

$$E = E_0 \sin(kx - \omega t) \tag{8.178}$$

or

$$E = E_0 \cos(kx - \omega t) \tag{8.179}$$

which is a plane wave of amplitude E_0 with wave vector $k(= 2\pi/\lambda)$ and angular frequency $\omega(= 2\pi\nu)$. Similar equations for the magnetic field vector **B** can be obtained, which is left as an exercise for the student.

8.5 Problems

8.1 When a spring with ends A and B is in a *vertical* direction such that end A is fixed and a mass of 0.05 kg is suspended at end B, the spring is stretched by 0.2 m. The same spring and mass are then placed on a smooth *horizontal* table with end A fixed and the mass at end B displaced by 0.2 m beyond the equilibrium position and then released.

(a)(i)calculate the force constant of the spring
(ii) if the horizontal direction is along the x-axis, show that

$$\frac{d^2x}{dt^2} + 49x = 0$$

(b)(i)find the position of the mass at any time t
(ii)hence find the values of the amplitude, angular frequency, frequency, and the period of the motion.

8.2 (a) Show that the equation of motion for the simple harmonic oscillator

$$m\frac{d^2u}{dt^2} + ku = 0$$

leads to the equation expressing energy conservation

$$\frac{1}{2}m\left(\frac{du}{dt}\right)^2 + \frac{1}{2}ku^2 = \text{Constant}$$

(b) A platform is executing simple harmonic motion in the vertical direction with an amplitude 5 cm and has an angular frequency of 20 vibrations per second. A block is placed on the platform when it is at its lowest position of its path.
(i) At what point above the equilibrium position will the block leave the platform?
(ii) How far will the block rise above the highest point reached by the platform? $[g = 9.8\text{ms}^{-2}]$

8.3 Two springs of force constants k_1 and k_2 are joined to a mass m either as in Figure 8.29(a) or Figure 8.29(b) below, and the system is on a smooth horizontal table.

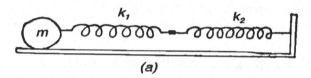

(a)

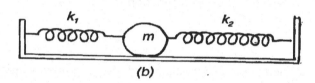

(b)

Figure 8.29: (a) and (b) Diagrams for problem 8. 3.

Show that the frequency of oscillation is given by:
(a)

$$\nu = \frac{1}{2\pi}\sqrt{\frac{k_1 k_2}{(k_1 + k_2)m}}$$

for figure (a) oscillations.
(b)

$$\nu = \frac{1}{2\pi}\sqrt{\frac{(k_1 + k_2)}{m}}$$

for figure (b) oscillations

8.4 (a) Show that a liquid covering a length l of a U-shaped tube exhibits simple harmonic oscillation when slightly displaced, and that the period of oscillation is given by

$$T = 2\pi\sqrt{\frac{l}{2g}}$$

(b) Nine kilograms of mercury are poured into a U-tube with a uniform bore of 1.2 cm diameter. The mercury oscillates freely up and down about its position of equilibrium. Ignoring surface tension effects, calculate
(i) the force constant responsible for oscillations, and
(ii) the period of oscillation
(Density of mercury = 13.6 gm.cm^{-3})

8.5 Time period of a simple pendulum is $2\,s$. How long will it take to move from the equilibrium position to the displacement equal to half the amplitude? What are the velocity, acceleration and force at this point?

8.6 The amplitude and frequency of a simple pendulum are $5\,cm$ and $2\,Hz$ respectively and at time $t = 0$ it starts oscillations from the equilibrium position. Determine :
(a) the displacement, velocity and acceleration of the pendulum at $1.25\,s$.
(b) velocity and acceleration of the pendulum at $2\,cm$, $3\,cm$ and $4\,cm$ displacements.

8.7 A $0.4\,kg$ mass attached to a spring executes SHM of amplitude $0.2\,m$, and has a maximum velocity of $5\,m\,s^{-1}$. Calculate the time period of the oscillator, and the force constant of the spring.

8.8 The displacement of an oscillatory motion as a function of time is given as:

$$x(t) = x_o \cos^2(\omega t + \phi)$$

Does this represent a simple harmonic motion? Explain and justify your conclusion.

8.9 The time period of a simple pendulum is $4.2\,s$. When its length is shortened by $1\,m$, its time period becomes $3.7\,s$. Calculate the acceleration due to gravity, and the original length of the pendulum.

8.10 The time period of a spring mass system, with a $0.2\,kg$ mass is $0.25\,s$, and the energy is $2\,J$. Determine the amplitude of oscillations.

8.11 A light cube of sides $0.1\,m$ and mass $200\,g$ floats upright in water. When the cube is depressed in water by $2\,cm$ from its equilibrium position, and let go, it executes simple harmonic motion. Setup the equation of motion for the cube in water, and determine the period of oscillations. (Density of water $= 1000\,kg\,m^{-3}$.

8.12 A $50\,g$ mass is attached to a spring of force constant $100\,N\,m^{-1}$. Another $100\,g$ mass is placed in contact with the first mass (not attached), and the two mass are displaced together on a smooth horizontal surface till the spring is compressed by $5\,cm$ and let go. Qualitatively describe the motion of the two masses, and determine the amplitude of the SHM of the spring-mass system.

8.13 (a) If $u_1 = A\cos\omega t$ and $u_2 = B\sin\omega t$ are solutions of a simple harmonic oscillator differential equation,
(i) show that $u_3 = A\cos\omega t + B\sin\omega t$ is also a solution.

(ii) Hence show that that u_3 can also be written in the form

$$u_3 = C \cos(\omega t - \phi)$$

where $C = \sqrt{A^2 + B^2}$ and $\phi = \tan^{-1}\left(\frac{B}{A}\right)$.

(b) Use a phase amplitude diagram to find the resultant amplitude and phase of the following sum of waves.

$$a \cos(kx - \omega t) + a \cos(kx - \omega t + \frac{\pi}{4}) + a \cos(kx - \omega t + \frac{\pi}{2})$$

8.14 An object of mass 0.2 kg is attached to a horizontal spring whose force constant is 80 N.m^{-1}. The object is subjected to a horizontal resistive force given by $-bv$ where b is the damping parameter and v is velocity in m.s^{-1}.
(a) Set up the differential equation of the system.
(b) If the frequency of the damped oscillations is $\frac{1}{2}\sqrt{3}$ of the resonant frequency, what is the value of b and its units?

8.15 After an electric current to be detected by a ballistic galvanometer has ceased, the angular displacement $\phi(t)$ of the galvanometer coil from its equilibrium position satisfies the following differential equation:

$$J\frac{d\omega}{dt} + B\omega + k\phi(t) = 0$$

where J is the moment of inertia of the coil about an axis through its centre, B is a parameter representing the resistive force, k is the force constant of the torsion suspension and $\omega = \frac{d\phi}{dt}$ is the angular velocity of the coil.
(a) Sketch a graph of $\phi(t)$ vs t as would be expected.
(b) Derive an expression for $\phi(t)$ and give the frequency of the underdamped oscillations.

8.16(a) (i) Show that the ratio of two successive maxima in the displacements of a damped harmonic oscillator is a constant.
(ii) Hence show that the ratio r of any maxima of damped oscillations to the next maxima is related to the damping constant Γ by

$$\Gamma = \frac{2 \ln r}{T}$$

where T is the period.
(b) An object of mass 0.4 kg is hooked to a spring of force constant 100 N.m^{-1} and is made to perform horizontal oscillations experiencing a resistive force of $-8v$ Newtons with v in m.s^{-1}. If one of the maxima has an amplitude of 0.5m, calculate the amplitude of the next two maxima.

8.17(a) After n cycles of oscillations, the amplitude of a damped harmonic oscillator drops to $1/e$ of its initial value. Show that the period T of this oscillation is related to the period T_0 of the same oscillator with no damping by

$$\frac{T}{T_0} = \left(1 + \frac{1}{4\pi^2 n^2}\right)^{\frac{1}{2}} \approx \left(1 + \frac{1}{8\pi^2 n^2}\right)$$

(b) For a vibrator with $m = 0.01$ kg and $k = 36$ N.m^{-1}, what value of b would make the amplitude decrease to $1/e$ in one second?

8.18 A vibrator with $m = 0.2$ kg, $k = 0.8$ N.m^{-1} and $b = 0.8$ kg.s^{-1} has an initial amplitude of 0.2m in the x-direction. (a) Write down the differential equation of the system.
(b) Show that
(i) the motion is critically damped
(ii) the solution

$$x = 0.2e^{-2t}(1 + 2t)$$

gives the position of the object at anytime t, and hence illustrate with a sketch a graph of x vs t.

8.19 For a damped spring-mass oscillator, $m = 0.2\,kg$, $k = 80\,N\,m^{-1}$, and $x_o = 2\,cm$. After 5 cycles of oscillations, the amplitude decreases to $\frac{1}{e}$ of its initial amplitude.
(a) Calculate the damping constant of the oscillator, and time period of oscillations.
(b) After how many oscillations, shall the energy have deceyed to $\frac{1}{e}$ of its maximum energy.
(c) What is the ratio of the amplitude after 2 and 4 complete oscillations to the maximum amplitude.
(c) What is the ratio of energy after 2 and 4 complete oscillations to the maximum energy.

8.20 In a LCR-circuit, $L = 0.308\mu\,H$, and $C = 10\,pF$. If the energy of the circuit decays to $\frac{1}{e}$ of its original energy in $1.5\,s$.
(a) Calculate the value of R, and the frequency of the oscillator.
(b) What is the frequancy of the broadcast signal which this LCR-circuit shall be able to recieve when included in a radio reciever set.

8.21 Prove that the rate of decay of energy of a damped oscillator is given by:

$$\frac{d\,E}{dt} = -b\,v^2$$

where the symbols have the usual meanings

8.22 A $30\,kg$ is suspended freely with $1.5m$ long light string, and is driven by a periodic force $3\cos(2\,t)$. What is the amplitude of oscillations?

8.23 A diatomic molecule can be modelled as in Figure (8.30) below, where the two masses m_1 and m_2 are coupled by a spring of force constant k. (a) Show that the natural angular frequency of oscillation of a diatomic molecule is given by

$$\omega = \sqrt{\frac{k}{\mu}}$$

when μ is the reduced mass of the system such that

$$\frac{1}{\mu} = \frac{1}{m_1} + \frac{1}{m_2}$$

(b) The natural angular frequency of a sodium chloride molecule is 1.14×10^{13} s^{-1}. Calculate the interatomic force constant if masses of the Na atom and Cl atom are given as 23 a.m.u. and 35 a.m.u. respectively.
(1 a.m.u. $= 1.67 \times 10^{-27}$ kg)

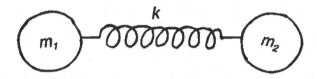

Figure 8.30: A diatomic molecule

8.24 A linear triatomic molecule may be considered as a system of three masses m_1, m_2 and $m_3 = m_1$ which are coupled by springs, each of force constant k as illustrated in Figure (8.31).

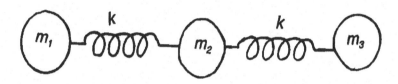

Figure 8.31: A triatomic molecule

Consider oscillations in which the masses are vibrating along the line joining their centres.
(a) Find the normal mode angular frequencies and illustrate with suitable diagrams the relative displacements of the masses for the normal modes.
(b) The CO_2 molecule can be likened to the model considered in this problem, such that $m_1 = m_3 = 16$ units and $m_2 = 12$ units. What is the ratio of the higher frequency to the lower frequency for CO_2 assuming this classical description were applicable?

8.25 A mass m_1 is connected to a fixed support by a spring of force constant k and slides on a horizontal plane without friction. Another mass m_2 is supported by a string of length l hangs as a simple pendulum from m_1 as illustrated in Figure (8.32).

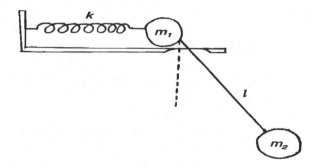

Figure 8.32: A coupled mass-spring and simple pendulum.

Assuming that $m_1 = m_2 = m$, show that the normal mode angular frequencies ω_1 and ω_2 satisfy

$$\omega_1^2 = (\omega_s^2 + \omega_g^2) + \sqrt{\omega_s^4 + \omega_g^4}$$

and

$$\omega_2^2 = (\omega_s^2 + \omega_g^2) - \sqrt{\omega_s^4 + \omega_g^4}$$

where $\omega_s^2 = \frac{k}{2m}$ and $\omega_g^2 = \frac{g}{l}$.

8.26 Two bodies, each of mass m are on a horizontal frictionless surface and are coupled by identical springs of force constant k such that the end of one spring is fixed as illustrated in Figure (8.33). Show that the two normal mode angular frequencies of oscillation are

$$\omega_1 = \frac{\sqrt{5}-1}{2}\sqrt{\frac{k}{m}}$$

and

$$\omega_2 = \frac{\sqrt{5}+1}{2}\sqrt{\frac{k}{m}}$$

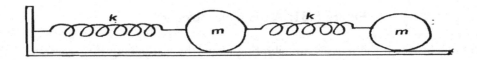

Figure 8.33: Coupled mass-spring system.

8.27 A double pendulum consists of two pendulums attached to each other by suspending a pendulum of mass m_2 and length l_2 to another of mass m_1 and length l_1 as illustrated in Figure (10.5). Show that the normal mode angular frequencies satisfy

$$\omega^4 - \left(\frac{m_1 + m_2}{m_1}\right)\left(\frac{g}{l_1} + \frac{g}{l_2}\right)\omega^2 + \left(\frac{m_1 + m_2}{m_1}\right)\frac{g^2}{l_1 l_2} = 0$$

8.28 If a function $\psi(x,t)$ given by

$$\psi(x,t) = Ae^{-(2x+2t)^2}$$

represents a progressive wave. Verify that the function satisfies the wave equation, and hence find the speed of the wave.

8.29 A linear chain of oscillators each of mass m with equilibrium separation a has a dispersion relation given by

$$\omega = 2\sqrt{\frac{T}{ma}}\sin\frac{qa}{2}$$

where T is the tension in the coupling springs.

(a) Plot the dispersion curve of the system.

(b) If Phase velocity is defined as $v_p = \frac{\omega}{q}$, derive an expression for the phase velocity in the continuous limit.

8.30 The velocity v_s of longitudinal vibrations in a medium of density ρ is given by

$$v_s = \sqrt{\frac{dP}{d\rho}}$$

where p is pressure

(a) Show that the velocity of sound in a solid is given by

$$v_s = \sqrt{\frac{B}{\rho}}$$

where B is the Bulk modulus.

(b) A longitudinal disturbance generated by an earthquake is observed to travel 5×10^3 km in quarter of an hour. Estimate the Bulk modulus of the rock through which the disturbance travels assuming that its average density is 2.7×10^3 kg.m^{-3}.

8.31 The velocity of longitudinal vibrations in a medium of density ρ is given by

$$v_s = \sqrt{\frac{dP}{d\rho}}$$

where P is pressure.

(a) Show that the velocity of sound in a gas is given by

$$v_s = \sqrt{\frac{\gamma P}{\rho}}$$

where $\gamma = \frac{C_p}{C_v}$ is the ratio of specific heats.

(b) Calculate the speed of sound in air at 0^0C and pressure of 760 mm of mercury. (Density of air at S.T.P.= 1.293×10^{-3}gcm^{-3}, Density of mercury = 13.6 gcm^{-3}, and $\gamma = 1.40$).

8.32 Suppose that a travelling wave pulse is described by the equation

$$y(x, t) = \frac{b^3}{b^2 + (x - vt)^2}$$

(a) If $b = 5$ cm, $v = 2.5$ cm.s^{-1}, sketch the profile of the pulse (y vs x) for $t = 0$.

(b) Show that the transverse velocity at $t = 0$ is given by

$$v_y(x, t) = \frac{2b^3 vx}{(b^2 + x^2)^2}$$

and sketch the velocity distribution (v_y vs x) using the values of b and v given in (a) above.

(c) Does the pulse move to the right or left? Explain.

Chapter 9

Rigid Body Dynamics

In preceding chapters on kinematics and rotational motion, we treated bodies as point masses, having only the mass but no dimensions, whereas bodies invariably have finite size. Representing a finite sized body as a point mass can be used only if the motion is purely translational. One only need to identify a point at which the entire mass of the body may be assumed to be concentrated, and the forces resulting in a translational motion are presumed to act at that point. However, the forces acting on the body may result not only in a translational motion but also in a rotational motion. From the rotational motion of a point mass we observed that the velocity, and acceleration of the mass and the torque acting on it, and consequently its momentum, angular momentum and kinetic energy depend on the distance of the mass from the axis of rotation. The point-mass approximation to rotational motion can only be applied to a mass whose size is very small compared to its distance from the axis of rotation, and to a finite number of discrete masses each one of which can individually be approximated as a point mass. If the axis of rotation either passes through a finite sized body, or if the axis is at a distance comparable to the dimensions of the body, the body can not be represented as a point mass. Such a body can be visualized as a collection of an infinite number of point masses located at different distances from the axis of rotation each of which has a different velocity, acceleration, momentum, torque, angular momentum, kinetic energy etc. It would be a daunting task for anyone to analyze the motion of a finite body by summing up the motion of an infinite number of point masses of which it comprises. The rotational motion of a finite sized body is dealt with by the *rigid body mechanics* which is the focus of this chapter. A rigid body is one in which the separation between its constituent elemental masses remains fixed. The basic approach involves defining certain physical quantities and properties of the rigid body, namely centre of mass, centre of gravity, moment of inertia about the axis of rotation and then analyze the rotational motion in terms of these quantities and properties. For the rotational motion of a point mass, the moment of inertia, rotational kinetic energy, and rotational (angular) momentum have been defined in terms of the mass and its distance from the axis of rotation in an earlier chapter. For finite bodies, while these definitions are essentially the same, they are expressed in terms of integral expressions. In this chapter we begin with the review of the rotational parameters for the rotational motion of a point mass, which lead to the definition of the same parameters for a (finite sized) rigid body. It is then followed by a number of examples of applications of rigid body dynamics. The chapter is concluded with further applications of rigid body mechanics to engineering problems where it finds its most applications.

9.1   Rotational parameters of a point mass

In this section we review the rotational parameters of a point mass presented in an earlier chapter to make an easy transition to the rigid body rotation. Furthermore, consequent to this review, this chapter can be studied independently without having to first cover the chapter on the rotational motion of a point mass. Consider a point mass m at a radial distance r from a fixed axis, rotating with an angular velocity ω as shown in Figure (9.1). Let $\mathbf{F}$ be the force acting on the mass whose tangential and radial components are F_t and F_r respectively. The tangential and radial components of acceleration $\mathbf{a}$ are a_t and a_r respectively.

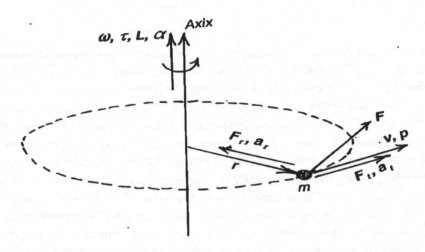

Figure 9.1: Rotational motion of a point mass about a fixed axis.

The linear velocity $\mathbf{v}$ of the mass in a pure rotational motion is tangential to the trajectory, referred to as the tangential velocity v_t, lies in the plane of rotation, and is given by:

$$\mathbf{v} = \omega \times \mathbf{r} \Rightarrow v_t = r\,\omega \tag{9.1}$$

From equation (9.1), we note that the angular velocity vector must lie along the axis of rotation Fas shown in igure (9.1).

The linear momentum $\mathbf{p}$ of the mass is:

$$\mathbf{p} = m\,\mathbf{v} = m\,\omega \times \mathbf{r} \tag{9.2}$$

The torque of the applied force is:

$$\tau = \mathbf{r} \times \mathbf{F} \Rightarrow r\,F_t \tag{9.3}$$

From equation (9.3), the direction of the torque is also along the axis of rotation and its magnitude is: $r\,F_t$

The components of acceleration and the force are related by Newton's second law of motion as:

$$F_t = m\,a_t \tag{9.4}$$
$$F_r = m\,a_r \tag{9.5}$$

The tangential acceleration causes the change in the magnitude of the velocity, and is the same as the acceleration for the translational motion along a straight line. The tangential acceleration is related to the angular acceleration α as:

$$\mathbf{a_t} = \alpha \times \mathbf{r} \Rightarrow r\,\alpha \tag{9.6}$$

and the direction of the angular acceleration vector is along the axis of rotation. The radial acceleration is responsible for the change in the direction of the velocity resulting in the rotational motion of the mass. In fact, a_r is the centripetal acceleration, and the radial force is the centripetal force, *i.e,*

$$F_r = \frac{m\,v^2}{r} = m\,r\,\omega^2 = m\,a_r \tag{9.7}$$

From equations (9.3), (9.4) and (9.6) we can express the torque in terms of the angular acceleration as:

$$\tau = r\,F_t = r\,m\,a_t = (m\,r^2)\,\alpha \tag{9.8}$$

The angular momentum $\mathbf{L}$ is:

$$\mathbf{L} = \mathbf{r} \times \mathbf{p} = m\,\mathbf{r} \times \mathbf{v} \Rightarrow m\,r\,v_t = (m\,r^2)\,\omega \tag{9.9}$$

The angular momentum lies along the axis of rotation, (also see section 9.4.6), and its magnitude is $(m\,r^2)\,\omega$.

The (rotational) kinetic energy KE of the mass is:

$$KE = \frac{1}{2}\,m\,v^2 = \frac{1}{2}\,m\,(r\,\omega)^2 = \frac{1}{2}\,(m\,r^2)\,\omega^2 \tag{9.10}$$

From the above expressions we note two important points:

(i) Expressions for the torque (equation 9.8), the angular momentum (equation 9.9), and the kinetic energy (equation 9.10) involve a term $m\,r^2$ which is a constant of motion for the point mass at a fixed distance from the axis of rotation. This is known as the moment of inertia (MI), or the rotational inertia I, of the mass about the axis of rotation. Its SI unit is $kg\,m^2$.

$$MI \rightarrow I = m\,r^2 \tag{9.11}$$

For an assembly of n point masses, m_1, m_2, m_3, ..., m_i... located at normal distances r_1, r_2, r_3, ..., r_i... respectively from the axis of rotation, the moment of inertia is:

$$I = \sum_i m_i\, r_i^2 \tag{9.12}$$

(ii) Vector quantities ω, α, τ, and $\mathbf{L}$ lie along the same direction, the axis of rotation. In terms of the moment of inertia $\mathbf{L}$, τ and KE can be expressed as:

$$\mathbf{L} \;=\; I\,\omega \tag{9.13}$$

$$\tau \;=\; I\,\alpha = I\,\frac{d}{dt}\omega = \frac{d}{dt}(I\,\omega) = \frac{d}{dt}\mathbf{L} \tag{9.14}$$

$$KE \;=\; \frac{1}{2}I\,\omega^2 \tag{9.15}$$

Equation (9.14) for rotational motion is analogous to the Newton's second law for translational motion. The Newton's second law for rotational motion can be stated as: *When an external torque τ is applied to a body free to rotate about a fixed axis, it undergoes angular acceleration α parallel to the direction of the torque. The torque and the angular acceleration are related as $\tau = I\,\alpha$, where I is the moment of inertia of the body about the axis of rotation.* The law of conservation of angular momentum, and the second condition of equilibrium also known as the condition of rotational equilibrium follow from equation (9.14) which are discussed in detail in a later section.

We note that the moment of inertia $I = m\,r^2$ plays the same role in rotational motion as does the mass in translational motion. The MI of the mass presents resistance to change in rotational motion in the same way as does the mass for the translational motion. Its value depends on the mass and the location of the axis of rotation. Larger is the distance of the mass from the axis of rotation, larger is its MI. We can state the equivalent of the Newton's first law of motion for the rotational motion as: *A body free to rotate about an axis of rotation, and in a state of rest or uniform rotational motion shall continue in that state unless changed by an external torque.*

Example 9.1: Four point masses $m_1 = 2\,kg$, $m_2 = 4\,kg$, $m_3 = 6\,kg$, and $m_4 = 8\,kg$ are located in the $xy-$plane at points $(0, 0)$, $(3, 0)$, $(0, 4)$, and $(3, 4)$ respectively, where the distances are in m. Determine the moment of inertia of the assembly of masses about: *(i)* the $x-$axis, and *(ii)* the axis passing through masses m_1 and m_4.

Solution Figure (9.2) shows the location of the masses m_1, m_2, m_3, and m_4 in the $xy-$plane. Let d_1, d_2, d_3, and d_4 respectively be the perpendicular distances of the masses from the axis of rotation.

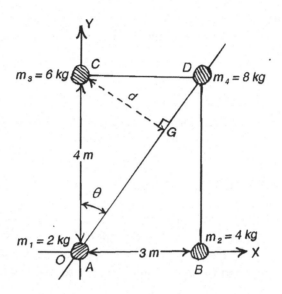

Figure 9.2: Moment of inertia of four point masses in the $xy-$plane.

(i) For this case $d_1 = 0$, $d_2 = 0$, $d_3 = 4\,m$, and $d_4 = 4\,m$. Hence the MI of the assembly of masses is:

$$I = 6 \times 4^2 + 8 \times 4^2 = 224\,kg\,m^2$$

(ii) For this case $d_1 = 0$, and $d_4 = 0$.

$$\theta = tan^{-1}\left(\frac{3}{4}\right) = 37^o$$
$$CG = d_3 = CA\,sin\,\theta = 4 \times sin\,37 = 2.4\,m = d_2$$

Hence the MI of the assembly of masses is:

$$I = 4 \times 2.4^2 + 6 \times 2.4^2 = 57.6\,kg\,m^2$$

Exercise 9.1: Four point masses $m_1 = 2\,kg$, $m_2 = 4\,kg$, $m_3 = 6\,kg$, and $m_4 = 8\,kg$ are located in the $xy-$plane at points $(0, 0)$, $(3, 0)$, $(0, 4)$, and $(3, 4)$ respectively, where the distances are in m. Determine the moment of inertia of the assembly of masses about: *(i)* the $y-$axis, *(ii)* the axis passing through masses m_2 and m_3, *(iii)* axis passing through the mid points of the lines joining masses $m_1\,m_2$ and $m_3\,m_4$ in the $xy-$plane, *(iv)* axis passing through the mid point of the line joining masses $m_1\,m_2$ and perpendicular to the $xy-$plane, *(v)* axis passing through the mid point of the line joining masses $m_1\,m_3$ and perpendicular to the $xy-$plane, and *(vi)* the $z-$axis.

Work and power: Let the point mass under goes an angular displacement $\Delta\theta$ in time interval Δt. The corresponding linear displacement is $\Delta d = r\,\Delta\theta$, and the work done is:

$$\Delta W = F_t \Delta d = (F_t \, r)\, \Delta\theta = \tau \, \Delta\theta \tag{9.16}$$

and $W = \tau\theta \Rightarrow \tau.\theta$ if the torque is constant during the angular displacement θ otherwise one must integrate equation (9.16) to determine the work done for a given angular displacement. The power by definition is the rate of work done given as:

$$P = \frac{\Delta W}{\Delta t} = \frac{\tau \, \Delta\theta}{\Delta t} = \tau\,\omega \Rightarrow \tau.\omega \tag{9.17}$$

Compare these expressions to the corresponding expressions for work and power in translational motion which are given as: $W = \mathbf{F}.\mathbf{d}$ and $P = \mathbf{F}.\mathbf{v}$ respectively.

9.2   Centre of mass and centre of gravity

For an assembly of discrete point masses or a finite sized rigid body, (system) there are two points, the centre of mass and the centre of gravity which play an important role in expressing the equilibrium and the motion of the system under applied external forces. Very often, the centre of mass and the centre of gravity of a system are assumed to be the same, an approximation which is valid only under certain assumptions that are examined in this section.

9.2.1   Centre of mass

Definition 9.1: Centre of Mass *of an assembly of point masses or a rigid body is a point at which the entire mass of the system is taken to be concentrated as a point mass, and the external forces on the system act as if they were applied to the point mass placed at the centre of mass.*

Centre of mass of an assembly of point masses:

Consider a system of point masses $m_1, m_2, m_3, ..., m_i...$ located at $\mathbf{r_1}, \mathbf{r_2}, \mathbf{r_3}, ..., (\mathbf{r_i} = (x_i, \, y_i, \, z_i)), ...$ from the origin of the Cartesian coordinates axes, (Figure 9.3 a).
The radius vector of the centre of mass, $\mathbf{r} = (x, \, y \, z)$, of the system is given by:

$$\mathbf{r} = \frac{\sum_i m_i \, \mathbf{r_i}}{\sum_i m_i} = \frac{\sum_i m_i \, \mathbf{r_i}}{M} = \frac{1}{M} \sum_i m_i \, \mathbf{r_i} \tag{9.18}$$

where M is the total mass of the system. The Cartesian coordinates of the centre of mass are given by:

$$
\begin{aligned}
x &= \frac{\sum_i m_i \, x_i}{\sum_i m_i} = \frac{1}{M} \sum_i m_i \, x_i \\
y &= \frac{\sum_i m_i \, y_i}{\sum_i m_i} = \frac{1}{M} \sum_i m_i \, y_i \\
z &= \frac{\sum_i m_i \, z_i}{\sum_i m_i} = \frac{1}{M} \sum_i m_i \, z_i
\end{aligned}
\tag{9.19}
$$

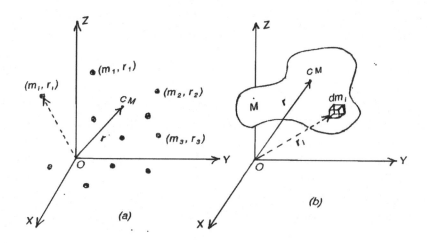

Figure 9.3: Centre of mass of an assembly of (a) point masses, and (b) a rigid body.

Centre of mass of a rigid body:

Consider an elemental mass $dm_i\,(\rightarrow 0)$ located at $\mathbf{r_i} = (x_i,\, y_i,\, z_i)$ as illustrated in Figure 9.3b . The centre of mass $\mathbf{r} = (x,\, y,\, z)$ of the body is obtained by integration over the entire mass, and is given by the following expressions:

$$
\begin{aligned}
\mathbf{r} &= \frac{\int_M \mathbf{r_i}\, dm_i}{\int_M dm_i} = \frac{1}{M}\int_M \mathbf{r_i}\, dm_i \\[2mm]
x &= \frac{\int_M x_i\, dm_i}{\int_M dm_i} = \frac{1}{M}\int_M x_i\, dm_i \\[2mm]
y &= \frac{\int_M y_i\, dm_i}{\int_M dm_i} = \frac{1}{M}\int_M y_i\, dm_i \\[2mm]
z &= \frac{\int_M z_i\, dm_i}{\int_M dm_i} = \frac{1}{M}\int_M z_i\, dm_i
\end{aligned}
$$

$$(9.20)$$

For a symmetric and homogeneous body the centre of mass is the geometrical centre of the body, for example the centre of mass of a uniform, homogeneous sphere is the centre of the sphere. The centre of mass of a body may or may not lie physically within the mass of the body. For example, the centre of mass of a uniform spherical shell is the centre of the shell, which is a point located in space and not within the shell-material. Likewise, the centre of mass of an assembly of point masses lies at a point in space within the assembly and not necessarily at any one of the constituent masses.

9.2.2 Centre of gravity

Definition 9.2: The Centre of gravity *of an assembly of point masses or a rigid body is the point at which if the entire weight of the system is assumed to be placed as a point weight. The gravitational force at the point weight produces the same effect as the total gravitational force acting on the system.*

Two forces are equivalent if they have *(i)* the same resultant and *(ii)* the same net torque about some fixed point in space. Consider a system of point masses $m_1, m_2, m_3, ..., m_i...$ located at $\mathbf{r_1}, \mathbf{r_2}, \mathbf{r_3}, ..., (\mathbf{r_i} = (x_i, y_i, z_i))...$ from a fixed point in space, which in our case is the origin of the Cartesian coordinates axes, (Figure 9.4a). $\mathbf{g_1}, \mathbf{g_2}, \mathbf{g_3}, ..., \mathbf{g_i}, ...$ are the gravitational field strengths (acceleration due to gravity) at the locations of the point masses. Let $\mathbf{r} = (x, y, z)$ be the radius vector of the centre of gravity, and $\mathbf{g}$ the acceleration due to gravity at this location. Then by definition the gravitational force and the gravitational torque at the centre of gravity must be the same as the net force and torque at the assembly of masses. Thus the location of the centre of gravity must satisfy the following conditions:

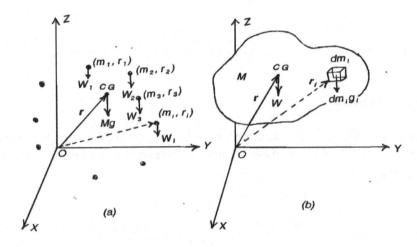

Figure 9.4: Centre of gravity of an assembly of (a) point masses, and (b) a rigid body.

$$\left(\sum_i m_i\right)\mathbf{g} = \sum_i (m_i\,\mathbf{g_i}) = M\,\mathbf{g}$$

$$\text{and} \quad \tau_\mathbf{G} = \mathbf{r} \times \left(\sum_i m_i\right)\mathbf{g} = \sum_i (\mathbf{r_i} \times m_i\,\mathbf{g_i}) = \mathbf{r} \times (M\,\mathbf{g}) \qquad (9.21)$$

Confining ourselves to the centre of gravity of masses in the gravitational field of the earth, the force of gravity on a mass acts towards the centre of the earth. Therefore, the gravitational forces on individual masses are not strictly parallel, but converge to the centre of the earth. Furthermore, the acceleration due to gravity varies with location. Therefore, locating the precise centre of gravity of a system of masses using equations (9.21) is a complex problem to solve. The problem can, however, be simplified under two justifiable assumptions:

(i) The separation between the masses are small compared to the radius of the earth, so that the gravitational forces acting on the individual masses are taken to be parallel to each other. Without going into the mathematical details, the centre of gravity $\mathbf{r} = (x,\, y\, z)$ of the point masses under this assumption is given by:

$$\mathbf{r} = \frac{\sum_i m_i\, g_i\, \mathbf{r_i}}{M\, g} = \frac{1}{M\, g} \sum_i m_i\, g_i\, \mathbf{r_i}$$

$$x = \frac{1}{M\, g} \sum_i m_i\, g_i\, x_i$$

$$y = \frac{1}{M\, g} \sum_i m_i\, g_i\, y_i$$

$$z = \frac{1}{M\, g} \sum_i m_i\, g_i\, z_i \tag{9.22}$$

Likewise, the centre of gravity of a body of finite size (Figure 9.4 (b)) is given by:

$$\mathbf{r} = \frac{\int_M g_i\, \mathbf{r_i}\, dm_i}{\int_M g_i\, dm_i} = \frac{1}{M\, g} \int_M g_i\, \mathbf{r_i}\, dm_i$$

$$x = \frac{\int_M g_i\, x_i\, dm_i}{\int_M g_i\, dm_i} = \frac{1}{M\, g} \int_M g_i\, x_i\, dm_i$$

$$y = \frac{\int_M g_i\, y_i\, dm_i}{\int_M g_i\, dm_i} = \frac{1}{M\, g} \int_M g_i\, y_i\, dm_i$$

$$z = \frac{\int_M g_i\, z_i\, dm_i}{\int_M g_i\, dm_i} = \frac{1}{M\, g} \int_M g_i\, z_i\, dm_i$$

$$\tag{9.23}$$

(ii) The value of the acceleration due to gravity is the same over the entire body, *i.e.*, $g_1 = g_2 = g_3 = ... = g$. By applying this condition to equations (9.22) and (9.23), we observe that the centre of gravity coincides with the centre of mass. Thus the approximation that the centre of mass and the centre of gravity are the same point applies only for a rigid body whose size, and the distances involved are small compared to the radius of the earth, and the acceleration due to gravity over the entire body is taken to be constant. These conditions are satisfied in the case of all terrestrial objects, and we can take the centre of mass and the centre of gravity of such bodies to be coincident with negligibly small error. However, for bodies separated by large distances comparable to the radius of the earth, for example the extraterrestrial bodies in our solar system, the centre of mass and the centre of gravity are two separate points. In this chapter, with the exception of some simple exercises and examples for the purpose of illustration, we shall regard the centre of mass and the centre of gravity to be coincident.

Experimental determination of the centre of gravity of a rigid body:

Suspend the body freely from a point. In equilibrium a vertical line from the pivot from which the body is suspended passes through the centre of gravity of the body, (explain why). Mark the vertical straight line through the body in this position. Now suspend the body freely from another point, and again mark the vertical line through the body in equilibrium. The intersection of the lines gives the centre of gravity of the body. If the body is suspended freely from this point, it shall always be in equilibrium no matter in what orientation the body is rotated, (why?).

9.2.3   Centre of mass motion

The centre of mass plays an important role in the dynamics of a rigid body, or an assembly of point masses, because most dynamical properties of the system can be represented in terms of the dynamics of the centre of mass. Consider a system of n point masses. Then from the definition of the centre of mass (equation 9.18):

$$M\,\mathbf{r} = m_1\,\mathbf{r_1} + m_2\,\mathbf{r_2} + m_3\,\mathbf{r_3} + ... + m_i\,\mathbf{r_i} + ... \tag{9.24}$$

Taking the time derivative of equation (9.24), we get:

$$M\,\mathbf{v} = m_1\,\mathbf{v_1} + m_2\,\mathbf{v_2} + m_3\,\mathbf{v_3} + ... + m_i\,\mathbf{v_i} + ... \tag{9.25}$$

where $\mathbf{v}$ is the velocity of the centre of mass, and $\mathbf{v_1}, \mathbf{v_2}, \mathbf{v_3}, ..., \mathbf{v_i}, ...$ are the velocities of the individual masses. In terms of the momentum, equation (9.25) becomes:

$$\mathbf{P_{cm}} = \mathbf{p_1} + \mathbf{p_2} + \mathbf{p_3} + ... + \mathbf{p_i} + ... \tag{9.26}$$

Thus the momentum of the centre of mass $\mathbf{P_{cm}}$ is the same as the total momentum of the individual masses. Taking further time derivative of equation (9.25):

$$M\,\mathbf{a} = m_1\,\mathbf{a_1} + m_2\,\mathbf{a_2} + m_3\,\mathbf{a_3} + ... + m_i\,\mathbf{a_i} + ... \tag{9.27}$$

where $\mathbf{a_{cm}}$ is the acceleration of the centre of mass, and $\mathbf{a_1}, \mathbf{a_2}, \mathbf{a_3}, ..., \mathbf{a_i}, ...$ are the accelerations of the individual masses. In terms of the forces, equation (9.27) yields:

$$\mathbf{F_{cm}} = \mathbf{F_1} + \mathbf{F_2} + \mathbf{F_3} + ... + \mathbf{F_i} + ... \tag{9.28}$$

The net external force $\mathbf{F_{cm}}$ on the centre of mass is the resultant of the external forces acting on the individual masses.

The motion of the centre of mass of a rigid body can now be described as follows:

The centre of mass of a system of particles or of a rigid body moves in such a way as if all the mass of the system were located at the centre of mass carrying the entire momentum of the system, and resultant of all the external forces was applied at the centre of mass.

Note that this principle applies to motion under gravitational field only if the centre of mass and the centre of gravity are taken to be coincident.

9.3 Rotational parameters of a rigid body

We now extend the definitions of rotational parameters for a point mass reviewed in section (9.1), and determine corresponding parameters for the rotational motion of a rigid body. Consider a rigid body of mass M free to rotate about a fixed axis. Let ω be the angular velocity and α be the angular acceleration at a given instant. Consider an elemental mass dm of the body at radius vector (normal distance) $\mathbf{r}$ from the axis of rotation (Figure 9.5)

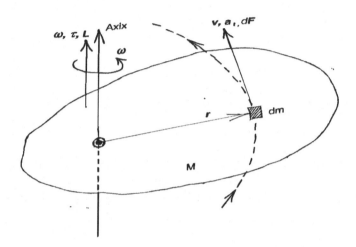

Figure 9.5: Rotational motion of a rigid body.

9.3.1 Angular momentum

The linear momentum of the elemental mass is: $d\mathbf{p} = \mathbf{v}\,dm$, from which the angular momentum $d\mathbf{L}$ of the elemental mass is obtained as:

$$d\mathbf{L} = \mathbf{r} \times d\mathbf{p} = (\mathbf{r} \times \mathbf{v})\,dm = \mathbf{r} \times (\omega \times \mathbf{r})\,dm \tag{9.29}$$

From the properties of triple vector product: $\mathbf{r} \times (\omega \times \mathbf{r}) = (\mathbf{r}.\mathbf{r})\,\omega - (\mathbf{r}.\omega)\,\mathbf{r} = r^2\,\omega$, $((\mathbf{r}.\omega)\,\mathbf{r} = 0$, because r and ω are perpendicular to each other), and equation (9.29) reduces to:

$$d\mathbf{L} = r^2\,\omega\,dm \tag{9.30}$$

The total angular momentum of the rigid body is obtained by integrating equation (9.30) over the entire mass of the body.

$$\mathbf{L} = \int_M d\mathbf{L} = \int_M (r^2\,dm)\,\omega = I\,\omega \tag{9.31}$$

where

$$I = \int_M (r^2\,dm) \tag{9.32}$$

is the moment of inertia of the rigid body about the axis of rotation. Compare this expression to the expression for the moment of inertia of discrete point masses (equation 9.12), and note how the summation changes to integration when one makes a transition from a discrete system to a continuous system which is a standard mathematical technique. In fact the integrand $r^2\, dm$ by definition is the moment of inertia of the elemental mass dm and its integral over the entire mass leads to the moment of inertia of the rigid body.

The angular momentum lies along the direction of the angular velocity ω which is the same as the direction of the axis of rotation, (also see section 9.4.6).

9.3.2   Torque

Let $\mathbf{a_t}$ be the tangential acceleration of the elemental mass dm. Then the tangential force on it is $d\mathbf{F_t} = dm\, \mathbf{a_t}$, and the torque on the elemental mass is:

$$d\tau = \mathbf{r} \times d\mathbf{F_t} = \mathbf{r} \times \mathbf{a_t}\, dm = \mathbf{r} \times (\alpha \times \mathbf{r})\, dm = r^2\, dm\, \alpha \qquad (9.33)$$

where α is the angular acceleration, and we have again used the property of the vector triple product, (prove equation 9.33!). The total torque on the body is obtained by integrating equation (9.33) over the entire mass:

$$\tau = \int_M d\tau = \int_M (r^2\, dm)\, \alpha = I\, \alpha \qquad (9.34)$$

The direction of the torque is also along the axis of rotation, the same as the direction of the angular acceleration. Note that the radial force is along the radius vector and its torque about the axis of rotation is zero, $\mathbf{r} \times \mathbf{F_r} = 0$

9.3.3   Kinetic energy

The (rotational) kinetic energy $d\,(KE)$ of the elemental mass is:

$$d\,(KE) = \frac{1}{2}\, dm\, v^2 = \frac{1}{2}\, dm\, (r\,\omega)^2 = \frac{1}{2}\, (r^2\, dm)\, \omega^2 = \frac{1}{2} dI \omega^2 \qquad (9.35)$$

and the total rotational kinetic energy of the body is:

$$KE = \int_M d\,(KE) = \frac{1}{2} \int_M (r^2\, dm)\, \omega^2 = \frac{1}{2}\omega^2 \int_M dI = \frac{1}{2}\, I\, \omega^2 \qquad (9.36)$$

From equations (9.31), (9.34) and (9.36) we note that the expressions for the angular momentum, the torque and the kinetic energy for a rigid body are identical to the corresponding expressions for an assembly of point masses (equations 9.13, 9.14 and 9.15), and involve the moment of inertia of the system about the axis of rotation. The calculation of the moment of inertia of rigid bodies, its properties and the theorems governing it are treated in a separate section.

9.3.4 Work and power

Let a torque τ be applied to a rigid body that is free to rotate about an axis such that it under goes an angular displacement $\Delta\theta$ in a time interval Δt. Let r be the distance of the point on the rigid body from the axis of rotation to which the force is applied. The linear displacement of the point is $\Delta d = r\,\Delta\theta$, and let F_t be the tangential component of the applied force. The work done in interval Δt is:

$$\Delta W = F_t\,\Delta d = (F_t\,r)\,\Delta\theta = \tau\,\Delta\theta \Rightarrow \tau.\Delta\theta \tag{9.37}$$

The total work done is obtained by integrating over the entire angular displacement as:

$$W = \int_\theta dW = \int_\theta \tau.d\theta \tag{9.38}$$

If the torque is constant, equation (9.38) gives $W = \tau.\theta$.

The power by definition is the rate of work done given as:

$$P = \frac{\Delta W}{\Delta t} = \frac{\tau\,\Delta\theta}{\Delta t} = \tau\,\omega \Rightarrow \tau.\omega \tag{9.39}$$

Compare these expressions to the corresponding expressions for work and power for the rotational motion of a point mass, (equations 9.16 and 9.17).

9.3.5 Angular momentum: Related laws

Recall equation (9.34). It can be expressed in terms of angular momentum as:

$$\tau = I\,\alpha = I\,\frac{d}{dt}\omega = \frac{d}{dt}(I\,\omega) = \frac{d}{dt}\mathbf{L} \tag{9.40}$$

This is an important equation for the analysis of the rotational motion, and interpretation of the equation leads to a number of interesting laws, conclusions and applications as discussed below.

(i) Conservation of Angular Momentum: If the total torque on the body is zero, $(\tau = 0)$, then the rate of change of angular momentum is also zero, *i.e.*

$$\frac{d}{dt}\mathbf{L} = 0 \tag{9.41}$$

This implies that the angular momentum of the system is constant, which is the law of conservation of angular momentum stated as follows:

Law of Conservation of angular momentum: *If the net external torque on a rotating body about the axis of rotation is zero, its angular momentum about the axis of rotation is conserved.*

Consider a rigid body (or an assembly of point masses) rotating about a fixed axis of rotation, for which by some internal means the distribution of mass about the axis of rotation can be changed without applying external torque. The redistribution of mass results in the change in the moment of inertia of the body about the axis of rotation. Let I_1 and I_2 be the MI of the body at two instants, and let ω_1 and ω_2 be the corresponding angular velocities. Since no external torque is applied, the angular momentum is conserved, *i.e.*,

$$L_1 = I_1 \omega_1 = I_2 \omega_2 = L_2 \tag{9.42}$$

In equation (9.42), if $I_1 > I_2$ then $\omega_2 > \omega_1$. A typical class-room demonstration of this principle involves a student seated on an almost frictionless turn-table holding heavy dumb-bells in both hands. Another student sets the student on the turn-table in rotation by giving it a push. While rotating the student on the turn-table alternately stretches out, and pulls-in his arms with dumb-bells in his hands. When the arms are stretched out, his moment of inertia is large (explain why) and he rotates at a slower rate, and when he pulls his arms close to his body, the moment of inertia decreases, and his speed of rotation increases.

This simple demonstration of the principle finds practical application in gymnastics, diving and circus events. Consider a diver from a high board in the swimming pool below. The diver starts the dive with a push to start the spinning of his body, pulls his body close together so that his moment of inertia is small, and he rotates at a faster rate. In this way he can complete many rotations (somersaults) in the air before hitting the water. When the swimmer is about to hit the water, he stretchable out his body completely, his moment of inertia increases, the rate of rotation decreases, and he enters the water at a safe low speed. In the same way one can explain the rotation of a skater with arms stretched out and pulled close to the body, somersault by gymnasts in sports events and in the circus while performing from the high bars.

(ii) Second condition of (rotational) equilibrium: In an earlier chapter we discussed the first condition of equilibrium, also known as the condition of translational equilibrium which basically is the consequence of the law of conservation of linear momentum. Likewise, the conservation of angular momentum leads to the second condition of equilibrium, or the condition of rotational equilibrium. Consider a body rotating about a fixed axis of rotation, and the net torque on the body is zero. Then the angular momentum of the body is conserved, which implies that the body continues to rotate with a constant angular velocity, *i.e.*, it is in a state of rotational equilibrium. Alternatively, the body could also be at rest, and in the absence of external torque it shall continue to be in static equilibrium. From these observations one can state the law of rotational equilibrium as:

Second Condition (Law) of (Rotational) equilibrium: *If the net torque on a body free to rotate about an axis is zero, the body remains in state of rotational (dynamic or static) equilibrium.*

The net torque on the system is determined using the sign convention: *A clockwise torque is positive and an anticlockwise torque is negative, i.e.,* a torque that tends to rotate the body in clockwise direction is positive, and a torque that tend to cause an anticlockwise rotation is negative. The net torque on the system is the algebraic sum of the torque of all the forces, which is zero if the system is in rotational equilibrium. The second condition of equilibrium can also be stated in terms of clockwise and anticlockwise torques as: *For a system in rotational equilibrium, total clockwise torque is equal to the total anticlockwise torque.*

(iii) Impulsive torque: Consider a torque τ applied to a rotating body for a small interval of time Δt, and it produces a change of $\Delta \mathbf{L}$ in the angular momentum. From equation (9.40) these are related as:

$$\tau = \frac{\Delta \mathbf{L}}{\Delta t}, \quad \text{or} \quad \tau \Delta t = \Delta \mathbf{L} \tag{9.43}$$

where $\tau \Delta t$ is known as the impulsive torque, its SI unit is $N\,m\,s$, and it is equal to the change in the angular momentum of the body. Thus when an impulsive torque is applied to a rotating body, the angular momentum of the body changes by equal to the applied impulse. Recall the impulsive force and its relation to the change in (linear) momentum discussed in an earlier chapter, and compare the two results.

9.4 Moment of Inertia (MI)

In sections (9.1) and (9.3.1). we defined moment of inertia by the following equations:

MI of an assembly of point masses: $\quad I = \sum_i m_i r_i^2 \, ,$

and MI of a rigid body of mass M: $\quad I = \int_M (r^2 \, dm)$

From these expressions alone, without even going any further, we can infer the following key properties of the moment of inertia.

- The MI depends on the mass of the system, and on the location of the axis of rotation. For a given system and the axis of rotation it is a constant of (rotational) motion.

- For systems of the same mass, MI is large if the mass is distributed at a large distance from the axis of rotation and it is small if the mass is distributed closer to the axis of rotation. For example MI of a long cylinder about the axis of the cylinder is smaller than its MI if the axis of rotation is perpendicular to the cylindrical axis. Likewise, if we consider a number of axes perpendicular to the cylindrical axis, the MI about the axis through the mid point of the cylinder shall be the least as compared to any other perpendicular axis along the length of the cylinder. Between a solid sphere and a spherical shell of the same mass M and same radius R which one has a larger moment of inertia about an axis through the centre of the sphere/ shell? (Explain why)

Within the scope of our discussion of rotational motion so far, MI can justifiably be termed as a scalar quantity. The moment of inertia as defined by the above equations is adequate for most applications. That is how it is also presented in most of the undergraduate books on mechanics. However, from advanced treatment of rotational motion it turns out that the moment of inertia in fact are the diagonal elements of the inertia tensor of second order (rank) with 9 components expressed as a 3×3 array of elements. Nevertheless, each of the component can still be regarded as a scalar quantity as it lacks the *'directional feature'* of a vector. We shall take an advanced insight into the angular momentum, and the moment of inertia tensor in a later subsection.

9.4.1 Radius of gyration

Consider a mass M with its moment of inertia I about an axis of rotation. One can locate a point at a perpendicular distance k from the axis of rotation at which if the entire mass of the body is placed as a point mass, then the MI of the point mass so placed is the same as that of the original body, (Figure 9.6), *i.e.,*.

$$I = M k^2 \Rightarrow k = \sqrt{\frac{I}{M}} \tag{9.44}$$

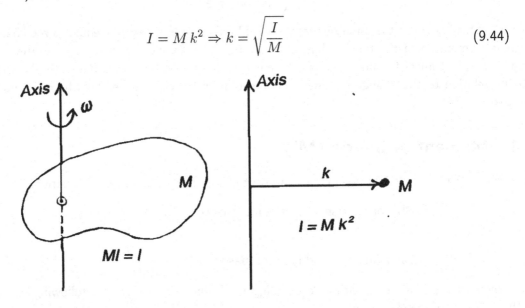

Figure 9.6: Radius of gyration of a rigid body about an axis of rotation.

The distance k is called the radius of gyration of the body. Clearly enough, the value of k depends on the axis of rotation with respect to the body.

9.4.2 Addition of moment of inertia

Consider n number of rigid bodies of masses M_1, M_2, M_3, ... rotating about a common axis, and their MI are: I_1, I_2, I_3, ... The moment of inertia of the entire system about the same axis is the sum of the MI of individual bodies, *i.e.,*

$$I_{System} = I_1 + I_2 + I_3 ... = \sum_i I_i \tag{9.45}$$

Prove equation (9.45)! From equation (9.45) one can also define the effective radius of gyration for the entire system (k_{sys}), as:

$$I_{sys} = \left(\sum_i M_i \right) k_{sys}^2 = \sum_i I_i$$

$$\text{or,} \quad k_{sys} = \left(\frac{\sum_i I_i}{\sum_i M_i} \right)^{\frac{1}{2}} \tag{9.46}$$

The addition theorem of the moment of inertia is particularly useful when one needs to calculate the MI of a body from which a part has been removed. The MI of the remainder of the body about an axis of rotation is equal to the MI of the whole body minus the MI of the part that has been removed about the same axis of rotation. For example:

(i) MI of a thick spherical shell of uniform density about an axis of rotation is equal to the MI of a large solid sphere of radius equal to the outer radius of the shell minus the MI of a small sphere of radius equal to the inner radius of the shell about the same axis of rotation where both spheres have the same density as that of the shell. If there is a variation of density of the shell, the same variation of density is applied to the inner and the outer spheres.

(ii) MI of a thick hollow cylinder is equal to the difference of the MI of an outer solid cylinder and the MI of an inner solid cylinder about the same axis of rotation.

9.4.3 Parallel axis theorem

This theorem gives the moment of inertia of a rigid body about an axis in terms of its MI about a parallel axis through the centre of mass of the body. The theorem states:

Parallel axis Theorem: *The moment of inertia (I) of a rigid body of mass M about an axis A is equal to the sum of the MI of the body about a parallel axis through the centre of mass of the body (I_{cm}) and the MI about axis A of a point mass M placed at the CM of the body.*

Referring to Figure (9.7 (a)), A_{cm} is an axis through the centre of mass, and A is the axis parallel to the A_{cm} axis at a distance d from it. From the parallel axis theorem:

$$I = I_{cm} + M d^2 \qquad (9.47)$$

Proof: Choose the Cartesian coordinates axis such that the axis of rotation A_{cm} through the centre of mass (CM) is along the $z-$axis, and the origin is at the centre of mass. Consider a mass element dm at point (x, y) in the $xy-$plane,(Figure 9.7(b)). Coordinates of the parallel axis A at a distance d in the $xy-$plane from the A_{cm} axis are (a, b). From the figure:

Separation between the two axes: $d = \sqrt{a^2 + b^2}$
Perpendicular distance of dm from the A_{cm}-axis: $r_{cm} = \sqrt{(x^2 + y^2)}$
Coordinates of dm from the foot of the $A-$axis in the $xy-$plane: $((x - a), (y - b))$
Perpendicular distance of dm from the A axis: $r = \sqrt{(x - a)^2 + (y - b)^2}$

MI of dm about A_{cm}-axis: $dI_{cm} = r_{cm}^2 dm = (x^2 + y^2) dm$
MI of dm about A-axis: $dI = r^2 dm = ((x^2 + y^2) + (a^2 + b^2) - 2 x a - 2 y b)dm = (r_{cm}^2 + d^2 - 2 x a - 2 y b)dm$

The total moment of inertia about axis A is:

$$\int_M dI = I = \int_M r_{cm}^2 \, dm + d^2 \int_M dm - 2\,a \int_M x \, dm - 2\,b \int_M y \, dm$$

$$= I_{cm} + M\,d^2 - 2\,a \int_M x \, dm - 2\,b \int_M y \, dm \qquad (9.48)$$

The last two integrals in equation (9.48) by definition are the coordinates of the centre of mass, which from the choice of our coordinates axes are $(0, 0)$. Hence:

$$I = I_{cm} + M\,d^2$$

which proves the parallel axis theorem.

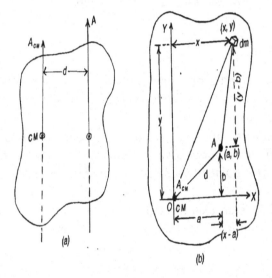

Figure 9.7: The parallel axis theorem for the moment of inertia of a rigid body.

9.4.4 Perpendicular axis theorem

The perpendicular axis theorem applies to a plane, 2-D body, and relates the MI of the plane about an axis perpendicular to the plane to its MI about two perpendicular axis in the plane of the body through a common point of intersection. This theorem, although not found in many of the text books can be used quite effectively for thin plane bodies. The statement of the theorem is as follows.

Parallel axis Theorem: *The moment of inertia of a thin, plane body about an axis normal to the plane is equal to the sum of its MI about two perpendicular axes in the plane of the body which pass through the same point as the normal axis, i.e.,*

$$I_z = I_x + I_y \qquad (9.49)$$

where I_z is the MI about the normal $z-$axis, and I_x and I_y are MI about two perpendicular $x-$ and $y-$axes respectively in the plane of the body.

Proof: Consider a thin, plane body of mass M. The Cartesian coordinates axes are chosen such that the $z-$ axis is perpendicular to the plane of the body, and $x-$ and $y-$ axes are in the plane of the body, (Figure 9.8).

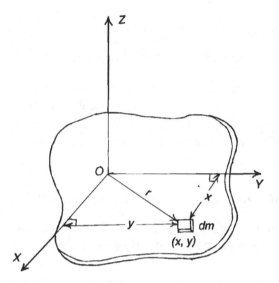

Figure 9.8: The perpendicular axis theorem for the moment of inertia of a thin, plane body.

Choose a mass element dm at point (x, y). The normal distance of the mass element from the $z-$axis is: $r = \sqrt{(x^2 + y^2)}$.

MI of the body about the $x-$axis: $I_x = \int_M y^2 \, dm$
MI of the body about the $y-$axis: $I_y = \int_M x^2 \, dm$
and MI about the $z-$axis is:

$$I_z = \int_M r^2 \, dm = \int_M (x^2 + y^2) \, dm = I_x + I_y \qquad (9.50)$$

which is the perpendicular axis theorem.

9.4.5 Calculation of the moment of inertia.

Recipe 9.1: MI of a symmetric body.

- Choose an appropriate set of coordinates axes with at least one of the axis along the axis of symmetry.

- Use the set of coordinates that are compatible with the symmetry of the body. As a basic guideline for bodies with rectangular, spherical or cylindrical symmetry one uses, rectangular, spherical and cylindrical coordinates system respectively.

- Choose a volume (area or length depending on the geometry being considered) element of the body, and express it in terms of the chosen set of coordinates.

- Express the mass of the element as the product of the volume element and the volume or area or length density of the body as the case may be.

- Write down the moment of inertia for the mass element in terms of the chosen coordinates.

- Determine the limits of each coordinate to include the entire volume of the rigid body.

- Integrate over the volume of the body to find the MI of the body.

- The recipe in the stated form can be used most conveniently to determine MI about an axis of symmetry. MI about axes which are not symmetrically located often can not be easily evaluated no matter how symmetrical the body is. In that case one makes use of the parallel and perpendicular axes theorems. The addition theorem of the moment of inertia is another useful tool that can be invoked.

Application/ Example: MI of a thin disc of mass M and radius R:

This example of a thin, solid, circular disc of mass M and radius R demonstrates the use of various tools and technique for the calculation of MI about differently located axes of rotation.

(a) MI, (I_1), about the axis through the centre and perpendicular to the plane of the disc, (Figure 9.9(a)).
We choose the $z-$axis along the axis of rotation, and the $x-$ and $y-$ axes are along two perpendicular diameters in the plane of the disc. The radial-polar coordinates are used to choose the area element of the disc between points (r, θ) and $((r + dr), (\theta + d\theta))$ as shown in the figure.

Area element: $dA = r\, d\theta\, dr$
Mass per unit area of the dis. Area density $= \frac{M}{\pi R^2}$
Mass of the area element dA: $dm = \frac{M}{\pi R^2} dA = \frac{M}{\pi R^2} r\, d\theta\, dr$
MI of the surface element: $dI_1 = r^2\, dm = r^2 \frac{M}{\pi R^2} r\, d\theta\, dr = \left(\frac{M}{\pi R^2} \right) r^3\, dr\, d\theta$
The limits of integration are: $r = 0 \,\text{to}\, R$ and $\theta = 0 \,\text{to}\, 2\pi$

The MI of the disc is obtained by integrating dI_1 over these limits as:

$$I_1 = \int_M dI_1 = \left(\frac{M}{\pi R^2} \right) \int_0^R r^3\, dr \int_0^{2\pi} d\theta = \frac{1}{2} M R^2 \qquad (9.51)$$

(b) MI, (I_2), about one of the diameters of the disc, *i.e.*, $x-$ or the $y-$axis, (Figure 9.9(b)).

From symmetry, MI of the disc about any of its diameters is the same, *i.e.*, $I_x = I_y = I_2$. Applying the perpendicular axis theorem:

$$I_1 = I_x + I_y = 2I_2, \quad \text{or} \quad I_2 = \frac{1}{2} I_1 = \frac{1}{4} M R^2 \qquad (9.52)$$

(c) MI, (I_3), about an axis along the edge of the disc, and in the plane of the disc, (Figure 9.9(c)).

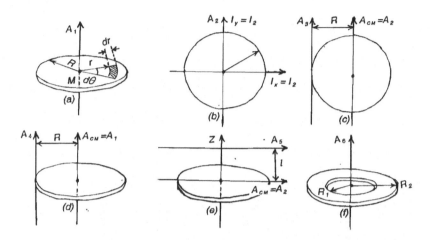

Figure 9.9: Moment of Inertia of a thin circular disc.

Use parallel axis theorem, the parallel axis through the CM being the diameter of the disc at a distance $d = R$.

$$I_3 = I_2 + M d^2 = \frac{1}{4} M R^2 + M R^2 = \frac{5}{4} M R^2 \tag{9.53}$$

(d) MI, (I_4), about an axis along the edge of the disc and perpendicular to its plane, (Figure 9.9(d)).

Apply parallel axis theorem, the parallel axis being the $z-$axis through the centre of the disc at a distance $d = R$.

$$I_4 = I_1 + M d^2 = \frac{1}{2} M R^2 + M R^2 = \frac{3}{2} M R^2 \tag{9.54}$$

(e) MI, (I_5), about an axis perpendicular to the $z-$ axis at a distance l from the disc, (Figure 9.9(e)).

Apply parallel axis theorem, the parallel axis being the one through the diameter of the disc at a distance l.

$$I_5 = I_2 + M l^2 = \frac{1}{4} M R^2 + M l^2 \tag{9.55}$$

(f) MI,(I_6), of a wide ring of mass M and inner and outer radii R_1 and R_2 respectively about the axis through the centre and perpendicular to the plane of the ring, (Figure 9.9(f)).

We can determine the MI in two ways: (i) By taking the difference of MI of a large and a small disc of radii R_2 and R_1 respectively, or (ii) by integration of the MI of a mass element of the ring. We shall deal with the problem concisely, leaving the mathematical details for students to work out.

(i) By differencing the MI of a large and a small dis.

Surface density of the dis. $\sigma = \frac{M}{\pi\left(R_2^2 - R_1^2\right)}$

MI of the inner, small dis. $I_{R1} = \frac{1}{2}\left(\pi R_1^2 \sigma\right) R_1^2$

MI of the outer, large dis. $I_{R2} = \frac{1}{2}\left(\pi R_2^2 \sigma\right) R_2^2$

MI of the ring:

$$I_6 = \left(I_{R2} - I_{R1}\right) = \frac{1}{2} M \left(R_1^2 + R_2^2\right) \tag{9.56}$$

(ii) By integration:

MI of the mass element: $dI_6 = \sigma\, r\, dr\, d\theta\, r^2 = \sigma\, r^3\, dr\, d\theta$. By integrating:

$$I_6 = \int_M dI_6 = \sigma \int_{R_1}^{R_2} r^3\, dr \int_0^{2\pi} d\theta = \frac{1}{2} M \left(R_1^2 + R_2^2\right) \tag{9.57}$$

(g) MI, (I_7), of a thin ring of mass M and radius R about the axis through the centre and perpendicular to the plane of the ring.

For a thin ring, put $R_1 = R_2 = R$ in the expression for MI for a wide ring (I_6), which gives:

$$I_7 = M\, R^2 \tag{9.58}$$

(h) MI, (I_8), of a solid cylinder of radius R; MI, (I_9), of a hollow cylinder of inner and outer radii R_1 and R_2; and MI, (I_{10}), of a thin cylindrical tube of radius R, each of mass M and length L about the cylindrical axis, (the $z-$axis).

These can be regarded as thin discs; thin wide rings; and a thin rings respectively stacked on top of each other to make up for the length. Their MI is the sum of the MI of the corresponding thin components, and are given by:

$$
\begin{aligned}
I_8 &= \frac{1}{2} M\, R^2 & (9.59)\\
I_9 &= \frac{1}{2} M \left(R_1^2 + R_2^2\right) & (9.60)\\
I_{10} &= M\, R^2 & (9.61)
\end{aligned}
$$

(i) We can continue with the same example a long way further to calculate the MI of the thick and the thin rings about various axes as in the case of the solid disc, and the MI of the solid, and hollow cylinders and thin tube along different axes. These are left for students to try.

Table 9.1: *Moment of inertia of some homogeneous, symmetrical solids about axes of symmetry, (excluding the cases covered by the example).*

Solid	Mass	Dimensions	Axis	MI
Solid sphere	M	Radius R	Through the centre	$\frac{2}{5} M R^2$
Thin spherical shell	M	Radius R	Through the centre	$\frac{2}{3} M R^2$
Thick spherical shell	M	Radii R_1 and R_2	Through the centre	$\frac{2}{5} M \frac{(R_2^5 - R_1^5)}{(R_2^3 - R_1^3)}$
Thin rectangular plate	M	Width a, breadth b	Through the centre and Perpendicular to the area	$\frac{1}{12} M (a^2 + b^2)$
Parallelepiped	M	Width a, breadth b and height c	Through the centre and parallel to the edge - c	$\frac{1}{12} M (a^2 + b^2)$
Cylinder	M	Radius R, length L	Perpendicular to cylindrical axis and through the centre	$\frac{1}{4} M R^2 + \frac{1}{12} M L^2$
Thin long wire	M	Length L	Perpendicular to the length through the mid point	$\frac{1}{12} M L^2$
Thin long wire	M	Length L	Perpendicular to the length at one end of the wire	$\frac{1}{3} M L^2$

9.4.6 Inertia tensor

NOTE: This section on advanced treatment of the angular momentum and moment of inertia may only be covered if students are familiar with tensors.

In equations (9.13) and (9.31) we expressed angular momentum in terms of angular velocity as $\mathbf{L} = I \omega$. If ω_x, ω_y, and ω_z are the Cartesian components of the angular velocity, then the corresponding components of the angular momentum according to this expression are:

$$L_x = I_1 \omega_x, \quad L_y = I_2 \omega_y, \quad \text{and} \quad L_z = I_3 \omega_z \Rightarrow \mathbf{L} = I_1 \omega_{\mathbf{x}} + I_2 \omega_{\mathbf{y}} + I_3 \omega_{\mathbf{z}} \qquad (9.62)$$

where I_1, I_2 and I_3 are the moment of inertia of the body about $x-$, $y-$, and $z-$ axes respectively. The equation (9.62) applies to the angular momentum and the angular velocity vectors only when the axes of rotation are the principal axes determined from the symmetry of the system, in which case the angular momentum vector is along the same direction as the angular velocity vector along the axis of rotation. However, if the axis of rotation is not the principal axis, then the angular momentum vector is not along the direction of the angular velocity vector, and equation (9.62) does not apply. In that case we have to treat the problem more generally. The general relationship between the angular momentum and the angular velocity vectors at our disposal is:

$$\mathbf{L} = M \left(\mathbf{r} \times (\omega \times \mathbf{r}) \right) \qquad (9.63)$$

From equation (9.63), one can draw the following conclusions: *(i)* The angular momentum vector is perpendicular to the radius vector **r**, and it is not necessarily along the axis of rotation which is the direction of the angular velocity vector. *(ii)* The components of angular momentum are the linear functions of the components of the angular velocity vector. These are expressed as:

$$L_x \;=\; = I_{xx}\,\omega_x + I_{xy}\,\omega_y + I_{xz}\,\omega_z$$
$$L_y \;=\; = I_{yx}\,\omega_x + I_{yy}\,\omega_y + I_{yz}\,\omega_z$$
$$L_z \;=\; = I_{zx}\,\omega_x + I_{zy}\,\omega_y + I_{zz}\,\omega_z \tag{9.64}$$

Equation (9.64) can be expressed in matrix notation as:

$$
\begin{pmatrix} L_x \\ L_y \\ L_z \end{pmatrix} =
\begin{bmatrix} I_{xx} & I_{xy} & I_{xz} \\ I_{yx} & I_{yy} & I_{yz} \\ I_{zx} & I_{zy} & I_{zz} \end{bmatrix}
\begin{pmatrix} \omega_x \\ \omega_y \\ \omega_z \end{pmatrix}
\Rightarrow \mathbf{L} = \mathcal{I}\,\omega \tag{9.65}
$$

where $\mathbf{L}$ and ω are the usual column vectors of the respective $x-$, $y-$ and $z-$ components, and $\mathcal{I}$ is known as the inertia tensor. It is a tensor of second rank (order), and in a 3-D space has nine components that are expressed as a 3×3 array of elements. We have devoted a few paragraphs to an elementary introduction to tensors at the end of this section. Here we continue with our discussion of the moment of inertia tensor.

The diagonal elements of the inertia tensor in the Cartesian coordinates are given by:

$$I_{xx} = \int_M (y^2 + z^2)\,dM$$

$$I_{yy} = \int_M (z^2 + x^2)\,dM$$

$$I_{zz} = \int_M (x^2 + y^2)\,dM \tag{9.66}$$

Comparing equation (9.66) to (9.32) (and 9.12), we recognize that the diagonal elements of the inertia tensor are the moment of inertia of the body about the $x-$, $y-$ and $z-$ axes respectively.

The off-diagonal elements are known as the product of inertia, defined as:

$$I_{xy} = \int_M x\,y\,dM$$

$$I_{yz} = \int_M y\,z\,dM$$

$$I_{zx} = \int_M z\,x\,dM \tag{9.67}$$

The products of inertia have the same units $(kg\,m^2)$ as the moment of inertia, and by definition, (equation 9.67), they are symmetric, *i.e.*, $I_{xy} = I_{yx}$, $I_{yz} = I_{zy}$, and $I_{zx} = I_{xz}$. A tensor with this property of the off-diagonal elements is known as the symmetric tensor. Thus the inertia tensor is a symmetric tensor of order two of which only six components, three diagonal and three off-diagonal, are independent.

For symmetric bodies it is possible to choose the coordinates axes in such a way that all or some of the off-diagonal elements of the inertia tensor (products of inertia) become zero. This is possible if the axis of rotation is the axis of symmetry, known as the principal axis.. For example if the $x-$axis is a principal axis, then $L_x = I_{xx} \omega_x$, and $I_{xy} = I_{xz} = 0$. Likewise if $y-$axis is principal axis then $I_{yx} = I_{yz} = 0$, and if $z-$axis is principal axis then $I_{zx} = I_{zy} = 0$. Thus in a coordinate system in which all the three axes are the principal (symmetry) axes, the inertia tensor is diagonal, *i.e.*, the only non-vanishing elements of the inertia tensor are the diagonal elements I_{xx}, I_{yy} and I_{zz} with all off-diagonal elements being zero. In principal axes equation (9.65) reduces to:

$$\begin{pmatrix} L_x \\ L_y \\ L_z \end{pmatrix} = \begin{bmatrix} I_{xx} & 0 & 0 \\ 0 & I_{yy} & 0 \\ 0 & 0 & I_{zz} \end{bmatrix} \begin{pmatrix} \omega_x \\ \omega_y \\ \omega_z \end{pmatrix} \Rightarrow \mathbf{L} = I_{xx}\,\omega_{\mathbf{x}} + I_{yy}\,\omega_{\mathbf{y}} + I_{zz}\,\omega_{\mathbf{z}} \qquad (9.68)$$

On comparison we find that equation (9.67) is the same as equation (9.62) with $I_{xx} = I_1$, $I_{yy} = I_2$, and $I_{zz} = I_3$

An elementary introduction to tensors:

Tensor is a set of mathematical quantities, called components whose properties (not the magnitude) are independent of the coordinate system used to describe them. A tensor is used most commonly when a vector $\mathbf{v}$ is expressed as a linear function of the components of another vector $\mathbf{u}$. For example in a 3-D space these can be expressed as:

$$\begin{aligned} v_x &= T_{xx}\,u_x + T_{xy}\,u_y + T_{xz}\,u_z \\ v_y &= T_{yx}\,u_x + T_{yy}\,u_y + T_{yz}\,u_z \\ v_z &= T_{zx}\,u_x + T_{zy}\,u_y + T_{zz}\,u_z \end{aligned} \qquad (9.69)$$

where

$$\mathcal{T} = \begin{bmatrix} T_{xx} & T_{xy} & T_{xz} \\ T_{yx} & T_{yy} & T_{yz} \\ T_{zx} & T_{zy} & T_{zz} \end{bmatrix} \qquad (9.70)$$

is a covariant tensor of the second rank (order). The rank r of a tensor is the number of indices that appear as subscript in case of the components of a covariant tensor, and as superscript for the components of a contravariant tensor. For example, T_i are the components of a convariant tensor of rank one, T^j are the components of a contravariant tensor of rank one, T_{ij} are the components of a covariant tensor of rank two, T^{ij}_{klm} are the components of a mixed tensor of rank three. A tensor A of rank zero is a scalar, and tensors B^i and C_j of rank one are vectors. A tensor of rank r in a $n-$dimensional space has n^r components. In our example (equation 9.70), the tensor $\mathcal{T}$ of rank two with components T_{ij} in 3-D space has $3^2 = 9$ components.

The components of a tensor in two coordinates system are related by transformation laws of tensors. The addition, subtraction, product, contraction and other algebraic operations on a tensor are governed by the laws of tensor algebra. These are beyond the scope of this book.

9.5   Applications and examples of rigid body rotation

In this section, we demonstrate the applications of the laws of rigid body rotation through some examples. Systems in static as well as in dynamic equilibrium are considered. Although each of the application presented is unique in some aspects, but solution to most problems can be approached by following the general recipe given below.

Recipe 9.2: General approach to solving rigid body rotation problems

- Identify all the forces acting on the system including the normal reaction and the frictional forces. If necessary, resolve the forces in an appropriate set of coordinates axes.

- If the system is in translational equilibrium, apply the first condition of equilibrium to the forces, *i.e.*, the net resultant force on the system is zero.

- Take the torque of all the forces about an appropriate point. Generally, the point through which maximum number of forces pass is considered to be the best choice, because the torque of all such forces vanishes, leaving fewer number of forces to deal with.

- If the system is in rotational equilibrium, apply the second condition of equilibrium, *i.e.*, the net torque on the system is zero.

- Determine the moment of inertia of the body, its rotational kinetic energy, and angular momentum if needed. The laws of conservation of energy and angular momentum may be used.

- Other relations that are useful include are the relations between various translational and rotational quantities, laws of motion, and the equations of motion.

9.5.1   Balancing on knife edge

An object balanced on a knife edge is in a state of translational and rotational equilibrium, *i.e.* $\sum \mathbf{F} = 0$ and $\sum \tau = 0$. This is illustrated by a solved example below.

Example 9.2: A uniform meter rule of $100\,g$ mass is balanced on a knife edge at its $40\,cm$ mark. A $50\,g$ mass is suspended from the $30\,cm$ mark and a $30\,g$ mass is suspended from the $80\,cm$ mark. In order to balance the meter rule on the knife edge a vertical force F is applied to one of the ends of the meter rule. Calculate the magnitude of the force, and to which end of the meter rule should it be applied. What is the normal reaction force of the knife edge?

Solution: Figure (9.10) shows the meter rule balanced on the knife edge, with the masses suspended from the given points. Since the meter rule is uniform, the mass of its $40\,cm$ length is $40\,g$ which acts at its centre of mass at the $20\,cm$ mark, and the mass of the $60\,cm$ length is $60\,g$ which acts at the centre of mass for this portion of the meter rule at $70\,cm$ mark. The normal reaction force N of the knife edge acts vertically up. Taking the torque of the weights about the knife edge:

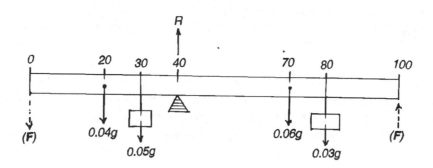

Figure 9.10: A meter rule balanced on a knife edge.

Total clockwise torque: $(0.06\,g \times 0.3) + (0.03\,g \times 0.4) = 0.03\,g\,kg\,m$
Total anticlockwise torque: $(0.05\,g \times 0.1) + (0.04\,g \times 0.2) = 0.013\,g\,kg\,m$
where g is the acceleration due to gravity.

In order to balance the meter rule, the net torque on it must be zero. Therefore, the external force must produce an anticlockwise torque of $(0.03\,g - 0.013\,g) = 0.027\,g\,kg\,m$. This can be achieved by applying a vertical force to either one of the two ends of the meter rule as follows:

(i) The force is applied vertically downwards to the $0\,cm$ mark of the meter rule. Its magnitude is: $F_1 = (0.027\,g/0.4) = 0.0675\,g\,N$. The normal reaction force at the knife edge from the first condition of equilibrium is: $N = (0.0675 + 0.04 + 0.05 + 0.06 + 0.03)g = 0.2375\,g\,N$

(ii) The force is applied vertically upwards to the $100\,cm$ mark of the meter rule. Its magnitude is: $F_2 = (0.027\,g/0.6) = 0.045\,g\,N$. The normal reaction force at the knife edge in this case is: $N = (-0.045 + 0.04 + 0.05 + 0.06 + 0.03)g = 0.135\,g\,N$

Discussion: In this example we have split the total weight of the meter rule into two parts on either side of the knife edge to demonstrate an important point: *Although the total forces due to weight on either side of the knife edge are equal, (0.09 g), the meter rule is not balanced because these forces produce unequal clockwise and anticlockwise torques.* In order to equalize the clockwise and anticlockwise torques an additional force is needed, so that the net torque on the meter rule is zero. The problem can also be solved in a more simple way by having the entire weight $0.1\,g\,N$ of the meter rule acting at its centre of mass at the $50\,cm$ mark. How does it work? On calculation one finds that the net torque due the weight of the meter rule about the knife edge is the same whether one places the entire weight at the $50,\,cm$ mark or distributes it on two sides of the knife edge as we have done in this example. Furthermore, we have taken the torque about the knife edge, because our primary aim was to determine the external force. This also eliminated the normal reaction force from the equation which is determined later once the external force has been determined. We can also take torque about any other point along the meter rule, provided only one unknown force appears in the resulting equation. For example, we have imposed a restriction in the problem that the external force must be applied to one of the ends of the meter rule. Therefore, by taking torque about either of the two ends we can first determine the normal reaction force and then the external force. Students are assigned to workout the problem first taking torque about the $0\,cm$ mark, and then about the $100\,cm$ mark, and compere the results obtained in both case to the results which are obtained above when we took torque about the knife edge.

Exercise 9.4: A meter rule of mass $100\,g$ is balanced on a knife edge at the $30\,cm$ mark by applying a force F to one of its end at an angle of 45° from the horizontal. Calculate the magnitude of the force F and the reaction force at the knife edge if the force is applied: *(i)* to the $0\,cm$ mark, and *(ii)* to the $100\,cm$ mark. (Hint: The reaction force at the knife edge is not normal to the meter rule. (Explain why.)

9.5.2   A ladder leaning against the wall

A person climbing a ladder leaning against a wall is another typical example of translational and rotational equilibrium. Figure (9.11) shows a uniform ladder of mass m and length l leaning against a vertical wall, making an angle θ from the horizontal floor. The coefficients of static friction for the floor and the wall are μ_1 and μ_2 respectively. A person of mass M climbs up the ladder upto a length x at which point the ladder is just about to slip. One may be interested to determine x as a fraction of the length l of the ladder.

As shown in the figure, the forces acting on the ladder-person system are: weight $w = m\,g$ of the ladder acting vertically downwards at its mid point, weight $W = M\,g$ of the person also acting vertically downwards at the point where (s)he is standing on the ladder, normal reaction forces R_1 and R_2, and the friction forces $F_1 = \mu_1\,R_1$ and $F_2 = \mu_2\,R_2$ at the floor and the wall respectively. As the system is in both translational and rotational equilibrium, the corresponding conditions of equilibrium to the forces and torques apply.

From the first condition of (translational) equilibrium:

$$\sum F_x = R_2 - F_1 = 0 \Rightarrow R_2 = F_1 = \mu_1\,R_1 \tag{9.71}$$

$$\sum F_y = R_1 + F_2 - w - W = 0 \Rightarrow R_1 + \mu_2\,R_2 = (m+M)\,g \tag{9.72}$$

Combining equations (9.71) and (9.72) and solving we get:

$$R_1 = \frac{(m+M)\,g}{(1+\mu_1\,\mu_2)} \tag{9.73}$$

$$R_2 = \mu_1\frac{(m+M)\,g}{(1+\mu_1\,\mu_2)} \tag{9.74}$$

From these values of the normal reaction forces, one can determine the corresponding friction forces. Now consider the torque of the forces about some point along the ladder, and apply the second condition of equilibrium, *i.e.*, the total clockwise torque is equal to the total anticlockwise torque.

As stated earlier the point through which maximum number of forces pass is the best choice for a point about which the torque should be taken. In the present case the top and the bottom ends of the ladder are two such points through each of which two forces, the normal reaction and the friction forces pass. Without going into the mathematical details for taking the torque, we have:

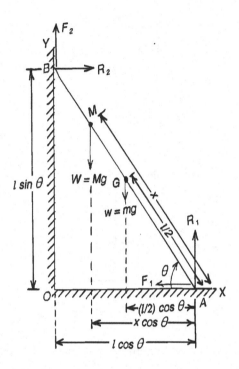

Figure 9.11: A person on a leaning ladder in equilibrium.

If the torques are taken about the bottom end of the ladder:

$$m\,g\,\frac{l}{2}\cos\theta + M\,g\,x\cos\theta = R_2\,l\,sin\,\theta + \mu_2\,R_2\,l\,cos\,\theta \qquad (9.75)$$

and if the torques are taken about the top end of the ladder:

$$m\,g\,\frac{l}{2}\cos\theta + M\,g\,(l-x)\cos\theta + \mu_1\,R_1\,l\,sin\,\theta = R_1\,l\,cos\,\theta \qquad (9.76)$$

Any of these equations (9.75) or (9.76) along with equations (9.73) and (9.74) can be used to determine the maximum length x of the ladder to which the person can climb without slipping, or any other parameter could be determined provided all other parameters in the equation are known.

9.5.3 Pulleys of finite mass

In an earlier chapter we considered ideal mass pulley systems, in which one of the assumptions was that the pulley is massless and frictionless. Here we consider the pulley to have a mass, but the assumption about being frictionless still remains in place. The consequence of the finite mass of the pulley is that it has a finite moment of inertia about the axis of rotation, and in order for it to rotate there must be net torque applied to the pulley. Hence the tension in the two portions of the string passing over it are not the same.

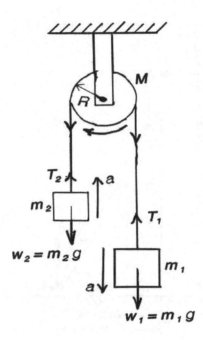

Figure 9.12: A mass-pulley system with a pulley of finite mass.

Figure (9.12) shows a pulley of mass M, radius R, and moment of inertia about the axle I. Two masses m_1 and m_2 attached with a massless, inextensible string are freely suspended over the pulley as shown in the figure. Let $m_1 > m_2$. When released, mass m_1 descends with an acceleration a and mass m_2 ascends with the same acceleration. α is the angular acceleration of the pulley. The weights of the masses act vertically downwards. Let T_1 and T_2 be the tensions in the two portions of the string as shown in the figure.

Moment of inertia of the pulley, treating it as a thin dis. $I = \frac{1}{2} M R^2$

The net torque on the pulley is: $(T_1 - T_2) R$

The linear acceleration of the masses and the angular acceleration of the pulley are related as: $a = R\alpha$

Applying Newton's second law of motion to the two masses:

$$m_1 g - T_1 = m_1 a \tag{9.77}$$
$$T_2 - m_2 g = m_2 a \tag{9.78}$$

and relating angular acceleration of the pulley to the net torque on it:

$$(T_1 - T_2)\,R = I\,\alpha = I\,\frac{a}{R} \tag{9.79}$$

We have three unknown parameters, namely T_1, T_2 and a, and three equation (9.77), (9.78) and (9.79), which can be solved to obtain the following expressions for them:

$$a = \frac{(m_1 - m_2)\,g}{(m_1 + m_2) + \frac{I}{R^2}} \tag{9.80}$$

$$T_1 = m_1\,g\,\frac{2\,m_2 + \frac{I}{R^2}}{(m_1 + m_2) + \frac{I}{R^2}} \tag{9.81}$$

$$T_2 = m_2\,g\,\frac{2\,m_1 + \frac{I}{R^2}}{(m_1 + m_2) + \frac{I}{R^2}} \tag{9.82}$$

For a massless pulley, $I = 0$, and the expressions for acceleration and tension in string reduce to the ones obtained earlier, *i.e.*,

$$a = \frac{m_1 - m_2}{m_1 + m_2}\,g\,,$$

and

$$T_1 = T_2 = T = \frac{2\,m_1\,m_2}{m_1 + m_2}\,g$$

9.5.4 Rolling down an inclined plane

When a rigid body with circular symmetry, (a cylinder, or a sphere, or a spherical shell, or a disc, or a ring) rolls down an inclined plane, its motion comprises two components: rotational motion about an axis through the centre and parallel to the horizontal edge of the plane, and the translational motion of the centre of mass of the body parallel to the plane. We are interested in finding the velocity with which the rigid body arrives at the foot of the inclined plane. Because different bodies of same mass and radius have different moment of inertia about the identical axes of rotation, it turns out that they have different velocities when they reach the bottom of the plane, and their times of travel along the length of the plane differ. This provides an experimental mean to distinguish between identical solid and hollow bodies.

Let M be the mass, R be the radius, and $I = M\,k^2$ be the moment of inertia of the body about the horizontal axis through the centre of the body, where k is the radius of gyration. l is the length of the inclined plane, h is the vertical height of the raised end of the plane, and θ is the angle of inclination above horizontal, (Figure 9.13). There are two methods by which the motion of the body can be analyzed: the conservation of energy method, and the dynamical method.

(i) Analysis by energy conservation method: Since the motion is under gravitational field, mechanical energy of the body is conserved, *i.e.*, as the body rolls down it gains kinetic energy at the expense of the gravitational potential energy. Although, there is rolling friction, the work done by it is zero, as the instantaneous linear displacement of the point of contact between the rolling object and the plane is zero. Let us assume that the body starts from rest at $t = 0$ from the top of the plane, and

arrives at the foot of the plane at $t = T$, (Figure 9.13 (a)). Let the angular velocity of the body about the axis through the centre be ω, and linear velocity of the centre of mass of the body parallel to the plane be v, where $v = \omega R$.

At the top of the plane the mechanical energy of the body comprises only of the gravitational potential energy:

$$E_i = PE + (KE = 0) = M g h \qquad (9.83)$$

and when the body reaches the foot of the incline, the total energy is just the kinetic energy that comprises the rotational and translational kinetic energies, *i.e.*:

$$E_f = KE_r + KE_t = \frac{1}{2} I \omega^2 + \frac{1}{2} M v^2 = \frac{1}{2} M k^2 \frac{v^2}{R^2} + \frac{1}{2} M v^2 = \frac{1}{2} M v^2 \left(1 + \frac{k^2}{R^2}\right) \qquad (9.84)$$

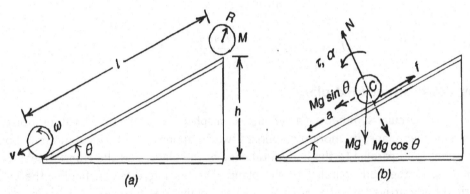

Figure 9.13: Rolling down and inclined plane: (a) Energy conservation, and (b) Dynamical analysis.

Conservation of energy gives:

$$E_i = E_f \Rightarrow v^2 = \frac{2 g h}{1 + \frac{k^2}{R^2}} \qquad (9.85)$$

(i) Analysis by dynamical method: Figure (9.13 (b)) shows the forces acting on the rigid body rolling down the plane at time t. The forces are: weight of the body $M g$ acting vertically downwards with its components $M g \sin\theta$ and $M g \cos\theta$ parallel and perpendicular to he plane respectively, normal reaction force N normal to the plane, and the rolling friction force f parallel to the plane, opposite to the motion. Let a and α be the linear and angular accelerations of the body, where $a = R \alpha$. From the first condition of equilibrium:

$$\begin{aligned} N - M g \cos\theta &= 0 \\ M g \sin\theta - f &= M a \end{aligned} \qquad (9.86)$$

The frictional force provides the torque $\tau = fR$ which is responsible for the rotational motion of the body, and is related to the angular acceleration by $\tau = I\alpha$. From this the force of friction is obtained as:

$$\tau = I\alpha = Mk^2\frac{a}{R} \Rightarrow f = \frac{Mk^2a}{R^2} \tag{9.87}$$

Note that the rolling friction is not related to the normal reaction force in the same way as the sliding friction. Substituting the expression for f in equation (9.86) and solving for the linear acceleration we get:

$$a = \frac{g\sin\theta}{1 + \frac{k^2}{R^2}} \tag{9.88}$$

Finally, from the equation of translational motion: $v^2 = u^2 + 2as$, where the symbols have the usual meanings as discussed in an earlier chapter, the velocity of the body on reaching the foot of the incline is obtained as:

$$v^2 = \frac{2gl\sin\theta}{1 + \frac{k^2}{R^2}} = v^2 = \frac{2gh}{1 + \frac{k^2}{R^2}} \tag{9.89}$$

This is the same as equation (9.85) obtained from energy conservation principles.

Discussion: From equation (9.85) or (9.89) one notes that the velocity of the rolling object on reaching the foot of the incline plane depends on its radius of gyration k. If k is large the velocity is small, and vice versa. Let us consider different rolling objects of the same mass M and radius R. From their moment of inertia about axes through their centres of mass given in Table (9.1), the corresponding radii of gyration are:

Solid sphere: $k_1^2 = \frac{2}{5}R^2$
Solid cylinder: $k_2^2 = \frac{1}{2}R^2$
Solid dis. $k_3^2 = \frac{1}{2}R^2$
Thin spherical shell: $k_4^2 = \frac{2}{3}R^2$
Thin ring: $k_5^2 = R^2$

From these expressions we note that $k_1 < k_2 = k_3 < k_4 < k_5$. Therefore, if these five objects are rolled down the same inclined plane, their velocities on reaching the foot of the incline shall be in the order: $v_1 > v_2 = v_3 > v_4 > v_5$, *i.e.*, the sphere shall have the largest velocity, the disc and the solid cylinder shall have the same velocity, and the ring shall have the least velocity. In this way one shall also be able to distinguish between the solid sphere and the spherical shell, if they were otherwise indistinguishable. Finally, if there is no rolling, and the objects simply slide down the plane, and the plane is frictionless, then the velocity in all cases from conservation of energy is the same: $v_{slide} = \sqrt{2gh}$, which is larger than linear velocity in the case of rolling (equation 9.89). This is so because the same potential energy at the top of the incline is converted to rotational and translational kinetic energy at the bottom of the incline. Hence, if the rotational kinetic energy is small, being zero in case of sliding, the translation kinetic energy becomes large, imparting large translational velocity to the body.

9.5.5 Compound pendulum

A compound pendulum, also known as the physical pendulum consists of a rigid body, which may be regularly shaped whose moment of inertia about the axis of oscillation can be evaluated from its mass and dimensions, or irregularly shaped for which the moment of inertia can not be determined by mathematical means. When the body is suspended freely from a fixed axis that does not pass through the centre of mass of the body, displaced by a small angle from its equilibrium position, and let go, it executes a simple harmonic motion. The period of oscillation depends on the moment of inertia of the body about the axis of rotation, and the value of the acceleration due to gravity. Experimentally the simple harmonic oscillations of a compound pendulum can be used either to determine the acceleration due to gravity or the moment of inertia of an irregularly shaped rigid body about the axis of rotation.

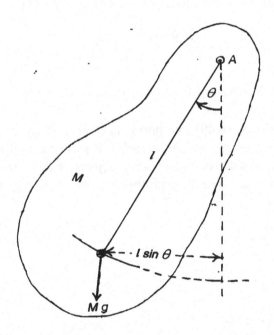

Figure 9.14: A compound (physical) pendulum.

Figure (9.14) shows a compound pendulum of mass M suspended freely from a fixed axis through A. The weight $W = mg$ of the body acts vertically downwards on the centre of mass of the body, at a distance l from the axis of rotation. The body is displaced by a small angle θ from its equilibrium position. The force of gravity $M g$ acting vertically downwards through the centre of mass constitutes the restoring torque on the body which is responsible for the oscillatory motion of the body when release. From the figure, the restoring torque τ is:

$$\tau = (-) M g l \sin\theta \approx (-) M g l \theta \tag{9.90}$$

where for small angle of displacement θ, $\sin\theta \approx \theta\,(radians)$, and the negative sign signifies the restoring nature of the torque acting opposite to the direction of the angular displacement. The torque can also be expressed in terms of the moment of inertia and the angular acceleration as:

$$\tau = I\,\alpha = I\,\frac{d^2\,\theta}{dt^2} \tag{9.91}$$

The two expressions for the torque from equations (9.90) and (9.91)are equal. Therefore,

$$I\,\frac{d^2\,\theta}{dt^2} = (-)\,M\,g\,l\,\theta$$

$$\text{or} \quad \frac{d^2\,\theta}{dt^2} + \frac{M\,g\,l}{I}\,\theta = 0 \tag{9.92}$$

Equation (9.67) is the equation of simple harmonic motion, with angular frequency ω and the time period T given by:

$$\omega = \sqrt{\frac{M\,g\,l}{I}}, \quad \text{and} \quad T = 2\,\pi\,\sqrt{\frac{I}{M\,g\,l}} \tag{9.93}$$

Let $I_{cm} = M\,k^2$ be the MI of the pendulum about an axis through the centre of mass of the body, parallel to the axis of oscillation where k is the radius of gyration about the centre of mass axis. From the parallel axis theorem: $I = I_{cm} + M\,l^2 = M\,(k^2 + l^2)$, and equation (9.93) becomes:

$$T = 2\,\pi\,\sqrt{\frac{(k^2 + l^2)}{g\,l}} \tag{9.94}$$

Note that for a simple pendulum, consisting of a massless string of length l, and a small bob of mass m, the entire mass is concentrated at the centre of mass which is the centre of the bob. Thus for a simple pendulum, $k = 0$, and equation (9.94) reduces to the equation for the time period of a simple pendulum.

Equation (9.93) can be used in two ways: *(i)* To determine the MI of the body about an axis parallel to the axis of oscillation through the centre of mass by measuring the time period of small oscillations. *(ii)* To determine the acceleration due to gravity. For this, one needs the value of the moment of inertia (radius of gyration) of the pendulum about the centre of mass axis. For a regularly shaped, homogenous body one can use the mathematical expression for the MI, but if the pendulum is irregularly shaped then one needs to find an alternate way either to determine the MI first (not from small oscillations), or find ways to eliminate the need to know MI explicitly. On further consideration of equation (9.94), one finds that the need to know MI of the body is bypassed if the data for the period of oscillation is analyzed graphically, as discussed below.

Determination of 'g' from compound pendulum:

Equation (9.94) for the time period of a compound pendulum can be expresses in terms of l, as follows:

$$l^2 - \frac{T^2\,g}{4\,\pi^2}\,l + k^2 = 0 \tag{9.95}$$

Equation (9.95) is a quadratic equation in l, and for a given value of the period of oscillation, there are two values l_1 and l_2 of l, the distance of the axis of oscillation from the centre of mass. These two roots are related (prove!) as:

$$(l_1 + l_2) = \frac{T^2 g}{4\pi^2} \tag{9.96}$$

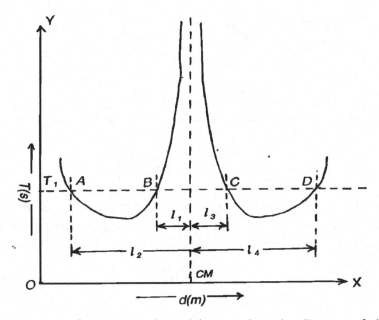

Figure 9.15: Time period of a compound pendulum against the distance of the axis of rotation from one end of the pendulum.

In order to determine g, one takes a number of points on a straight line along the length of the pendulum, passing through the centre of mass, and measures the time period of oscillation about axes passing through these points. A graph of T versus the distance d of the axis from one end of the body (not from the centre of mass) is plotted. The graph has a general shape as shown in Figure (9.15). The two branches of the graph are for the two set of points on either side of the centre of mass, and the vertical line drawn symmetrically between the two branches of the graph represent the location of the centre of mass. Theoretically from equation (9.94), if the axis of oscillation passes through the centre of mass ($l = 0$), the period of oscillation $T = \infty$, as is seen from the graph. Draw a horizontal line corresponding to time period T_1 of the pendulum, which intersects the two branches of the graph at four points, and gives four values of l as shown in the figure. Using these values in equation (9.94), we get:

$$g_1 = \frac{4\pi^2}{T_1^2}(l_1 + l_2) \quad ,\text{and} \quad g_2 = \frac{4\pi^2}{T_1^2}(l_3 + l_4) \tag{9.97}$$

Thus one horizontal line gives two values of g. In this way by taking a number of horizontal lines at different time periods, a number of values of g are obtained the average of which give the acceleration due to gravity at the location of the experiment. In this way we have not only eliminated the need to know the MI of the pendulum about the centre of mass axis, but the need for the precise location of the centre of mass of the pendulum has also been eliminated.

9.5.6 Collision and explosion in rotational motion

The principle of conservation of angular momentum can be applied to the collision and explosion of rigid bodies rotating about a fixed axis in the same way as the conservation of linear momentum applies to the collision and explosion of bodies in translational motion. Here we shall consider only the perfectly inelastic collision and the explosion for which the kinetic energy is not conserved as was the case with the inelastic collision and explosion of bodies in translational motion. The problem of perfectly elastic collision of bodies in rotational motion is rather more complex, because that involves the moving axes of rotation, and one must consider both the linear and angular momentum, and the rotational and the translational kinetic energies.

Consider a rigid body of moment of inertia I_i rotating with an angular velocity ω_i. The angular momentum of the body is $L_i = I_i \omega_i$. Let us first consider the inelastic collision. A point mass m initially at rest is placed at a distance r from the axis of rotation on to the rotating rigid body. The new moment of inertia of the system now is: $I_f = (I_i + m r^2)$, and the angular velocity changes to ω_f, which is determined from the conservation of angular momentum, *i.e.*,

$$L_i = L_f \Rightarrow I_i \omega_i = (I_i + m r^2) \omega_f \tag{9.98}$$

Next let us consider spontaneous explosion of the same rigid body as above. A mass m which we shall treat as a point mass spontaneously breaks off the rigid body from a point at a distance r from the axis of rotation and flies off with a tangential velocity v. The moment of inertia of the remainder of the rigid body is $I_1 = (I_i - m r^2)$, and let its new angular velocity be ω_1. The moment of inertia of the broken piece at the time of explosion is $I_2 = m r^2$ and its angular velocity is $\omega_2 = \frac{v}{r}$. Then the final angular momentum after explosion is:

$$L_f = L_1 + L_2 = I_1 \omega_1 + I_2 \omega_2 = (I_i - m r^2) \omega_1 + m r^2 \frac{v}{r} \tag{9.99}$$

and from the conservation of angular momentum:

$$L_i = L_f \Rightarrow I_i \omega_i = (I_i - m r^2) \omega_1 + m r^2 \frac{v}{r} \tag{9.100}$$

From equation (9.99) one can determine the new angular velocity of the rigid body if other parameters are known.

9.5.7 Hinged beams and doors

This category of problems have wide range of engineering and design applications such as in heavy hinged trap doors opened with cables attached to them, the large cranes used to lift heavy weights, suspension of large billboards with cables and hinges, supporting of the sections of suspension bridges etc. In analyzing such problems one is interested in determining the strength (tension) that the cable should be able to withstand in order to support or lift a given weight, the maximum weight that can be lifted by a crane without tipping it over, the stress on the hinges produced and whether they shall be able to withstand the stress etc. The range of problems in this category are so vast, that one needs to devote a full chapter to them alone to do a full justice to the subject. Here we shall consider only a representative example to illustrate such applications. The method remains the same: identify all the forces, determine the torque about suitable points on the system, and apply the first and second conditions of equilibrium to determine the unknown forces.

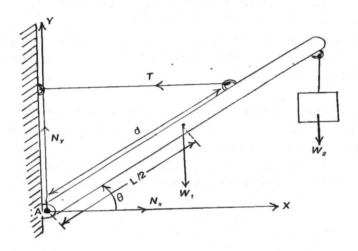

Figure 9.16: A hinged beam supported with a single cable.

Figure (9.16) shows a large, uniform hinged beam of weight W_1 and length L supported at an angle θ above horizontal with a horizontal cable attached to the beam at a distance d from the hinged end. A weight W_2 is suspended from the far end of the beam. (This model could be applied to represent a hinged trap door opened with a cable, or a cable of a crane lifting a heavy weight, or a large hinged bill board suspended with a cable). Find the expressions for the tension T in the cable, and the reaction force N at the hinge.

Let N_x and N_y be the horizontal and vertical components of the reaction force at the hinge. From the condition of translational equilibrium:

$$N_x = T, \quad \text{and} \quad N_y = W_1 + W_2 \tag{9.101}$$

Taking torque about the hinge:

$$T\,d\sin\theta = W_1 \frac{L}{2}\cos\theta + W_2 L \cos\theta \tag{9.102}$$

From equations (9.101) and (9.102) one can determine up to three unknown parameters, provided the remaining others are known.

9.6 Problems

9.1. A 10 N force is applied tangentially to a 5 kg point mass free to rotate about an axis at 30 cm from it. Initially, at time $t = 0$, the mass is at rest. Determine the angular acceleration of the mass, and its angular momentum after 5 s without using the equations of uniform rotational motion. What is the kinetic energy of the mass, and its angular and translational velocities at this moment? If the mass is joined to the axis of rotation with a thin, rigid rod of negligible mass, what is the compressional force in the rod at $t = 5\,s$?

9.2. Eight point masses of 1 kg each are placed at eight corners of a cube of side 10 cm, and joined with thin rods. Ignore the masses of the rods, and determine the moment of inertia of the cube about: *(i)* an axis along one of the edges of the cube, *(ii)* an axis along one of the face diagonals of the cube, *(iii)* an axis along one of the body diagonals of the cube, *(iv)* an axis through the centre of two opposite faces.

9.3. Four point masses of 6, 8, 10, and 12 kg mass are placed in $xy-$plane at points (3, -6), (-4, 6), (-5, -4) and (x, y) respectively. If the centre of mass of the masses lies at the origin, determine the location (x, y) of the fourth mass.

9.4 The nine planets in the solar system are aligned along a straight line. Using the planetary data given in the chapter on Gravitation, and treating them as point masses determine the distance of the centre of mass of the planets from the sun.

9.5. Find the moment of inertia of the planets when aligned in a straight line about an axis through their centre of mass and perpendicular to the line joining them. Determine their moment of inertia about an axis through the sun directly from the first principles, and by applying the parallel axis theorem.

9.6. Determine the centre of mass of a thin long rod of mass M and length L by integrating over the length of the rod.

9.7. Determine the centre of mass of a thin disc of mass M and radius R by integrating over the area of the disc.

9.8. The nine planets in the solar system are aligned along a straight line. Using the planetary data given in the chapter on Gravitation, and treating them as point masses determine the distance of the centre of gravity of the planets from the sun in the gravitational field of the sun.

9.9. An object is projected with velocity v, at an angle θ above horizontal from the ground. The object spontaneously explodes in mid air. Qualitatively describe the motion of the centre of mass of the fragments after explosion before any of the fragments hits the ground.

9.10. Four forces $\mathbf{F_1} = (5\,N, 37^o)$, $\mathbf{F_2} = (8\,N, 120^o)$, $\mathbf{F_3} = (7\,N, 210^o)$ and $\mathbf{F_2} = (10\,N, -45^o)$ are applied to four masses 3 kg 5 kg, 4 kg, and 6 kg respectively. The masses are respectively placed at points $(3, 4)$, $(-4, 5)$,$(-5, 4)$, and $(-4, -3)$ in the xy-plane where the distances are in m. Determine the acceleration of the centre of mass of the system.

9.11. A thin long rod of mass $2\,kg$ and length $75\,cm$ is pivoted at one end and is held horizontally at $t = 0$. The far end of the rod is pushed up with a force $F_1 = 10\,N$ applied at an angle of 45° below horizontal, and a force $F_2 = 15\,N$ pushes the rod vertically down at the mid point. Calculate the net torque on the rod, its angular acceleration, its angular momentum and kinetic energy at $t = 10\,s$ assuming that the torque remains constant. Calculate the work done and hence the angular displacement in $10\,s$, and the average power supplied to the rod.

9.12. A uniform rod of mass M and length L is pivoted at one end, and released from the horizontal position. Calculate the linear velocity of the centre of mass of the rod when it passes the vertical position *(i)* from the torque and angular displacement considerations and *(ii)* from energy conservation principles.

9.13. A $50\,kg$ rectangular homogeneous cupboard of base $75\,cm \times 1\,m$ and height $2\,m$ is placed on a horizontal floor. What minimum force should be applied horizontally, perpendicular to the top edge of the cup-board to tilt it on the *(i)* long base edge, and *(ii)* short base edge?

9.14. A child walks back and forth from one end to another along one of the diameters on a rotating merry-go-round . Qualitatively describe how the angular velocity of the merry-go-round changes during one complete trip of the child?

9.15. Derive all the expressions for moment of inertia for symmetrical solids given in Table (9.1). Prove that the moment of inertia for a thick spherical shell reduces to the MI of thin spherical shell in the limit $R_1 = R_2 = R$.

9.16. Calculate the products of inertia of a rectangular plate of area $a \times b$ and mass M. The origin of the coordinate axes is at one of the corners of the plate, and the $x-$ and $y-axes$ are along the two edges of the plate.

9.17. Calculate the products of inertia of a circular plate of radius R and mass M. The centre of the plate is the origin of the coordinates axes and the $z-axis$ is perpendicular to the plate.

9.18. A meter rule of $150\,g$ mass is balanced on a knife edge at the $50\,cm$ mark. A $50\,g$ mass is suspended from its $20\,cm$ mark, and a $25\,g$ mass is suspended from its $40\,cm$ mark. *(i)* What mass should be suspended from the $80\,cm$ mark, and *(ii)* from what point a $75\,g$ mass should be suspended to balance the meter rule? Calculate the reaction force of the knife edge in both cases.

9.19. A $4\,m$ long uniform ladder of $20\,kg$ mass is leaning against a smooth wall on a rough floor. The ladder makes an angle of 37° from the floor. A painter of $80\,kg$ mass climbs the ladder three-fourth of the way up. What is the minimum coefficient of static friction that shall keep the ladder from slipping? If the wall is also rough and the coefficients of friction for the wall and the floor are equal, what should be its minimum value to keep the ladder from slipping?

9.20. Find the moment of inertia of the solid plates, each of mass M, shown in Figure 9.17 below about all the axes indicated. Axis 3 in each case is perpendicular to the plane of the plates, and the shaded portions in each case are the cut out portions removed from the plates. Also find expressions for the radius of gyration for each case.

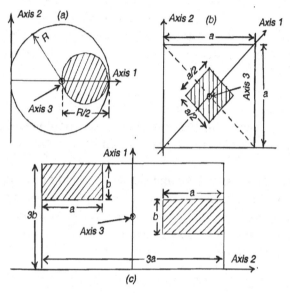

Figure 9.17: (i) Circular plate, (ii) A square plate, (iii) A rectangular plate all with cut out shaded portions as shown.

9.21. A $5\,kg$ mass placed on a rough horizontal plane is attached to a $4\,kg$ mass with a massless, unstreachable string. The string passes over a uniform pulley of mass $500\,g$ and radius $5\,cm$. The coefficient of kinetic friction between the mass and the plane is 0.2. Calculate the tension in the string and acceleration of the masses.

9.22. In Exercise (9.21) if the plane is inclined at 30^0 above horizontal, calculate the tension and acceleration of the masses.

9.23. A massless, unstreachable string is wound around a large uniform pulley of mass $2\,kg$ and radius $15\,cm$. The other end of the string is attached to a $5\,kg$ mass, which is released from rest from a height of $2\,m$ above ground. With what speed does the mass hit the ground, and how long does it take for the mass to reach the ground? Compare this to the same calculations if the pulley was massless, and explain any differences in the velocities and the times in the two cases.

9.24. A hollow cylinder and a solid cylinder, both of mass $5\,kg$, radius $15\,cm$ roll down without slipping a $2\,m$ long plane inclined at 30° above horizontal. Calculate the time taken by both to reach the foot of the plane, and the velocity with which they reach the bottom. If a rectangular block of the same mass slides down the plane (ignore friction), how long will it take, and what shall be its velocity on reaching the ground? Compare the results to those for the hollow and solid cylinders, and explain the differences.

9.25. A uniform rod of mass M and length L oscillates about a pivot *(i)* at one end of the rod, and *(ii)* at a distance of $\frac{L}{4}$ from one end. Determine the period of oscillation of the rod for both the cases.

9.26. A uniform disc of $10\,kg$ mass and $25\,cm$ radius is rotating at $\pi/4$ radian per second about a vertical axis through its centre. Another disc of mass $3\,kg$ and radius $15\,cm$ initially at rest is lowered on to the rotating disc such that the axis of rotation passes through its centre, and both the discs rotate as one. Calculate the new angular speed of the two discs combined. What fraction of the total energy is lost in the process?

9.27. A potters wheel of $50\,kg$ mass and $40\,cm$ radius is rotating at 5 revolutions per second. A $5\,kg$ clump of clay placed at $20\,cm$ distance from the axis of rotation spontaneously flies off the wheel tangentially with a speed of $2\,m\,s^{-1}$. Calculate the new speed of rotation of the wheel.

9.28. A uniform, square trap door of mass $200\,kg$ and side $2\,m$ is held horizontally by hinges on one of the side. The door is lifted open by a cable attached to the mid point of the opposite side of the door, and the cable makes an angle of 45° from the horizontal when the door is closed. Calculate the minimum tension that the string should be able to withstand when the door begins to lift. What are the forces on the hinges at this moment?

Chapter 10

Lagrangian Dynamics

The focus of this book so far has been to solve mechanics problems using Newtonian mechanics. The objective of this chapter is to introduce an alternative method of solving mechanics problems using a more elegant and sophisticated but simple technique known as Lagrangian dynamics. The technique developed by a French Mathematician, Joseph Louis Lagrange uses a Lagrangian function of the system to define the equations of motion of the system, which are then solved to determine the desired unknown parameters and properties of the system. The technique can be applied to many body problems of connected and constrained particles with much more ease as compared to the Newtonian mechanics. To illustrate this, a number of examples that have earlier been solved using Newtonian mechanics are revisited, and solved by Lagrangian dynamics. The chapter concludes by a brief introduction to yet another technique, Hamilton's equations due to Scottish mathematician Sir William R Hamilton that uses a Hamiltonian function of the system to formulate the equations of motion.

In Newtonian mechanics, one identifies and deals with the external forces acting on the system, which are then used to formulate the equations of motion using Newton's laws of motion. Solution of the equations of motion in conjunction with the conservation laws, namely the conservation of energy, momentum and angular momentum then leads to the desired unknown quantities and properties of the mechanical system. The method is quite simple and straightforward so long as one is dealing with systems of not more than few particles, or a rigid body in which the interparticle separation is fixed, and the internal forces resulting from interparticle interactions are known not to influence the motion of the rigid body as such. However, many a manybody system may have constraints that restrict the independent motion of individual particles. The position coordinates r_i of particles in a constrained system are no longer independent. The equations of motions of such systems are not independent, and the use of Newtonian mechanics leads to coupled equations, as we saw from the examples of coupled oscillations in Chapter 8. Solving coupled equations is not always a simple task, and in some cases the equations may not even be solvable. Further difficulty with the Newtonian mechanics is presented by the forces of constraints that are unknown. These forces manifest themselves in terms of the motion of the system as a whole, and can only be determined from the solution of the equations of motions. Both these problems are dealt with by Lagrangian dynamics if the constraints are holonomic, *i.e.*, if they can be expressed as mathematical equations.

If the constraints are nonholonomic, *i.e.* they are not expressible as mathematical equations, another technique of Lagrange multipliers is used which is beyond the scope of this book.

In Lagrangian dynamics, the problem of constraints is overcome by defining a set of independent coordinates known as the *generalized* coordinates. For a system of N particles with no constraints there are $3N$ degrees of freedom, and hence there are $3N$ generalized coordinates. But a system of N particles with k number of constraints has $(3N - k)$ degrees of freedom, and there are only $(3N - k)$ generalized coordinates. The issue of the forces of constraints is tackled by formulating the problem in such a way that the forces of constraint disappear from the equations of motion. Thus one only deals with the applied known forces, and a set of independent generalized coordinates which give equations of motion in terms of independent coordinates. Other parameters that enter the Lagrange equations of motion through the generalized coordinates are the generalized velocity, and generalized forces as discussed below.

10.1 Lagrangian formulation

In this section, we define the generalized parameters, and present the mathematical formulation of Lagrangian dynamics leading to Lagrange's equations of motion.

10.1.1 Generalized Coordinates

In general, a system with n degrees of freedom requires a minimum of n independent coordinates to specify the system, denoted as:

$$q_1, q_2, \cdots q_n \quad \text{or} \quad (q_i, i = 1, 2, \cdots, n) \tag{10.1}$$

These are known as *generalized coordinates*. The Cartesian coordinates can be expressed as functions of the generalized coordinates as:

$$x = x(q) \quad \text{for one degree of freedom}$$

$$x = x(q_1, q_2) \quad \text{for two degrees of freedom}$$
$$y = y(q_1, q_2) \quad \text{for two degrees of freedom}$$

$$x = x(q_1, q_2, q_3) \quad \text{for three degrees of freedom}$$
$$y = y(q_1, q_2, q_3) \quad \text{for three degrees of freedom}$$
$$z = z(q_1, q_2, q_3) \quad \text{for three degrees of freedom}$$

In general for a system of N particles with k number of constraints, the position vector $\mathbf{r_i}$ of the ith particle can be expressed in terms of $n = (3N - k)$ number of generalized coordinates as:

$$\mathbf{r_i} = \mathbf{r_i}(q_1, q_2, \ldots, q_n, t) \Rightarrow \tag{10.2}$$

$$\begin{cases} x_i = x_i(q_1, q_2, \ldots, q_n, t) \\ y_i = y_i(q_1, q_2, \ldots, q_n, t) \\ z_i = z_i(q_1, q_2, \ldots, q_n, t) \end{cases}$$

If the generalized coordinates change from $q_1, q_2, \cdots q_n$ to neigbouring values $q_1+\delta q_1, q_2+\delta q_2, \cdots q_n+\delta q_n$, then the corresponding changes in Cartesian coordinates are such that the particle moves from x_i, y_i, z_i to $x_i + \delta x_i, y_i + \delta y_i, z_i + \delta z_i$, where

$$\left.\begin{aligned}
\delta x_i &= \frac{\partial x_i}{\partial q_1}\delta q_1 + \frac{\partial x_i}{\partial q_2}\delta q_2 + ... + \frac{\partial x_i}{\partial q_n}\delta q_n = \sum_{k=1}^{n}\frac{\partial x_i}{\partial q_k}\delta q_k \\
\delta y_i &= \frac{\partial y_i}{\partial q_1}\delta q_1 + \frac{\partial y_i}{\partial q_2}\delta q_2 + ... + \frac{\partial y_i}{\partial q_n}\delta q_n = \sum_{k=1}^{n}\frac{\partial y_i}{\partial q_k}\delta q_k \\
\delta z_i &= \frac{\partial z_i}{\partial q_1}\delta q_1 + \frac{\partial z_i}{\partial q_2}\delta q_2 + ... + \frac{\partial z_i}{\partial q_n}\delta q_n = \sum_{k=1}^{n}\frac{\partial z_i}{\partial q_k}\delta q_k
\end{aligned}\right\} \Rightarrow$$

$$\delta \mathbf{r_i} = \frac{\partial \mathbf{r_i}}{\partial q_1}\delta q_1 + \frac{\partial \mathbf{r_i}}{\partial q_2}\delta q_2 + ... + \frac{\partial \mathbf{r_i}}{\partial q_n}\delta q_n = \sum_{k=1}^{n}\frac{\partial \mathbf{r_i}}{\partial q_k}\delta q_k \tag{10.3}$$

As Cartesian coordinates are the most frequently used set of coordinates, our common conception of coordinates is that they are vectors and have dimension of length. In complete contrast to this, an important aspect of generalized coordinates is that they do not necessarily have the dimensions of length and may not be vectors. They need not be orthogonal as well. Any parameter of the system such as angles, energy, angular momentum, etc. can constitute generalized coordinates.

10.1.2 Generalized velocities

Corresponding to n number of generalized coordinates, we define n number of generalized velocities, the kth generalized velocity ($k = 1, 2, ..., n$) given as:

$$\dot{\mathsf{q}}_k = \frac{d\,q_k}{dt} \tag{10.4}$$

As in the case of generalized coordinates, this is not necessarily a velocity in the conventional sense.

10.1.3 Generalized Forces

If a particle undergoes a displacement $\delta \mathbf{r}$ under the action of a force $\mathbf{F}$, then the work done δW is given by

$$\begin{aligned}
\delta W &= \mathbf{F} \cdot \delta \mathbf{r} \\
&= \sum_i F_i \delta x_i
\end{aligned}$$

Likewise, for a system of N particles, if the jth particle undergoes a displacement $\delta \mathbf{r_j}$ under the action of force $\mathbf{F_j}$, the work done δW is given as:

$$\delta W = \sum_{j=1}^{N}\mathbf{F}_j \cdot \delta \mathbf{r}_j = \sum_{j=1}^{N}(F_{xj}\,\delta x_j + F_{yj}\,\delta y_j + F_{zj}\,\delta z_j) \tag{10.5}$$

Equation (10.5) comprises $3N$ terms corresponding to (x, y, z) coordinates of N number of particles. When one substitutes δx_j etc. in terms of generalized coordinates using equation (10.3), it leads to a mathematically long and cumbersome expressions with nesting of several summations. In order to simplify such expressions it is a common practice in mathematics to use a single notation x_i to represent all the three components (x, y, z) where the index i runs from 1 to $3N$ for the N particles system such that for $i = j$, $x_i = x_j$, for $i = j + 1$, $x_i = y_j$, and for $i = j + 2$, $x_i = z_j$ where $j = 1, 4, 7 \ldots, (3N - 2)$ takes N values for a N particle system. In terms of the new notation, equation (10.5) becomes:

$$\delta W = \sum_{i=1}^{3N} F_i\, \delta x_i = \sum_{i} F_i\, \delta x_i \tag{10.6}$$

From here onwards we use this notation, unless otherwise stated. Substituting δx_i from equation (10.3) in equation (10.6):

$$
\begin{aligned}
\delta W &= \sum_{i}\left(F_i \sum_{k} \frac{\partial x_i}{\partial q_k}\delta q_k\right) \\
&= \sum_{i}\left(\sum_{k} F_i \frac{\partial x_i}{\partial q_k}\delta q_k\right) \\
&= \sum_{k}\left(\sum_{i} F_i \frac{\partial x_i}{\partial q_k}\right)\delta q_k \quad \text{where order of summation has been reversed} \\
&= \sum_{k} Q_k \delta q_k \tag{10.7}
\end{aligned}
$$

where the quantity Q_k is given by

$$Q_k = \sum_{i}\left(F_i \frac{\partial x_i}{\partial q_k}\right) \tag{10.8}$$

and is known as the *generalized force*, where $k = 1$ to n for n number of generalized coordinates. Once again Q_k does not necessarily have the dimensions of force.

It can be recalled that in a conservative force field, the force is given as the gradient of the potential energy function, V, in the form

$$F_i = -\frac{\partial V}{\partial x_i} \tag{10.9}$$

and hence the generalized force given by equation (10.8) becomes

$$
\begin{aligned}
Q_k &= -\left(\sum_{i} \frac{\partial V}{\partial x_i}\frac{\partial x_i}{\partial q_k}\right) \\
&= -\frac{\partial V}{\partial q_k} \tag{10.10}
\end{aligned}
$$

In the case when part of the generalised forces *are not* conservative, say Q'_k, and part are derivable from the potential V then the total generalised force is

$$Q_k = Q'_k - \frac{\partial V}{\partial q_k} \tag{10.11}$$

10.1.4 Lagrange's Equations

Derivation of Lagrange's Equation from Newton's Second law

Let us find an equation of motion in terms of the generalized coordinates, q_i and their derivatives, the generalized velocities. From $\mathbf{r_i} = \mathbf{r_i}(q_1, q_2, \cdots, q_n, t)$, one can prove that

$$\frac{\partial \dot{\mathbf{r}}_{\mathbf{i}}}{\partial \dot{q}_k} = \frac{\partial \mathbf{r_i}}{\partial q_k} \tag{10.12}$$

Multiplying both sides by $\dot{\mathbf{r}}_{\mathbf{i}}$ and differentiating with respect to t, we obtain

$$\frac{d}{dt}\left(\dot{\mathbf{r}}_{\mathbf{i}}\frac{\partial \dot{\mathbf{r}}_{\mathbf{i}}}{\partial \dot{q}_k}\right) = \frac{d}{dt}\left(\dot{\mathbf{r}}_{\mathbf{i}}\frac{\partial \mathbf{r_i}}{\partial q_k}\right)$$

$$= \ddot{\mathbf{r}}_{\mathbf{i}}\frac{\partial \mathbf{r_i}}{\partial q_k} + \dot{\mathbf{r}}_{\mathbf{i}}\frac{\partial \dot{\mathbf{r}}_{\mathbf{i}}}{\partial q_k}$$

or

$$\frac{d}{dt}\left(\frac{\partial}{\partial \dot{q}_k}\frac{\dot{\mathbf{r}}_{\mathbf{i}}^2}{2}\right) = \ddot{\mathbf{r}}_{\mathbf{i}}\frac{\partial \mathbf{r_i}}{\partial q_k} + \frac{\partial}{\partial q_k}\left(\frac{\dot{\mathbf{r}}_{\mathbf{i}}^2}{2}\right)$$

$$\frac{d}{dt}\left(\frac{\partial}{\partial \dot{q}_k}\frac{m_i\dot{\mathbf{r}}_{\mathbf{i}}^2}{2}\right) = m_i\ddot{\mathbf{r}}_{\mathbf{i}}\frac{\partial \mathbf{r_i}}{\partial q_k} + \frac{\partial}{\partial q_k}\left(\frac{m_i\dot{\mathbf{r}}_{\mathbf{i}}^2}{2}\right) \tag{10.13}$$

where in the last equation, m_i has been multiplied throughout. But, $m_i\ddot{\mathbf{r}}_{\mathbf{i}} = F_i$, is the force, as is well known from Newtonian mechanics, and $\frac{1}{2}m_i\dot{\mathbf{r}}_{\mathbf{i}}^2$ is the kinetic energy. Summing over i, equation (10.13) becomes

$$\frac{d}{dt}\frac{\partial T}{\partial \dot{q}_k} = \sum_i\left(F_i\frac{\partial \mathbf{r_i}}{\partial q_k}\right) + \frac{\partial T}{\partial q_k} \tag{10.14}$$

$$= Q_k + \frac{\partial T}{\partial q_k}$$

$$= -\frac{\partial V}{\partial q_k} + \frac{\partial T}{\partial q_k} \quad \text{where equations (10.8) and (10.10) have been used}$$

$$= \frac{\partial T}{\partial q_k} - \frac{\partial V}{\partial q_k} \tag{10.15}$$

A function L, known as the Lagrangian, is defined as

$$L = T - V \tag{10.16}$$

where T is the kinetic energy and V is the potential energy, and these quantities are expressed in terms of generalized coordinates, that is

$$L(q_k, \dot{q}_k, t) = T(q_k, \dot{q}_k, t) - V(q_k, t)$$

Now, since $V = V(q_k) \Rightarrow \partial V / \partial \dot{q}_k = 0$, we obtain

$$\frac{\partial L}{\partial q_k} = \frac{\partial T}{\partial q_k} - \frac{\partial V}{\partial q_k}$$

$$\frac{\partial L}{\partial \dot{q}_k} = \frac{\partial T}{\partial \dot{q}_k} \qquad\qquad (10.17)$$

Using equations (10.15) and (10.16) in (10.17), we obtain

$$\frac{d}{dt}\frac{\partial L}{\partial \dot{q}_k} = \frac{\partial L}{\partial q_k} \quad \text{or}$$

$$\frac{d}{dt}\frac{\partial L}{\partial \dot{q}_k} - \frac{\partial L}{\partial q_k} = 0 \qquad\qquad (10.18)$$

which are known as *Lagrange's Equations*. They constitute a set of equations for n number of generalised coordinates.

In the case when there are non-conservative forces present, Lagrange's equations take the form

$$\frac{d}{dt}\frac{\partial L}{\partial \dot{q}_k} - \frac{\partial L}{\partial q_k} = Q'_k \qquad\qquad (10.19)$$

where Q'_k are non conservative generalised forces.

Derivation of Lagrange's Equations from Hamilton's Variational Principle

Hamilton's Variational principle (or principle of least action) states that *Every mechanical system is characterised by a definite function of position and velocity known as the Lagrangian, $L(q, \dot{q}, t) = T - V$, and any motion is such that the action, S, takes the least possible value, where the action is defined as*

$$S = \int_{t_1}^{t_2} L(q_k, \dot{q}_k, t)dt \qquad\qquad (10.20)$$

What Hamilton's principle says is that out of all possible ways a system can evolve in a given interval of time $(t_2 - t_1)$, the motion that will occur will be the one in which the action is a minimum. This can be expressed as:

$$\delta S = \delta \int_{t_1}^{t_2} L(q_k, \dot{q}_k, t)dt = 0 \qquad\qquad (10.21)$$

where δ refers to a small variation. Let us find the variation δS.

$$\delta S = \delta \int_{t_1}^{t_2} L(q_k, \dot{q}_k, t)dt$$

$$= \int_{t_1}^{t_2} \delta L(q_k, \dot{q}_k, t)dt$$

$$= \int_{t_1}^{t_2} \sum_k \left(\frac{\partial L}{\partial q_k}\delta q_k + \frac{\partial L}{\partial \dot{q}_k}\delta \dot{q}_k \right) dt$$

The second term on the right hand side in the above equation can be integrated by parts to obtain

$$\int_{t_1}^{t_2} \sum_k \frac{\partial L}{\partial \dot{q}_k}\delta \dot{q}_k dt = \left[\frac{\partial L}{\partial \dot{q}_k}\delta q_k \right]_{t_1}^{t_2} - \int_{t_1}^{t_2} \sum_k \frac{d}{dt}\frac{\partial L}{\partial \dot{q}_k}\delta q_k dt$$

$$= 0 - \int_{t_1}^{t_2} \sum_k \frac{d}{dt}\frac{\partial L}{\partial \dot{q}_k}\delta q_k dt$$

where the fact that the variation $\delta q_k = 0$ at t_1 and t_2 has been used to obtain the vanishing zero term in the above result. Inserting this result in the expression for δS, we obtain

$$\delta S = \int_{t_1}^{t_2} \sum_k \left[\frac{\partial L}{\partial q_k} - \frac{d}{dt}\frac{\partial L}{\partial \dot{q}_k} \right]\delta q_k dt = 0 \qquad (10.22)$$

which implies that

$$\frac{\partial L}{\partial q_k} - \frac{d}{dt}\frac{\partial L}{\partial \dot{q}_k} = 0$$

or

$$\frac{d}{dt}\frac{\partial L}{\partial \dot{q}_k} - \frac{\partial L}{\partial q_k} = 0$$

which are precisely the same as *Lagrange's Equations* obtained ealier in (10.18).

10.1.5 Generalized momentum

Although not directly involved in Lagrange's equations of motion or in the Lagrangian function, we introduce generalized momentum for the completeness sake of generalized parameters of a system, which also highlights an important property of the system. It is also used in Hamiltonian mechanics which has applications in quantum and statistical mechanics. The generalized momentum is defines as:

$$p_j = \frac{\partial L}{\partial \dot{q}_j} \qquad (10.23)$$

where we once again emphasize that generalized momentum is not necessarily momentum in the conventional sense, and may not have the same dimensions as that of momentum.

The Lagrangian L is a function of generalized coordinates and generalized velocities, *i.e.*, $L = L(q_k, \dot{q}_k, t)$. If any of the generalized coordinate, say q_j does not appear explicitly in the Lagrangian function, although it may contain the corresponding generalized velocity, $\dot{q}_j$, then such a generalized coordinate is known as the *cyclic coordinate*. For a cyclic coordinate:

$$\frac{\partial L}{\partial q_j} = 0 \qquad (10.24)$$

and the corresponding Lagrange's equation becomes:

$$\frac{d}{dt}\frac{\partial L}{\partial \dot{q}_j} = 0 = \frac{dp_j}{dt} \qquad (10.25)$$

Equation (10.25) implies:

$$p_j = \text{Constant} \qquad (10.26)$$

Thus the generalized momentum, corresponding to a cyclic coordinate is conserved.

10.2   Applications of Lagrange's Equations

In this section, Lagrange's equations are applied to several cases. Some of these problems will have been solved using Newton's laws of motion or energy principles in the earlier chapters. The point, however, is to show the versatility of Lagrangian dynamics. A generalised recipe, given below, is useful in solving problems of dynamics using the Lagrangian formulation.

Recipe 10.1

- Choose a suitable set of generalized coordinates to represent the system.

- Write down the transformation equations relating the conventional and generalized coordinates

- Obtain the kinetic energy, T, as a function of the generalized coordinates.

- If the system is conservative, obtain the potential energy, V as a function of the generalized coordinates.

- Obtain the Lagrangian, $L = T - V$

- Apply the Lagrange's Equations and obtain n number of equations of motion corresponding to each of the n generalized coordinates.

- Solve the resulting equations.

10.2.1   The Simple Atwood Machine

Consider the simple Atwood machine consisting of two particles of masses m_1 and m_2 connected by a light inextensible cord which passes over a pulley of radius a and moment of inertia I, as illustrated in Figure (10.1).

The simple Atwood machine is a three particle system which shall have 9 degrees of freedom, if all particles moved independently. But the system has constraints as described below:

- All particles move in the same vertical plane, there is no motion perpendicular to the plane $\rightarrow$ 3 constraints.

- The two masses move only along vertical, there is no horizontal motion $\rightarrow$ 2 constraints.

- The vertical displacement of the two masses are related through the length of the string, so that the displacement of one determines the displacement of the other $\rightarrow$ 1 constraint.

- The pulley does not have an independent displacement. Its angular rotation θ is related to the linear displacement x of the mass(es) through the radius of the pulley as: $x = a\theta \rightarrow 2$ constraints.

These add up to a total of 8 constraints, leaving us with only one dgree of freedon, and hence only one generalized coordinate. This we choose as the vertical displacement x of one of the masses as shown in Figure (10.1).

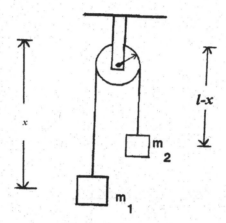

Figure 10.1: The Simple Atwood machine.

The corresponding generalized velocity is $\dot{x}$. The kinetic energy, T, potential energy, V, and the Lagrangian, L, of the system are given as

$$
\begin{aligned}
T &= \frac{1}{2}m_1\dot{x}^2 + \frac{1}{2}m_2\dot{x}^2 + \frac{1}{2}I\frac{\dot{x}^2}{a^2} \\
V &= -m_1gx - m_2g(l-x) \\
L &= T - V = \frac{1}{2}\left(m_1 + m_2 + \frac{I}{a^2}\right)\dot{x}^2 + g(m_1 - m_2)x + m_2gl
\end{aligned}
\tag{10.27}
$$

Hence

$$
\begin{aligned}
\frac{\partial L}{\partial x} &= g(m_1 - m_2) \\
\frac{\partial L}{\partial \dot{x}} &= \left(m_1 + m_2 + \frac{I}{a^2}\right)\dot{x} \\
\frac{d}{dt}\frac{\partial L}{\partial \dot{x}} &= \left(m_1 + m_2 + \frac{I}{a^2}\right)\ddot{x}
\end{aligned}
$$

Applying Lagrange's equation,

$$\frac{d}{dt}\frac{\partial L}{\partial \dot{x}} = \frac{\partial L}{\partial x}$$

$$\left(m_1 + m_2 + \frac{I}{a^2}\right)\ddot{x} = g(m_1 - m_2)$$

which gives

$$\ddot{x} = \frac{(m_1 - m_2)g}{m_1 + m_2 + I/a^2} \tag{10.28}$$

which is the acceleration of the masses. This is the same expression as was obtained in Chapter 9 by rigid body dynamics. Note that if $m_1 > m_2$, m_1 moves downwards and m_2 moves upwards, whereas if $m_1 < m_2$, m_1 moves upwards and m_2 moves downwards with constant equal accelerations. The inertial effect of the pulley shows up in the term I/a^2 in the denominator. In elementary Newtonian mechanics, this acceleration is obtained as $(m_1 - m_2)g/(m_1 + m_2)$ because the effect of the mass of the pulley is ignored.

10.2.2 The Simple Pendulum

Consider a simple pendulum of mass m and length l, as illustrated in Figure (10.2). Although we have considered this system before, we shall study it again using Lagrangian dynamics. The system has only one degree of freedom and the corresponding generalized coordinate is taken to be the angular displacement θ.

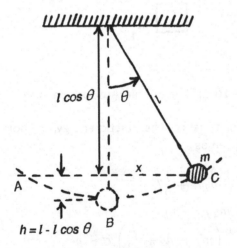

Figure 10.2: The simple pendulum.

The kinetic energy, T, potential energy, V, and the Lagrangian, L, are given by

$$T = \frac{1}{2}ml^2\dot{\theta}^2$$

$$V = mgl(1 - \cos\theta)$$

$$L = T - V = \frac{1}{2}ml^2\dot{\theta}^2 - mgl(1 - \cos\theta) \tag{10.29}$$

Hence

$$\frac{\partial L}{\partial \theta} = -mgl\sin\theta$$

$$\frac{\partial L}{\partial \dot\theta} = ml^2\dot\theta$$

$$\frac{d}{dt}\frac{\partial L}{\partial \dot\theta} = ml^2\ddot\theta$$

Applying Lagrange's equation,

$$\frac{d}{dt}\frac{\partial L}{\partial \dot\theta} - \frac{\partial L}{\partial \theta} = 0$$

$$ml^2\ddot\theta + mgl\sin\theta = 0$$

$$\ddot\theta + \frac{g}{l}\sin\theta = 0$$

$$\ddot\theta + \frac{g}{l}\theta = 0 \qquad (10.30)$$

where the approximation for small angles, $\sin\theta \approx \theta$, has been used. Equation (10.30) is a familiar second order differential equation for a simple pendulum, obtained ealier in Chapter 8 using Newtonian mechanics. These oscillations have an angular frequency $\omega = \sqrt{g/l}$.

10.2.3 The Mass-spring system: Simple Harmonic Oscillations

Consider a mass-spring system on a smooth horizontal plane, as illustrated in Figure (10.3). After considering the constraints of the system, we conclude that there is only one degree of freedom, and hence only one generalized coordinate is needed. This is chosen to be the linear displacement x of the mass from the equilibrium position.

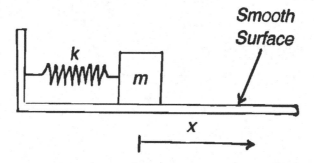

Figure 10.3: The mass-spring system exhibiting simple harmonic oscillations.

The kinetic energy, T, potential energy, V, and the Lagrangian, L, are given by

$$T = \frac{1}{2}m\dot x^2$$

$$V = \frac{1}{2}kx^2$$

$$L = T - V = \frac{1}{2}m\dot x^2 - \frac{1}{2}kx^2 \qquad (10.31)$$

Hence

$$\frac{\partial L}{\partial x} = -kx$$

$$\frac{\partial L}{\partial \dot{x}} = m\dot{x}$$

$$\frac{d}{dt}\frac{\partial L}{\partial \dot{x}} = m\ddot{x}$$

Applying Lagrange's equation,

$$\frac{d}{dt}\frac{\partial L}{\partial \dot{x}} - \frac{\partial L}{\partial x} = 0$$

$$m\ddot{x} + kx = 0 \qquad (10.32)$$

which is again the familiar second order differential equation for simple harmonic oscillations, obtained ealier in Chapter 8 using Newtonian mechanics. These oscillations have an angular frequency $\omega = \sqrt{k/m}$.

10.2.4   The Mass-spring system: Damped Oscillations

Consider a mass-spring system on a rough horizontal surface, as illustrated in Figure (10.4). Suppose that the damping force is proportional to velocity. This system differs fom the previous example in that it is *nonconservative*.

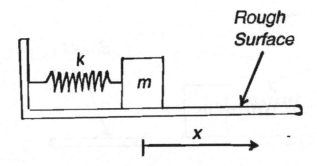

Figure 10.4: The Mass-spring system exhibiting damped oscillations.

The kinetic energy, T, potential energy, V, and the Lagrangian, L, are given by

$$T = \frac{1}{2}m\dot{x}^2$$

$$V = \frac{1}{2}kx^2$$

$$L = T - V = \frac{1}{2}m\dot{x}^2 - \frac{1}{2}kx^2 \qquad (10.33)$$

Hence

$$\frac{\partial L}{\partial x} = -kx$$

$$\frac{\partial L}{\partial \dot{x}} = m\dot{x}$$

$$\frac{d}{dt}\frac{\partial L}{\partial \dot{x}} = m\ddot{x}$$

Applying Lagrange's equation for nonconservative systems, noting that the nonconservative part of the generalised force is the damping force given as

$$Q'_k = -b\dot{x}$$

we obtain

$$\frac{d}{dt}\frac{\partial L}{\partial \dot{x}} - \frac{\partial L}{\partial x} = Q'_k$$

$$m\ddot{x} + kx = -b\dot{x}$$

$$m\ddot{x} + b\dot{x} + kx = 0 \qquad (10.34)$$

which is the familiar second order differential equation for damped oscillations, which was obtained ealier in Chapter 8 using Newtonian mechanics. These oscillations have a displacement, $x = x_0 e^{-\frac{b}{2m}t}\cos\omega_1 t$, and an angular frequency, $\omega_1 = \sqrt{\frac{k}{m} - \frac{b^2}{4m^2}}$ as was discussed in Chapter 8.

10.2.5 The Double Pendulum: Coupled Oscillations

Consider the double pendulum which is a system consisting of two coupled pendula of masses m_1, m_2 and lengths l_1, l_2, as illustrated in Figure (10.5). The system has two degrees of freedom represented by angles θ and ϕ which are angular displacements of each of the pendulum in the plane of oscillation.

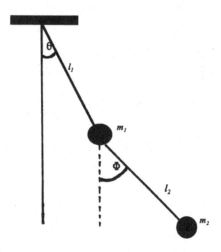

Figure 10.5: The double pendulum.

The kinetic energy, T, potential energy, V, and the Lagrangian, L, are given by

$$
\begin{aligned}
T &= \frac{1}{2}m_1 l_1^2 \dot{\theta}^2 + \frac{1}{2}m_2\left[l_1^2\dot{\theta}^2 + l_2^2\dot{\phi}^2 + 2l_1\dot{\theta}l_2\dot{\phi}\cos(\phi-\theta)\right] \\
&\approx \frac{1}{2}m_1 l_1^2 \dot{\theta}^2 + \frac{1}{2}m_2\left[l_1^2\dot{\theta}^2 + l_2^2\dot{\phi}^2 + 2l_1 l_2\dot{\theta}\dot{\phi}\right] \quad \text{for small oscillations,} \\
V &= -m_1 g l_1 \cos\theta - m_2 g(l_1\cos\theta + l_2\cos\phi) \\
&\approx -m_1 g l_1(1 - \frac{\theta^2}{2}) - m_2 g\left[l_1(1 - \frac{\theta^2}{2}) + l_2(1 - \frac{\phi^2}{2})\right] \quad \text{for small oscillations,} \\
L &= T - V \\
&= \frac{1}{2}m_1 l_1^2 \dot{\theta}^2 + \frac{1}{2}m_2\left[l_1^2\dot{\theta}^2 + l_2^2\dot{\phi}^2 + 2l_1 l_2\dot{\theta}\dot{\phi}\right] + m_1 g l_1 + m_2 g(l_1 + l_2) \\
&\quad - m_1 g l_1\frac{\theta^2}{2} - \frac{1}{2}m_2 g[l_1\theta^2 + l_2\phi^2] \quad\quad\quad\quad\quad\quad\quad\quad\quad (10.35)
\end{aligned}
$$

Hence, for the θ coordinate, we have

$$
\begin{aligned}
\frac{\partial L}{\partial \theta} &= -m_1 g l_1 \theta - m_2 g l_1 \theta \\
\frac{\partial L}{\partial \dot{\theta}} &= m_1 l_1^2 \dot{\theta} + m_2 l_1^2 \dot{\theta} + m_2 l_1 l_2 \dot{\phi} \\
\frac{d}{dt}\frac{\partial L}{\partial \dot{\theta}} &= (m_1 + m_2)l_1^2\ddot{\theta} + m_2 l_1 l_2\ddot{\phi}
\end{aligned}
$$

Applying Lagrange's equation,

$$
\begin{aligned}
\frac{d}{dt}\frac{\partial L}{\partial \dot{\theta}} - \frac{\partial L}{\partial \theta} &= 0 \\
(m_1 + m_2)l_1^2\ddot{\theta} + m_2 l_1 l_2\ddot{\phi} + (m_1 + m_2)g l_1\theta &= 0 \\
(m_1 + m_2)l_1\ddot{\theta} + m_2 l_2\ddot{\phi} + (m_1 + m_2)g\theta &= 0 \quad (10.36)
\end{aligned}
$$

For the ϕ coordinate, we have

$$
\begin{aligned}
\frac{\partial L}{\partial \phi} &= -m_2 g l_2 \phi \\
\frac{\partial L}{\partial \dot{\phi}} &= m_2 l_2^2 \dot{\phi} + m_2 l_1 l_2 \dot{\theta} \\
\frac{d}{dt}\frac{\partial L}{\partial \dot{\phi}} &= m_2 l_2^2\ddot{\phi} + m_2 l_1 l_2\ddot{\theta}
\end{aligned}
$$

Applying Lagrange's equation,

$$
\begin{aligned}
\frac{d}{dt}\frac{\partial L}{\partial \dot{\phi}} - \frac{\partial L}{\partial \phi} &= 0 \\
m_2 l_2^2\ddot{\phi} + m_2 l_1 l_2\ddot{\theta} + m_2 g l_2\phi &= \\
m_2 l_2\ddot{\phi} + m_2 l_1\ddot{\theta} + m_2 g\phi &= 0 \quad (10.37)
\end{aligned}
$$

Assuming normal mode solutions of the form $\theta = \theta_0 e^{-i\omega t}$ and $\phi = \phi_0 e^{-i\omega t}$ equations (10.36) and (10.37) can be expressed in a matrix form as below

$$\begin{bmatrix} (m_1 + m_2)(g - l_1\omega^2) & -m_2 l_2 \omega^2 \\ \\ -m_2 l_1 \omega^2 & m_2(g - l_2\omega^2) \end{bmatrix} \begin{bmatrix} \theta \\ \phi \end{bmatrix} = \begin{bmatrix} 0 \\ 0 \end{bmatrix} \tag{10.38}$$

A solution exists if the determinant of the 2×2 matrix vanishes, that is

$$(m_1 + m_2)(g - l_1\omega^2)m_2(g - l_2\omega^2) - m_2^2 l_1 l_2 \omega^4 = 0$$

which can be simplified to the form

$$\omega^4 - \left(\frac{m_1 + m_2}{m_1}\right)\left(\frac{g}{l_1} + \frac{g}{l_2}\right)\omega^2 + \left(\frac{m_1 + m_2}{m_1}\right)\frac{g^2}{l_1 l_2} = 0 \tag{10.39}$$

which is a quadratic equation in ω^2 and gives two normal mode frequencies as solutions of the equation (10.39). It is interesting, however, to consider three special cases:

Case (i): $m_1 = m_2 = m$ and $l_1 = l_2 = l$

When the double pendulum consists of two equal masses ($m_1 = m_2 = m$) and the two strings are of equal length ($l_1 = l_2 = l$), equation (10.39) reduces to

$$\omega^4 - 4\frac{g}{l}\omega^2 + 2\frac{g^2}{l^2} = 0 \tag{10.40}$$

which can be easily recognized as a quadratic equation in ω^2 with a solution

$$\omega^2 = \frac{g}{l}(2 \pm \sqrt{2})$$

and the system has two normal mode frequencies

$$\omega_+ = \sqrt{\frac{g}{l}(2 + \sqrt{2})} \quad \text{and} \quad \omega_- = \sqrt{\frac{g}{l}(2 - \sqrt{2})} \tag{10.41}$$

which have a ratio which is independent of the length l, given by

$$\frac{\omega_+}{\omega_-} = \left(\frac{2 + \sqrt{2}}{2 - \sqrt{2}}\right)^{1/2} = 2.414 \tag{10.42}$$

Case (ii): $m_1 \gg m_2$ and $l_1 \neq l_2$

When the double pendulum consists of two unequal masses ($m_1 \gg m_2$) so that $(m_1 + m_2) \approx m_1$ and the two strings are of unequal length ($l_1 \neq l_2$), equation (10.39) reduces to

$$\omega^4 - \left(\frac{g}{l_1} + \frac{g}{l_2}\right)\omega^2 + \frac{g^2}{l_1 l_2} = 0 \tag{10.43}$$

which can be easily recognized as a quadratic equation in ω^2 with a solution which has two normal mode frequencies

$$\omega_+ = \sqrt{\frac{g}{l_1}} \quad \text{and} \quad \omega_- = \sqrt{\frac{g}{l_2}} \tag{10.44}$$

Case (iii): $m_1 \ll m_2$ and $l_1 \neq l_2$

When the double pendulum consists of two unequal masses $(m_1 \ll m_2)$ so that $(m_1 + m_2) \approx m_2$ and the two strings are of unequal length $(l_1 \neq l_2)$, equation (10.39) reduces to

$$\omega^4 - \frac{m_2}{m_1}\left(\frac{g}{l_1} + \frac{g}{l_2}\right)\omega^2 + \frac{m_2}{m_1}\frac{g^2}{l_1 l_2} = 0 \tag{10.45}$$

which can be easily recognized as a quadratic equation in ω^2 with a solution which has two normal mode frequencies

$$\omega_\pm^2 = \frac{1}{2}\frac{m_2}{m_1}\left(\frac{g}{l_1} + \frac{g}{l_2}\right) \pm \frac{1}{2}\left\{\frac{m_2^2}{m_1^2}\left(\frac{g}{l_1} + \frac{g}{l_2}\right)^2 - 4\frac{m_2}{m_1}\frac{g^2}{l_1 l_2}\right\}^{1/2} \tag{10.46}$$

10.2.6 Mass-spring systems: Coupled Oscillations

Consider the longitudinal oscillations of a coupled mass-spring system as illustrated in Figure (10.6). u_a and u_b are the two generalized coordinates of the system.

The kinetic energy, T, potential energy, V, and the Lagrangian, L, are given by

$$\begin{aligned} T &= \frac{1}{2}m\dot{u_a}^2 + \frac{1}{2}m\dot{u_b}^2 \\ V &= \frac{1}{2}ku_a^2 + \frac{1}{2}k(u_b - u_a)^2 + \frac{1}{2}ku_b^2 \\ L &= T - V = \frac{1}{2}m\dot{u_a}^2 + \frac{1}{2}m\dot{u_b}^2 - k(u_a^2 + u_b^2 - 2u_a u_b) \end{aligned} \tag{10.47}$$

Figure 10.6: Mass-spring coupled oscillations.

Hence

$$\frac{\partial L}{\partial u_a} = -2ku_a + ku_b$$

$$\frac{\partial L}{\partial u_b} = -2ku_b + ku_a$$

$$\frac{\partial L}{\partial \dot{u}_a} = m\dot{u}_a$$

$$\frac{\partial L}{\partial \dot{u}_b} = m\dot{u}_b$$

$$\frac{d}{dt}\frac{\partial L}{\partial \dot{u}_a} = m\ddot{u}_a$$

$$\frac{d}{dt}\frac{\partial L}{\partial \dot{u}_b} = m\ddot{u}_b$$

Applying Lagrange's equation,

$$\frac{d}{dt}\frac{\partial L}{\partial \dot{u}_a} = \frac{\partial L}{\partial u_a}$$

$$\frac{d}{dt}\frac{\partial L}{\partial \dot{u}_b} = \frac{\partial L}{\partial u_b}$$

which gives

$$m\ddot{u}_a = -2ku_a + ku_b$$

$$m\ddot{u}_b = -2ku_b + ku_a$$

which upon considering normal mode solutions $u_a \sim e^{-i\omega t}$ and $u_b \sim e^{-i\omega t}$ leads to the following matrix equation

$$\begin{bmatrix} \left(\frac{2k}{m} - \omega^2\right) & -\frac{k}{m} \\ \\ -\frac{k}{m} & \left(\frac{2k}{m} - \omega^2\right) \end{bmatrix} \begin{bmatrix} u_a \\ \\ u_b \end{bmatrix} = \begin{bmatrix} 0 \\ \\ 0 \end{bmatrix} \tag{10.48}$$

which is exactly the same as equation (8.121) obtained by Newtonian mechanics in chapter 8. A discussion of the normal mode solutions was given earlier in chapter 8.

10.2.7 Central Force Motion

Consider a particle moving in a central force field as illustrated in Figure (10.7), for example, the motion of an electron in an atom. Since the motion is confined to a plane, there are two generalized coordinates which we choose to be the polar coordinates (r, θ). The particle's velocity is given by $\mathbf{v} = \dot{r}\hat{\mathbf{r}} + r\dot{\theta}\hat{\theta}$.

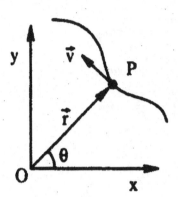

Figure 10.7: An orbit of a particle under the influence of a central force.

The kinetic energy, T, potential energy, V, and the Lagrangian, L, are given by

$$\begin{aligned}
T &= \frac{1}{2}mv^2 = \frac{1}{2}m(\dot{r}^2 + r^2\dot{\theta}^2) \\
V &= V(r) \\
L &= T - V = \frac{1}{2}mv^2 - V(r) = \frac{1}{2}m(\dot{r}^2 + r^2\dot{\theta}^2) - V(r)
\end{aligned} \qquad (10.49)$$

Hence, for the r coordinate, we have

$$\begin{aligned}
\frac{\partial L}{\partial r} &= mr\dot{\theta}^2 - \frac{\partial V}{\partial r} \\
\frac{\partial L}{\partial \dot{r}} &= m\dot{r} \\
\frac{d}{dt}\frac{\partial L}{\partial \dot{r}} &= m\ddot{r}
\end{aligned}$$

Applying Lagrange's equation,

$$\begin{aligned}
\frac{d}{dt}\frac{\partial L}{\partial \dot{r}} &= \frac{\partial L}{\partial r} \\
m\ddot{r} - mr\dot{\theta}^2 + \frac{\partial V}{\partial r} &= 0
\end{aligned} \qquad (10.50)$$

which can be solved if the potential energy function $V(r)$ is known.

For the θ coordinate, we have

$$\frac{\partial L}{\partial \theta} = 0$$

$$\frac{\partial L}{\partial \dot\theta} = mr^2\dot\theta$$

$$\frac{d}{dt}\frac{\partial L}{\partial \dot{x}} = \frac{d}{dt}mr^2\dot\theta \tag{10.51}$$

Applying Lagrange's equation,

$$\frac{d}{dt}\frac{\partial L}{\partial \dot\theta} = \frac{\partial L}{\partial \theta}$$

$$\frac{d}{dt}mr^2\dot\theta = 0 \tag{10.52}$$

which leads to $mr^2\dot\theta = l$, a constant. This constant l is the angular momentum, and thus this equation expresses conservation of angular momentum.

10.2.8 Charged Particle in an Electromagnetic Field

If a charge q of mass m is moving in an electromagnetic field with a velocity $\dot{\mathbf{r}}$, it experiences a force $\mathbf{F}$, and has kinetic energy T, given by

$$\mathbf{F} = q(\mathbf{E} + \dot{\mathbf{r}} \wedge \mathbf{B}) \tag{10.53}$$

$$T = \frac{1}{2}m\dot{\mathbf{r}}^2 \tag{10.54}$$

To obtain the Lagrangian, L, in addtion to the kinetic energy T we need to know the potential energy V. This is obtained using the following results from elctromagnetic theory

$$\mathbf{E} = -\nabla\phi - \frac{\partial A}{\partial t} \tag{10.55}$$

$$\mathbf{B} = \nabla \wedge \mathbf{A} \tag{10.56}$$

where ϕ is the scalar potential, $\mathbf{A}$ is the vector potential. Using equations (10.55) and (10.56) in (10.53), the x component of the force is obtained as

$$\begin{aligned}
F_x &= q(E_x + \dot{y}B_z - \dot{z}B_y) \\
&= q\left(-\frac{\partial\phi}{\partial x} - \frac{\partial A_x}{\partial t}\right) + q\dot{y}\left(\frac{\partial A_y}{\partial x} - \frac{\partial A_x}{\partial y}\right) - q\dot{z}\left(\frac{\partial A_x}{\partial z} - \frac{\partial A_z}{\partial x}\right) \\
&= -\frac{\partial}{\partial x}(q\phi - q\dot{x}A_x - q\dot{y}A_y - q\dot{z}A_z) - q\left(\frac{\partial A_x}{\partial t} + \frac{\partial A_x}{\partial x}\dot{x} + \frac{\partial A_x}{\partial y}\dot{y} + \frac{\partial A_x}{\partial z}\dot{z}\right) \\
&= -\frac{\partial}{\partial x}(q\phi - q\dot{\mathbf{r}}\cdot\mathbf{A}) - q\frac{dA_x}{dt} \\
&= -\frac{\partial V}{\partial x} - q\frac{dA_x}{dt} \tag{10.57}
\end{aligned}$$

from which the potential V can be identified as $(q\phi - q\dot{\mathbf{r}}\cdot\mathbf{A})$. From the second term, one can identify qA_x as the x-component of the electromagnetic momentum $q\mathbf{A}$ as is shown later in rquation (10.59).

An application of Lagrangian dynamics leads to:

$$
\begin{aligned}
T &= \frac{1}{2}m\dot{\mathbf{r}}^2 \\
V &= q\phi - q\dot{\mathbf{r}} \cdot \mathbf{A} \\
L &= T - V = \frac{1}{2}m\dot{\mathbf{r}}^2 - q\phi + q\dot{\mathbf{r}} \cdot \mathbf{A}
\end{aligned}
\tag{10.58}
$$

Hence, the generalised momentum can be found by $p_i = \partial L / \partial \dot{q}_i$, which gives

$$
\begin{aligned}
p_x &= \frac{\partial L}{\partial \dot{x}} = m\dot{x} + qA_x \\
p_y &= \frac{\partial L}{\partial \dot{y}} = m\dot{y} + qA_y \\
p_z &= \frac{\partial L}{\partial \dot{z}} = m\dot{z} + qA_z
\end{aligned}
$$

which gives the momentum as

$$
\begin{aligned}
\mathbf{p} &= p_x\mathbf{i} + p_y\mathbf{j} + p_z\mathbf{k} \\
&= m(\dot{x}\mathbf{i} + \dot{y}\mathbf{j} + \dot{z}\mathbf{k} + q(A_x\mathbf{i} + A_y\mathbf{j} + A_z\mathbf{k}) \\
&= m\dot{\mathbf{r}} + q\mathbf{A}
\end{aligned}
\tag{10.59}
$$

The interpretation of equation (10.59) is that the momentum of a charged particle in an electromagnetic field has two parts, with the first term describing the usual mechanical momentum $m\dot{\mathbf{r}}$ and the second term decribes an electromagnetic momentum $q\mathbf{A}$.

10.3   Hamilton's Equations

Another approach to solving mechanical problems is through the Hamilton's formulation which defines a Hamiltonian function in terms of which the equations of motion are expressed. It is not only in classical mechanics where the Hamiltonian approach provides an important tool, it also forms the basis of Hamilton-Jacobi theory, and perturbation techniques, and it is equally widely applied in statistical and quantum mechanics. An added advantage of Hamiltonian approach is that for a system defined in terms of n generalized coordinates, it leads to $2n$ first order differential equations, where as the Lagrangian approach gives n number of second order differential equations. We define the Hamiltonian function H in terms of generalized coordinates, and generalized momentum as:

$$
H = \sum_k \dot{q}_k\, p_k - L = \sum_k \dot{q}_k \frac{\partial L}{\partial \dot{q}_k} - L = H(q_k, p_k\, t)
\tag{10.60}
$$

Using $L = (T - V)$ in equation (10.60) we have:

$$
H = \sum_k \dot{q}_k \frac{\partial T}{\partial \dot{q}_k} - L = 2T - (T - V) = T + V
\tag{10.61}
$$

where only conservative forces are assumed to be acting on the system, which can be expressed as derivative of a potential energy function, and $\frac{\partial V}{\partial \dot{q}_k} = o$. We have used Euler's theorem for homogenous functions (students are referred to appropriate text in mathematics) to express:

$$\sum_k \dot{q}_k \frac{\partial T}{\partial \dot{q}_k} = 2T$$

Thus the *Hamiltonian function of a conservative system is defined as the sum of its kinetic and potential energy functions* expressed in terms of generalized coordinates and generalized momentum. Without going into the mathematical derivation, the Hamilton's equations of motion using the Hamiltonian function are given as:

$$\frac{\partial H}{\partial p_k} = \dot{q}_k \tag{10.62}$$

$$\frac{\partial H}{\partial q_k} = -\dot{p}_k \tag{10.63}$$

10.3.1 Application of Hamilton's equations of motion

Application of Hamilton's equations of motion is illustrated with a simple example of a one dimensional simple harmonic oscillator comprising a spring-mass system. The mass m is placed on a smooth horizontal surface, attached to a spring of force constant k. The motion of the system can be described by one generalized coordinate x, which represents the displacement of the mass, as well as the extension (compression) of the spring from equilibrium position. We have:

$$\text{Kinetic energy of the system: } T = \frac{1}{2}m\dot{x}^2$$

$$\text{Potential energy of the system: } V = \frac{1}{2}kx^2$$

$$\text{Lagrangian of the system: } L = \frac{1}{2}m\dot{x}^2 - \frac{1}{2}kx^2$$

$$\text{Generalized momentum: } p = \frac{\partial L}{\partial \dot{x}} = m\dot{x}$$

$$\text{Hamiltonian of the system: } H = T + V = \frac{1}{2}m\dot{x}^2 + \frac{1}{2}kx^2$$

$$= \frac{1}{2m}p^2 + \frac{1}{2}kx^2 \tag{10.64}$$

from equations (10.62) and (10.63), Hamilton's equations of motion for the oscillator are:

$$\frac{\partial H}{\partial p} = \frac{p}{m} = \dot{x} \Rightarrow p = m\dot{x} \tag{10.65}$$

$$\frac{\partial H}{\partial x} = kx = -\dot{p} \tag{10.66}$$

Combining equations (10.65) and (10.66) we get:

$$kx = -\frac{d}{dt}(m\dot{x}) = -m\ddot{x} \Rightarrow m\ddot{x} + kx = 0 \qquad (10.67)$$

This is the same second order differential equation of motion which we have encountered on many occasions earlier, and can be solved in the usual manner. The significance of this method is that we have not considered the restoring force explicitly, and the Newtonian mechanics has not been employed.

10.4 Problems

Unless stated otherwise, all problems in this chapter should be solved using Lagrangian dynamics.

10.1. An object of mass m_1 on a smooth inclined plane at an angle α is connected by a massless cord passing over a frictionless and massless pulley to an object of mass m_2 placed on another smooth inclined at an angle β as shown in the Figure (10.8).
(a) Write down the Lagrangian of the system, defining all your symbols.
(b) Hence, show by Lagrangian dynamics, that the acceleration of the system is given by

$$\frac{g(m_2 \sin \beta - m_1 \sin \alpha)}{m_2 + m_1}$$

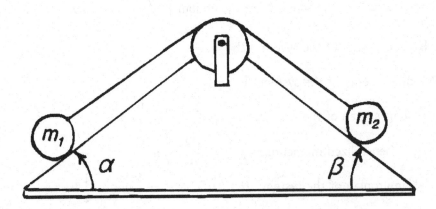

Figure 10.8: Diagram for problem (10. 1).

10.2. An object of mass m_1 on a smooth horizontal table is connected by a massless cord passing over a frictionless and massless pulley to a hanging mass m_2 as shown in the Figure (10.9).

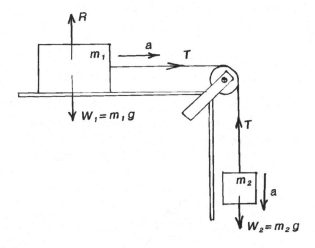

Figure 10.9: Diagram for problem (10.2).

(a) Write down the Lagrangian of the system, defining all your symbols.
(b) Hence, show, by Lagrangian dynamics, that the acceleration a and tension T are given by

$$a = \frac{m_2 g}{m_1 + m_2}$$

$$T = \frac{m_1 m_2 g}{m_1 + m_2}$$

10.3. A mass m_3 hangs at one end of a string which passes over a massless frictionless pulley A. At the other end of this string there is a frictionless pulley B of mass m_b over which passes a string carrying masses m_1 and m_2. The respective displacements in the system are as shown in Figure (10.10).

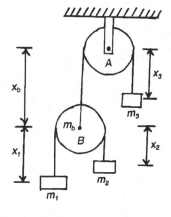

Figure 10.10: Diagram for problem (10.3).

(a) Set up the Lagrangian of the system.

(b) Show that the accelerations of mass m_1 and pulley B, given by $\ddot{x}_1$ and $\ddot{x}_b$ are given by

$$\ddot{x}_1 = \frac{2gm_3(m_1 - m_2)}{(m_b + m_3)(m_1 + m_2) + 4m_1m_2}$$

$$\ddot{x}_b = \frac{g\{(m_b - m_3)(m_1 + m_2) + 4m_1m_2\}}{(m_b + m_3)(m_1 + m_2) + 4m_1m_2}$$

10.4. A mass m_1 is connected to a fixed support by a spring of force constant k and slides on a horizontal plane without friction. Another mass m_2 is supported by a string of length l hangs as a simple pendulum from m_1 as illustrated in Figure (10.11).

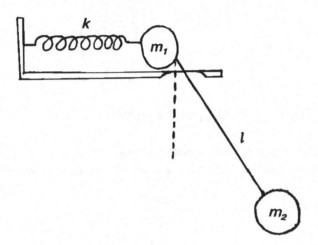

Figure 10.11: Diagram for problem (10.4).

(a) Set up the Lagrangian of the system, defining all your symbols.

(b) (i) Assuming that $m_1 = m_2 = m$, use Lagrangian dynamics to show that the normal mode frequencies satisfy the equation:

$$\omega^4 - \omega^2 \left[\frac{k}{m} + \frac{2g}{l}\right] + \frac{kg}{ml} = 0$$

(ii) Hence, show that the normal mode angular frequencies ω_1 and ω_2 satisfy

$$\omega_1^2 = (\omega_s^2 + \omega_g^2) + \sqrt{\omega_s^4 + \omega_g^4}$$

and

$$\omega_2^2 = (\omega_s^2 + \omega_g^2) - \sqrt{\omega_s^4 + \omega_g^4}$$

where $\omega_s^2 = \frac{k}{2m}$ and $\omega_g^2 = \frac{g}{l}$.

10.5. A particle of mass m is moving in a horizontal orbit in the gravitational field inside the inner surface of a cone of half angle α. The axis of the cone is z-axis and its apex is at the origin as shown in Figure (10.12). The velocity of the particle in cylindrical coordinators (r, θ, z) is

$$\mathbf{v} = \dot{r}\hat{\mathbf{r}} + r\dot{\theta}\hat{\theta} + \dot{z}\hat{\mathbf{r}}$$

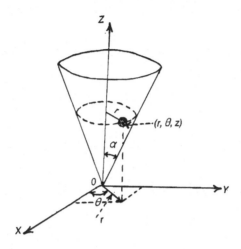

Figure 10.12: Diagram for problem 10.5.

(a) Show that the Lagrangian of the particle is

$$L = \frac{1}{2}m(\dot{r}^2 csc^2\alpha + r^2\dot{\theta}^2) - mgr\cot\alpha$$

(b) Hence, show that
(i) angular momentum is conserved.
(ii) the following equation is satisfied:

$$\ddot{r} - r\dot{\theta}^2\sin^2\alpha + g\sin\alpha\cos\alpha = 0$$

10.6. A particle of mass m on a smooth horizontal table is attached to a string passing through a small hole in the table and the other end of the string carries a particle of equal mass hanging vertically under the table. At time t, the particle is at a distance r from the hole and the string makes an angle θ with some fixed line on the table.
(a) Set up the Lagrangian of the system.
(b) If the particle on the table is initially projected at right angles to the string with a velocity $\sqrt{2gh}$ when at a distance a from the hole, show that

$$\left(\frac{dr}{dt}\right)^2 = g(a - r)\left\{1 - \frac{h(r + a)}{r^2}\right\}$$

Appendix A

Some Useful Constants etc

Table A.1: Some Fundamental Physical Constants

Name	Symbol	Value	Units
Acceleration due to gravity	g	$9.807^\dagger$	ms^{-2}
Universal gravitational constant	G	6.673×10^{-11}	Nm^2kg^{-2}
Electronic charge	e	1.602×10^{-19}	C
Rest mass of electron	m_e	9.109×10^{-31}	kg
Rest mass of proton	m_p	1.673×10^{-27}	kg
Rest mass of neutron	m_n	1.675×10^{-27}	kg
Mass of hydrogen atom	m_H	1.673×10^{-27}	kg
Bohr radius	a_o	0.529×10^{-10}	m
Binding energy of hydrogen	E_o	13.606	eV
Speed of light in vacuum	c	2.998×10^8	ms^{-1}
Permittivity of free space	$\epsilon_o = \frac{1}{\mu_o c^2}$	8.854×10^{-12}	$C^2(Nm^2)^{-1}$
Permeability of free space	μ_o	$4\pi \times 10^{-7}$	NA^{-2}
Coulomb's constant for free space	$k = \frac{1}{4\pi\epsilon_o}$	8.988×10^9	Nm^2C^{-2}
Planck's constant	h	6.626×10^{-34}	Js
	$\hbar = \frac{h}{2\pi}$	1.055×10^{-34}	Js
Stefan-Boltzmann constant	σ	5.67×10^{-8}	$Wm^{-2}K^{-4}$
Boltzmann's constant	$k = \frac{R}{N_A}$	1.381×10^{-23}	JK^{-1}
Avogadro's number	N_A	6.022×10^{26}	$(kmol)^{-1}$
Universal gas constant	R	8.315	$J(mol-K)^{-1}$
Standard temperature	T	273.15	K
Standard pressure	P	1	$atmosphere$

$\dagger$: Average value at sea level; decreases to $9.598\ ms^{-2}$ at an altitude of $100,000\ m$. Varies with latitude from $9.780\ ms^{-2}$ at the equator to $9.832\ ms^{-2}$ at the poles.

Table A.2: Some common physical properties and parameters

Name	Symbol	Value	Units
Volume of ideal gas at STP	V	22.415×10^3	$l\,(kmol)^{-1}$
Density of air at STP	ρ_{air}	1.293	kgm^{-3}
Average molecular weight of air	m_{air}	28.970	$kg(kmol)^{-1}$
Density of water at 4^oC	ρ_w	1000	kgm^{-3}
Specific heat capacity of water	C_w	4200	Jkg^{-1}
Latent heat of fusion of ice	L_f	3.360×10^5	Jkg^{-1}
Latent heat of vaporization of water	L_v	2.260×10^6	Jkg^{-1}
Speed of sound in air at STP	$v_{s,air}$	331	ms^{-1}
Absolute zero (0 K) temperature	K	-273.15	^oC

Table A.3: Earth, Sun and Moon data

Earth	Mass	$5.98 \times 10^{24}kg$
	Mean radius	$6.378 \times 10^6 m$
	Period of revolution (sun)	$1y = 365d\ 5h\ 48m\ 45.97s$
	Period of rotation (axial)	$86164.1s = 23h\ 56m\ 4.1s$
	Mean Earth - sun distance	$1.496 \times 10^{11}m$
	Mean Earth - moon distance	$3.844 \times 10^8 m$
Moon	Mass	$7.35 \times 10^{22}kg$
	Mean radius	$1.738 \times 10^6 m$
	Period of revolution (earth)	$27.32\ days$
Sun	Mass	$1.99 \times 10^{30}kg$
	Mean radius	$6.96 \times 10^8 m$
	Period of rotation (axial)	$25 - 35\ days$

Table A.4: Some conversion factors

Fundamental units:

$$1\ slug\ =\ 14.59\ kg$$
$$1\ slug\ =\ 32.17\ lb\ \text{mass where}\ g = 9.81\ ms^{-2}$$
$$1\ kg\ =\ 2.20\ lb\ \text{mass where}\ g = 9.81\ ms^{-2}$$
$$1\ year\ (y)\ =\ 365.242(19878)\ days = 3.156 \times 10^7 s$$
$$1\ in\ =\ 2.54\ cm$$
$$1\ mi\ =\ 5280\ ft = 1.61\ km$$
$$1\ nautical\ mile\ (U.S.)\ =\ 1.151\ mi = 6074\ ft = 1.852\ km$$

Secondary units:

Speed	$1\ knot$ =	$1.852\ kmh^{-1} = 1.151\ mih^{-1}$
Volume	$1\ gallon\ (= 4qt)\ (U.S.)$ =	$3.78\ l = 0.83\ gal\ (Imperial)$

Atomic mass unit (u)	$1\ u$ =	$1.661 \times 10^{-27} kg = 931.434\ MeV\ c^{-2}$
Electron volt (eV)	$1\ eV$ =	$1.602 \times 10^{-19} J$
Mechanical equivalent of heat (J)	J =	$4.185\ J\ (cal)^{-1} = 3.97 \times 10^{-3}\ Btu\ (cal)^{-1}$
Angstrom $(\mathring{A})$	$1\ \mathring{A}$ =	$1 \times 10^{-10} m$
Fermi (fm)	$1\ fm$ =	$1 \times 10^{-15} m$
Light-year (ly)	$1\ ly$ =	$9.462 \times 10^{15} m$
Atmospheric pressure (atm)	$1\ atm$ =	$1.013 \times 10^5 Nm^{-2} = 1.013\ bar = 760\ torr$
	$1\ radian\ (rad)$ =	$57.30^{\circ} = 57^{\circ}\ 18'$

Table A.5: The Greek Alphabet

Alpha	A	α		Nu	N	ν	
Beta	B	β		Xi	Ξ	ξ	
Gamma	Γ	γ		Omicorn	O	o	
Delta	Δ	δ		Pi	Π	π, ϖ	
Epsilon	E	ϵ, ε		Rho	P	ρ, ϱ	
Zeta	Z	ζ		Sigma	Σ	σ, ς	
Eta	H	η		Tau	T	τ	
Theta	Θ	θ, ϑ		Upsilon	Υ	υ	
Iota	I	ι		Phi	Φ	Φ, φ	
Kappa	K	κ		Chi	X	χ	
Lambda	Λ	λ		Psi	Ψ	ψ	
Mu	M	μ		Omega	Ω	ω	

Appendix B

Mathematical Relations

B.1. Mathematical Symbols and Constants

Table B.1.1. Mathematical symbols and numerical constants

Expression	Explanation
$a \propto b$	a is prpopotional to b
$a \approx b$	a is approximately equal to b
$a \sim b$	a is same order of magnitude (*i.e.* same power of 10) as b
$a \neq b$	a is not equal to b
$a \perp b$	a is perpendicular to b
$a \parallel b$	a is parallel to b
$a > b$	a is greater than b
$a < b$	a is less than b
$a \geq b$	a is greater than or equal to b OR a is not less than b
$a \leq b$	a is less than or equal to b OR a is not greater than b
$a \gg b$	a is much larger than b
$a \ll b$	a is much less (smaller) than b
$\sum_i a_i$	Summation: $a_1 + a_2 + a_3 + ...$
$\prod_i a_i$	Product: $a_1 \times a_2 \times a_3 \times ...$
n!	Factorial n(integer): $n \times (n-1) \times (n-2)... \times 3 \times 2 \times 1.$
Δx	Small change in x
$\Delta x \to 0$	Δx approaches zero
$\bar{x}$	Average value of x
∞	Infinity

Constant	Value
π(rad)	3.14159 26535 89793... rad $= 180^o$
1 rad	$57.2957795^o = 57^o17'44.8''$
e	2.7182 818...
ln 2	0.6931472...
ln 10	2.3025851...
$\log_{10} e$	0.4342945...

B.2. Algebra

NOTE: In this section some of the common algebraic mistakes made by students are also included as a cautionary measure. These are expressed using the symbol $\neq$.

Quadratic equation

The quadratic equation of the form: $ax^2 + bx + c = 0$ has two roots given by:

$$x = \frac{(-)b \pm \sqrt{b^2 - 4ac}}{2a}$$

Ratios, fractions, and inversion

$$\text{If } \frac{a}{b} = \frac{c}{d}, \text{ then}$$
$$\frac{a}{c} = \frac{b}{d},$$
$$\frac{b}{a} = \frac{d}{c}, \text{ and}$$
$$\frac{c}{a} = \frac{d}{b}$$

$$\text{If } a = (b+c+d), \text{ then}$$
$$\frac{1}{a} = \frac{1}{(b+c+d)}, \text{ and}$$
$$\frac{1}{a} \neq \frac{1}{a} + \frac{1}{b} + \frac{1}{c}$$

$$\text{If } \frac{a}{b} = \frac{c}{d} + \frac{e}{f}, \text{ then}$$
$$\frac{b}{a} = \frac{1}{\{\frac{c}{d} + \frac{e}{f}\}}, \text{ and}$$
$$\frac{b}{a} \neq \frac{d}{c} + \frac{f}{e}$$

Algebraic expansions and factorization

$$(x \pm a)^2 = x^2 \pm 2ax + a^2$$
$$(x \pm a)^3 = x^3 \pm 3ax^2 + 3a^2x \pm a^3$$
$$(x \pm a)^n = x^n + C_1 a n^{n-1} + C_2 a^2 x^{n-2} + \ldots + C_j a^j x^{(n-j)} + \ldots C_n a^n$$
$$\text{where} C_j = (\pm 1)^j \frac{\{n(n-1)(n-2)\ldots(n-j+1)\}}{j!}$$

$$(x^2 - a^2) \;=\; (x+a)(x-a)$$
$$(x^3 \pm a^3) \;=\; (x \pm a)(x^2 \mp ax + a^2)$$
$$(x^4 - a^4) \;=\; (x^2 + a^2)(x^2 - a^2) = (x^2 + a^2)(x+a)(x-a)$$

Manupulation of the powers and roots

$$x^0 \;=\; 1$$
$$x^1 \;=\; x$$
$$\frac{1}{x^n} \;=\; x^{-n}$$
$$x^{\frac{1}{n}} \;=\; \sqrt[n]{x}$$
$$x^n . x^m \;=\; x^{n+m}$$
$$(x^n)^m \;=\; x^{nm}$$
$$\sqrt{(a^2 + b^2 + c^2)} \;\neq\; (a + b + c)$$

Logrithmic and exponential functions

$$\ln x \;=\; \ln_e x$$
$$\text{If}\quad \ln x \;=\; y, \quad \text{then}$$
$$x \;=\; e^y$$
$$\ln e \;=\; 1$$
$$\ln 1 \;=\; 0$$
$$\ln 10 \;=\; 2.302585...$$

$$\ln(ab) \;=\; \ln a + \ln b$$
$$\ln \frac{a}{b} \;=\; \ln a - \ln b$$
$$\ln a^b \;=\; b \ln a$$
$$\text{If}\quad z \;=\; ae^{bx}, \quad \text{then}$$
$$\ln z \;=\; b\,x + \ln a$$

$$\log x \;=\; \log_{10} x$$
$$\text{If}\quad \log x \;=\; y, \quad \text{then}$$
$$y \;=\; 10^x$$
$$\log 1 \;=\; 0$$
$$\log 10 \;=\; 1$$
$$\log 100 \;=\; 2$$

$$\log 10^n = n$$
$$\ln f(x) = \ln 10 \times \log f(x)$$
$$= (2.302585) \times \log f(x)$$

$$e^{\pm ax} = 1 \pm ax + \frac{(ax)^2}{2!} \pm \frac{(ax)^3}{3!} + ...(\pm 1)^j \frac{(ax)^j}{j!}..., \qquad (j = 0, 1, 2, 3, ...\infty)$$

For $(-)1 < x < 1$,

$$\ln(1 + x) = x - \frac{1}{2}x^2 + \frac{1}{3}x^3 - \frac{1}{4}x^4 + \frac{1}{5}x^5 - ...$$
$$\ln(1 - x) = -x - \frac{1}{2}x^2 - \frac{1}{3}x^3 - \frac{1}{4}x^4 - \frac{1}{5}x^5 - ...$$

Approximations

For $x \ll 1$:

$$e^{\pm x} \cong (1 \pm x)$$
$$\ln(1 \pm x) \cong \pm x$$
$$(1 \pm x)^n \cong (1 \pm nx)$$
$$(1 \pm x)^{\frac{1}{2}} \cong 1 \pm \frac{1}{2}x$$
$$(1 \pm x)^{-1} \cong 1 \mp x$$
$$(1 \pm x)^{-\frac{1}{2}} \cong 1 \mp \frac{1}{2}x$$

B.3. Plane Geometry

Triangles

For a general triangle of sides *a*, *b*, and *c* and opposite angles *A*, *B*, and *C*:

$$A + B + C = 180^o = \pi \text{ radians}$$
$$c^2 = a^2 + b^2 - 2 a b \cos C$$

$$\text{Area of the triangle} = \frac{1}{2}a b \sin C = \frac{1}{2}b c \sin A = \frac{1}{2}c a \sin B$$
$$= \sqrt{s \times (s - a) \times (s - b) \times (s - c)}$$
$$\text{where } s = \frac{1}{2}(a + b + c)$$

$$\frac{a}{\sin A} = \frac{b}{\sin B} = \frac{c}{\sin C}$$

For a right angled trangle in which $\angle C = 90^o$, the opposite side *c* is called the hypotenous, and $c^2 = a^2 + b^2$. This relationship between the sides of a right angle triangle is called the *Pythagora's Theorem*.

Straight line:

General equation of a straight line is: $y = mx + c$.
where c is the $y-$ intercept of the line, *i.e.* the value of y when $x = 0$, and
m is the slope of the line given as: $m = \frac{y_2 - y_1}{x_2 - x_1} = \Delta y / \Delta x = \tan\theta$. The angle θ is measured from the positive $x-$ axis and is positive anticlockwise and negative clockwise.

Angle between two straight lines of slope m_1 and m_2 is given by:

$$\tan\alpha = \frac{m_2 - m_1}{1 + m_1 m_2}$$
$$\text{For parallel lines: } m_1 = m_2, \text{ and}$$
$$\text{for perpendicular lines: } m_1 \times m_2 = (-)1.$$

Equation of a straight line passing through points (x_1, y_1) and (x_2, y_2) is:
$y = m(x - x_1) + y_1 = m(x - x_2) + y_2$ where $m = \frac{(y_2 - y_1)}{(x_2 - x_1)}$

Distance d between two points (x_1, y_1) and (x_2, y_2) is given by: $[(x_2 - x_1)^2 + (y_2 - y_1)^2]^{\frac{1}{2}}$

Circle

The general equation of a circle is: $x^2 + y^2 + 2bx + 2cy + d = 0$
with center at $(-b, -c)$ and radius, $r = \sqrt{b^2 + c^2 - d}$

Equation of a circle with center at (x_c, y_c) and radius R is: $(x - x_c)^2 + (y - y_c)^2 = R^2$

Area of a circle of radius R is: πR^2, and its circumfrence is: $2\pi R$ where $\pi = 3.1415...$ radians.

Area of a sector of angle θ (radians) of a circle of radius R is: $\frac{1}{2}\theta R^2$ and the arc of the sector is: θR

Ellipse

The general equation of an ellipse is: $Ax^2 + Bxy + Cy^2 + Dx + Ey + F = 0$ where $(B^2 - 4AC) < 0$.

Equation of an ellipse with center at the origin, and the axes of the ellipse of lengths $2a$ and $2b, (a \neq b)$, along the $x-$ and $y-$ axes respectively is: $\frac{x^2}{a^2} + \frac{y^2}{b^2} = 1$.
Area of the ellipse is: $\pi a b$ and its circumfrence is $\approx 2\pi [\frac{1}{2}(a^2 + b^2)]^{\frac{1}{2}}$

Equation of an ellipse with center at (x_o, y_o) and axes of the ellipse parallel to the coordinate axes is: $\frac{(x - x_o)^2}{a^2} + \frac{(y - y_o)^2}{b^2} = 1$

Longer of the two axis is know as the major axis, and the shorter is the minor axis. Let $2a$ be the major axis and $2b$ be the minor axis, $\{(a/b) > 1\}$, then a and b are the semi-major and semi-minor axes respectively.

The eccentricity (e) of the ellipse is given by: $e = \frac{\sqrt{(a^2 - b^2)}}{a} < 1$

Circle is a special case of an ellipse for which $a = b, (a/b) = 1$, ecentricity $e = 0$, and the radius of the circle is $R = a = b$

Parabola

Eccentricity of a parabola: $e = 1$.

The general equation of a parabola is: $Ax^2 + Bxy + Cy^2 + Dx + Ey + F = 0$ where $(B^2 - 4AC) = 0$, and the axis of the parabola is oblique to the coordinate axes.

The equation of a parabola with its axis parallel to the $x-$ axis is: $x = ay^2 + by + c$.

The equation of a parabola with its axis parallel to the $y-$ axis is: $y = ax^2 + bx + c$.

The equation of a parabola with its axis along the $x-$ axis, vertex at the origin, and focus at $(p, 0)$ is: $y^2 = 4px$.

Hyperbola

Let $2a$ be the transverse axis, and $2b$ be the conjugate axis of the hyperbola. The eccentricity of a hyperbola is given by: $e = \frac{\sqrt{(a^2+b^2)}}{a} > 1$

For a rectangular hyperbola $a = b$ and $e = \sqrt{2}$, and the asymptotes are perpendicular.

Genaral equation of a hyperbola with axes oblique to the coordinate axes is: $Ax^2 + Bxy + Cy^2 + Dx + Ey + F = 0$ where $(B^2 - 4AC) > 1$

Equation of a hyperbola with center at the origin, and the transverse axis of length $2a$ and and the conjugate axis of length $2b$ along the $x-$ and $y-$ axes respectively is: $\frac{x^2}{a^2} - \frac{y^2}{b^2} = 1$.

Equation of an ellipse with center at (x_o, y_o), and the transverse axis $2a$ and the conjugate axis $2b$ parallel to the $x-$ and $y-$ axes respectively is: $\frac{(x-x_o)^2}{a^2} - \frac{(y-y_o)^2}{b^2} = 1$.

Transformation of coordinates

Let the coordinates of a point P in a 3-D space in cartesian, spherical and cylinderical coordinates system be: $P = (x, y, z) = (r, \theta, \phi) = (\rho, \phi, z)$ respectively. Then the spherical and the cylindrical coordinates are related to the cartesian coordinates by the following expressions.

Spherical and cartesian coordinates:

$$
\begin{aligned}
x &= r \sin\theta \cos\phi \\
y &= r \sin\theta \sin\phi \\
z &= r \cos\theta
\end{aligned}
$$

$$
\begin{aligned}
r &= \sqrt{x^2 + y^2 + z^2} \\
\theta &= \cos^{-1}\left(\frac{z}{\sqrt{x^2 + y^2 + z^2}}\right) = \cos^{-1}\left(\frac{z}{r}\right) \\
\phi &= \tan^{-1}\left(\frac{y}{x}\right)
\end{aligned}
$$

Cylindrical and cartesian coordinates:

$$
\begin{aligned}
x &= \rho \cos \phi \\
y &= \rho \sin \phi \\
z &= z
\end{aligned}
$$

$$
\begin{aligned}
\rho &= \sqrt{x^2 + y^2} \\
\phi &= \tan^{-1}\left(\frac{y}{x}\right) \\
z &= z
\end{aligned}
$$

Volume element in the three coordinate systems:
Cartesian coordinates: $dv = dx\,dy\,dz$
Spherical coordinates: $dv = r^2\,dr\,\sin\theta\,d\theta\,d\phi$
Cylindrical coordinates: $dv = \rho d\rho d\phi dz$

Surface element in the three coordinate systems:
Cartesian coordinates: $ds = dx\,dy$
Spherical coordinates: $ds = R^2\,\sin\theta\,d\theta\,d\phi$
Cylindrical coordinates: $ds = R\,d\phi\,dz$
where R is the radius of the spherical (cylinderical) surface.

B.4. Solid (3-D) geometry

The following symbols are used for expressions in this section:
V = volume,
T = total surface area,
S = lateral surface area, and
s = surface area of one of the face where all faces are equal.

Parallelepiped

Let the three edges of the parallelepiped are given by vectors **a, b,** and **c,** and the dihedral angles are $\alpha \neq \beta \neq \gamma$.

$V = \mathbf{a}.(\mathbf{b} \times \mathbf{c}) = \mathbf{b}.(\mathbf{c} \times \mathbf{a}) = \mathbf{c}.(\mathbf{a} \times \mathbf{b})$
$T = 2 \times |(\mathbf{a} \times \mathbf{b}) + (\mathbf{b} \times \mathbf{c}) + (\mathbf{c} \times \mathbf{a})| = 2(a\,b\,\sin\gamma + b\,c\,\sin\alpha + c\,a\,\sin\beta)$

For a rectangular parallelepiped:
$\alpha = \beta = \gamma = 90^o,$
$V = a \times b \times c,$
$T = 2(ab + bc + ca),$ and
body diagonal: $D = (a^2 + b^2 + c^2)^{\frac{1}{2}}.$

Tetrahedron

A solid bounded by four equalateral triangles of side a each.

Height: $h = a\sqrt{\frac{2}{3}}$, $s = \frac{\sqrt{3}}{4}a^2$, and $V = \frac{1}{6\sqrt{2}}a^3.$

Pyramid

$V = \frac{1}{3}$(area of base) $\times$ (altitude)
$T = $ (area of slant surfaces) $+$ (area of base) $= \frac{1}{2}$(perimeter of base) $\times$ (slant height) $+$ (area of base).

Cone

Cone is a regular pyramid with a circular base of radius R, and height h.

$V = \frac{1}{3}\pi R^2 h$
Stant height $h_s = (R^2 + h^2)^{\frac{1}{2}}$
$S = \pi R h_s = \pi R (R^2 + h^2)^{\frac{1}{2}}$.
$T = S + \pi R^2$.

Cylinder

For a right circular cylinder† of base radius R, and height h.

$V = \pi R^2 h$, cylinder surface area: $S = 2\pi R h$, and $T = \pi R \times (R + 2h)$

$\dagger$ Same relations apply to a disk of radius R and thickness t whereby h is replaced with t.

Sphere

$V = \frac{4}{3}\pi R^3$, and $T = 4\pi R^2$, where R is the radius of the sphere.

Ellipsoid

$V = \frac{4}{3}\pi \, a \, b \, c$, where a, b, and c are the semi axes of the ellipsoid.

Spheroid

Formed by rotating an ellipse of major and minor semiaxes a and b respectively and ecentricity e about one of the axes.

Oblate spheroid: Rotation about the minor axis: $V = \frac{4}{3}\pi a^2 b$, and $T = 2\pi a^2 + \pi \, \frac{b^2}{e} \, \ln \frac{(1+e)}{(1-e)}$

Prolate spheroid: Rotation about the major axis: $V = \frac{4}{3}\pi a b^2$, and $T = 2\pi b^2 + 2\pi \, \frac{ab}{e} \, \sin^{-1} e$

B.5. Trigonometry

Consider a right angle triangle ABC with sides a, b and c. $\angle C = 90^o$, $\angle A = \theta$ and $\angle B = (90 - \theta)^o$. The side c of the triangle is the hypotenuse, and with respect to $\angle A = \theta$, a is designated as the opposite side, and b is designated as the adjacent side. The three sides of the triangle are related by the Pythagora's theorem: $c^2 = (a^2 + b^2)$. Using this nomenclature, we define the following trigonometric functions for $\angle A = \theta$.

$$\text{sine } \theta \quad = \quad \sin\theta = \frac{a}{c} = \frac{\text{opposite}}{\text{hypotenuse}}$$

$$\text{cosine } \theta \;=\; \cos \theta = \frac{b}{c} = \frac{\text{adjacent}}{\text{hypotenuse}}$$

$$\text{tangent } \theta \;=\; \tan \theta = \frac{a}{b} = \frac{\text{opposite}}{\text{adjacent}}$$

and

$$\text{cosecant } \theta \;=\; \operatorname{cosec} \theta = \frac{1}{\sin \theta}$$

$$\text{secent } \theta \;=\; \sec \theta = \frac{1}{\cos \theta}$$

$$\text{cotangent } \theta \;=\; \cot \theta = \frac{1}{\tan \theta}$$

Table of trigonometric functions

θ^o	$\sin \theta$	$\cos \theta$	$\tan \theta$
0	0	1	0
30	$\frac{1}{2}$	$\frac{\sqrt{3}}{2}$	$\frac{1}{\sqrt{3}}$
45	$\frac{1}{\sqrt{2}}$	$\frac{1}{\sqrt{2}}$	1
60	$\frac{\sqrt{3}}{2}$	$\frac{1}{2}$	$\sqrt{3}$
90	1	0	∞
In the I^{st} quadrant	(+)	(+)	(+)
In the IInd quadrant	(+)	(-)	(-)
In the IIIrd quadrant	(-)	(-)	(+)
In the IVth quadrant	(-)	(+)	(-)
$(-)\alpha$	(-) $\sin \alpha$	(+) $\cos \alpha$	(-) $\tan \alpha$
$90 \pm \alpha$	(+) $\cos \alpha$	($\mp$) $\sin \alpha$	($\mp$) $\cot \alpha$
$180 \pm \alpha$	($\mp$) $\sin \alpha$	(-) $\cos \alpha$	($\pm$) $\tan \alpha$
$270 \pm \alpha$	(-) $\cos \alpha$	($\pm$) $\sin \alpha$	($\mp$) $\cot \alpha$
$360 \pm \alpha$	($\pm$) $\sin \alpha$	(+) $\cos \alpha$	($\pm$) $\tan \alpha$

Trigonometric identities

$$\sin^2 \theta + \cos^2 \theta \;=\; 1$$
$$\sec^2 \theta \;=\; 1 + \tan^2 \theta$$
$$\operatorname{cosec}^2 \theta \;=\; 1 + \cot^2 \theta$$

$$\sin(\alpha \pm \beta) \;=\; \sin \alpha \cos \beta \pm \cos \alpha \sin \beta$$
$$\cos(\alpha \pm \beta) \;=\; \cos \alpha \cos \beta \mp \sin \alpha \sin \beta$$
$$\tan(\alpha \pm \beta) \;=\; \frac{(\tan \alpha \pm \tan \beta)}{(1 \mp \tan \alpha \tan \beta)}$$

$$\sin 2\alpha \;=\; 2\sin\alpha\cos\alpha$$
$$\cos 2\alpha \;=\; \cos^2\alpha - \sin^2\alpha = 2\cos^2\alpha - 1 = 1 - 2\sin^2\alpha$$
$$\tan 2\alpha \;=\; \frac{2\tan\alpha}{(1-\tan^2\alpha)}$$

$$\sin 3\alpha \;=\; 3\sin\alpha - 4\sin^2\alpha$$
$$\cos 3\alpha \;=\; 4\cos^2\alpha - 3\cos\alpha$$
$$\tan 3\alpha \;=\; \frac{3\tan\alpha - \tan^3\alpha}{(1-3\tan^2\alpha)}$$

$$\sin\alpha + \sin\beta \;=\; 2\sin\left(\frac{\alpha+\beta}{2}\right)\cos\left(\frac{\alpha-\beta}{2}\right)$$

$$\sin\alpha - \sin\beta \;=\; 2\cos\left(\frac{\alpha+\beta}{2}\right)\sin\left(\frac{\alpha-\beta}{2}\right)$$

$$\cos\alpha + \cos\beta \;=\; 2\cos\left(\frac{\alpha+\beta}{2}\right)\cos\left(\frac{\alpha-\beta}{2}\right)$$

$$\cos\alpha - \cos\beta \;=\; (-)2\sin\left(\frac{\alpha+\beta}{2}\right)\sin\left(\frac{\alpha-\beta}{2}\right)$$

$$\tan\alpha + \tan\beta \;=\; \frac{\sin(\alpha+\beta)}{\cos\alpha\cos\beta}$$

$$\tan\alpha - \tan\beta \;=\; \frac{\sin(\alpha-\beta)}{\cos\alpha\cos\beta}$$

$$\sin\alpha\sin\beta \;=\; \frac{1}{2}\cos(\alpha-\beta) - \frac{1}{2}\cos(\alpha+\beta)$$

$$\cos\alpha\cos\beta \;=\; \frac{1}{2}\cos(\alpha-\beta) + \frac{1}{2}\cos(\alpha+\beta)$$

$$\sin\alpha\cos\beta \;=\; \frac{1}{2}\sin(\alpha+\beta) + \frac{1}{2}\sin(\alpha-\beta)$$

$$\cos\alpha\sin\beta \;=\; \frac{1}{2}\sin(\alpha+\beta) - \frac{1}{2}\sin(\alpha-\beta)$$

For angle α in radians:

$$\sin\alpha \;=\; \frac{e^{i\alpha} - e^{-i\alpha}}{2i}; \quad \text{where } i = \sqrt{-1}$$

$$\cos\alpha \;=\; \frac{e^{i\alpha} + e^{-i\alpha}}{2}$$

$$e^{\pm i\alpha} \;=\; \cos\alpha \pm i\sin\alpha$$

$$\sin\alpha \;=\; \alpha - \frac{\alpha^3}{3!} + \frac{\alpha^5}{5!} - \frac{\alpha^7}{7!} + \ldots + (-1)^{2j+1}\frac{\alpha^{2j+1}}{(2j+1)!} + \ldots, \quad \text{where } j = 0,1,2,3,\ldots$$

$$\cos\alpha \;=\; 1 - \frac{\alpha^2}{2!} + \frac{\alpha^4}{4!} - \frac{\alpha^6}{6!} + \ldots + (-1)^j\frac{\alpha^{2j}}{(2j)} + \ldots$$

For α (radians) $\ll 1$: $\sin \alpha = \alpha$; $\quad \cos \alpha = 1$; $\quad \tan \alpha = \alpha$

Hyperbolic functions

For a real argument u:

$$\text{hyperbolic sine of u} \quad = \quad \sinh u = \frac{e^u - e^{-u}}{2}$$

$$\text{hyperbolic cosine of u} \quad = \quad \cosh u = \frac{e^u + e^{-u}}{2}$$

$$\text{hyperbolic tangent of u} \quad = \quad \tanh u = \frac{\sinh u}{\cosh u}$$

$$\text{hyperbolic cosecant of u} \quad = \quad \operatorname{cosech} u = \frac{1}{\sinh u}$$

$$\text{hyperbolic secant of u} \quad = \quad \operatorname{sech} u = \frac{1}{\cosh u}$$

$$\text{hyperbolic cotangent of u} \quad = \quad \coth u = \frac{1}{\tanh u}$$

Relationship between the squares of functions

$$\cosh^2 u - \sinh^2 u = 1$$
$$\tanh^2 u + \operatorname{sech}^2 u = 1$$
$$\coth^2 u - \operatorname{cosech}^2 u = 1$$
$$\operatorname{cosech}^2 u - \operatorname{sech}^2 u = \operatorname{cosech}^2 u \operatorname{sech}^2 u$$

Symmetry and periodicity

$$\sinh(-u) = (-)\sinh u$$
$$\cosh(-u) = \cosh u$$
$$\tanh(-u) = (-)\tanh u$$
$$\operatorname{cosech}(-u) = (-)\operatorname{cosech} u$$
$$\operatorname{sech}(-u) = \operatorname{sech} u$$
$$\coth(-u) = (-)\coth u$$

Range of functions for real argument u

Function	Range of u	Range of function
sinh u	$(-\infty, +\infty)$	$(-\infty, +\infty)$
cosh u	$(-\infty, +\infty)$	$(1, +\infty)$
tanh u	$(-\infty, +\infty)$	$(-1, +1)$
cosech u	$(-\infty, 0)$	$(0, -\infty)$
	$(0, +\infty)$	$(+\infty, 0)$
sech u	$(-\infty, +\infty)$	$(0, 1)$
coth u	$(-\infty, 0)$	$(1, -\infty)$
	$(0, +\infty)$	$(+\infty, 1)$

Special values of hyperbolic functions

x	0	$\frac{\pi}{2} i$	πi	$\frac{3\pi}{2} i$	∞
sinh x	0	i	0	$-i$	∞
cosh x	1	0	-1	0	∞
tanh x	0	∞i	0	$-\infty i$	1
cosech x	∞	$-i$	∞	i	0
sech x	1	∞	-1	∞	0
coth x	∞	0	∞	0	1

B.6. Differential Calculus

NOTE: In the differential expressions given below, and in the integral expressions in the next section A, B and C are the functions of x, $f(A)$ is a function of A, and a, b, c, n, and m are real constants. The arguments of trigonometric functions are expressed in radians.

Definition:

$$\frac{d}{dx}A = Lim_{\Delta x \to 0} \frac{\Delta A}{\Delta x}$$

$$\frac{d}{dx}(A \pm B \mp C) = \frac{d}{dx}A \pm \frac{d}{dx}B \mp \frac{d}{dx}C$$

$$\frac{d}{dx}(ABC) = BC\frac{d}{dx}A + CA\frac{d}{dx}B + AB\frac{d}{dx}C$$

$$\frac{d}{dx}\left(\frac{1}{ABC}\right) = \frac{1}{BC}\frac{d}{dx}\frac{1}{A} + \frac{1}{CA}\frac{d}{dx}\frac{1}{B} + \frac{1}{AB}\frac{d}{dx}\frac{1}{C}$$

$$\frac{d}{dx}\left(\frac{A}{BC}\right) = \frac{1}{BC}\frac{d}{dx}A + A\frac{d}{dx}\frac{1}{BC}$$

$$\frac{d}{dx}f(A) = \frac{d}{dA}f(A)\frac{d}{dx}A$$

$$\frac{d}{dx}A^B = B\,A^{B-1}\frac{d}{dx}A + A^B \ln A\,\frac{d}{dx}B$$

Table of some common differential expressions

A(x)	$\frac{d}{dx}$ A
$a\,x^n$	$a\,nx^{n-1}$
a^x	$a^x \ln a$
e^x	e^x
$\ln x$	$\frac{1}{x}$
$\sin x$	$\cos x$
$\cos x$	$(-)\sin x$
$\tan x$	$\sec^2 x$
$\operatorname{cosec} x$	$(-)\cot x \operatorname{cosec} x$
$\sec x$	$\tan x \sec x$
$\cot x$	$(-)\operatorname{cosec}^2 x$

f(A)	$\frac{d}{dx}$ f(A)
$a\,A^n$	$a\,nA^{(n-1)}\frac{d}{dx}A$
a^A	$a^A \ln a \frac{d}{dx}A$
e^A	$e^A \frac{d}{dx}A$
$\ln A$	$\frac{1}{A}\frac{d}{dx}A$
$\sin A$	$\cos A \frac{d}{dx}A$
$\cos A$	$(-)\sin A \frac{d}{dx}A$
$\tan A$	$\sec^2 A \frac{d}{dx}A$

$$d(ABC) = BC\,dA + CA\,dB + AB\,dC$$

If $y = f(x)$, then $dy = \frac{d}{dx}f(x)\,dx$

B.7. Integral Calculus

Definition:

$$\sum_{Lim\,\Delta x_i \to 0}(A(x_i)\,\Delta x_i) = \int A(x)\,dx$$

Integral of the product of two functions AB is obtained by **integration by parts** following the expression given below:

$$\int (AB)\,dx = A\int B dx - \int\left(\frac{d}{dx}A\int B\,dx\right)dx$$

Table of some common integral expressions

$A(x)$	$\int A(x)\,dx$
$a\,x^n$	$\frac{a}{n+1}\,x^{n+1}$
e^x	e^x
$\ln ax$	$x \ln ax - x$
$\sin ax$	$-\frac{1}{a} \cos ax$
$\cos ax$	$\frac{1}{a} \sin ax$
$\tan ax$	$-\frac{1}{a} \ln (\cos ax)$
$\operatorname{cosec} ax$	$\frac{1}{a} \ln (\operatorname{cosec} ax - \cot ax)$
$\sec ax$	$\frac{1}{a} \ln (\sec ax + \tan ax)$
$\cot ax$	$\frac{1}{a} \ln (\sin ax)$

Bibliography

Arya, A. P., (1979). *Introductory College Physics*, McMillan Publishing Co., Inc., New York, USA.

Benson, H., (1996). *University Physics*, revised edition, John Wiley & Sons, Inc., New York, USA.

Bueche, F. J., (1986). *Introduction to Physics for Scientists and Engineers*, 4th Edition, McGraw-Hill Book Company, New York, USA.

Wilson, J. D. and A. J. Bufa, (2003). *College Physics*, 5th Edition, Prentica Hall, New Jersey, USA.

Chester, W., (1991). *Mechanics*, Chapman & Hall, London, UK.

Duncan, T., (1994). *Advanced Physics*, 4th Edition, John Murray (Publishers) Ltd, London, UK.

Fishbane, P. M. S., S. Gasiorowitz and S. T. Thorton, (1996). *Physics for Scientists and Engineers*, 2nd Edition, Prentics Hall, New Jersey, USA.

French, A. P., (1979). *Vibrations and Waves: M.I.T. Introductory Physics Series*, Thomas Nelson and Sons Ltd., UK.

Giancoli, D. C., (2000). *Physics for Scientists and Engineers*, 3rd Edition, Prentice Hall, New Jersey, USA.

Halliday, D., R. Resnick, (1978). *Physics: Part-I & Part-II Combined*, 3rd Edition, John Wiley & Sons, New York, USA.

Kibble, T. W. B., (1973). *Classical Mechanics*, 2nd Edition, Mcgraw-Hill Book Company (UK) Ltd, London, UK

Landau, L. D. and E. M. Liftshitz, (1960). *Electrodynamics of Continous Media*, Pergamon Press, Oxford, UK.

Mansfield, M. and C. O'Sullivan, (1998). *Understanding Physics*, John Wiley & Sons, New York, USA.

Nelkon, M. and P. Parker, (1990). *Advanced Level Physics*, 6th Edition, Heinemann Educational, Oxford, UK.

Ohanian, H. C., (1994). *Principles of Physics*, W. W. Norton & Co, New York, USA.

Ohanian, H. C.,(1989). *Principles of Physics*, 2nd Edition Expanded, W. W. Norton & Co, New York, USA.

Sears, F. W., M. W. Zemansky and H. D. Young, (1987). *University Physics*, 7th Edition, Addison-Wesley Publishing Company, Reading (MA), USA.

Serway, R. A. and J. W. Jewett Jr., (2004). *Physics for Scientista and Engineers* , 6th Edition, Thomson Brooks/ Cole, UK.

Smith, C. J., (1943). *Intermediate Physics*, Edward Arnold and Co., London, UK.

Tipler, P. A., (1999). *Physics for Scientists and Engineers*, 4th Edition, W. H. Freefan and Company/ Worth Publishers, New York, USA.

Urone, P. P., (2001). *College Physics*, 2nd Edition, Thomson Brooks/ Cole, UK.

Whelan, P. M. and M. J. Hodgson, (1987). *Essential Principles of Physics*, 2nd Edition, John Murray, UK.

Wolfson, R. and J. M. Pasachoff, (1999). *Physics*. Addison-Wesley, UK.

Young, H. D. and R. A. Freedman, (2000). *University Physics*, Addison-Wesley, UK.

Index

Printed in the United States
By Bookmasters